AF787207

Advances in Road Engineering: Innovation in Road Pavements and Materials

Advances in Road Engineering: Innovation in Road Pavements and Materials

Editors

Andrea Baliello
Di Wang

MDPI

Basel • Beijing • Wuhan • Barcelona • Belgrade • Novi Sad • Cluj • Manchester

Editors
Andrea Baliello
University of Padova
Padova
Italy

Di Wang
University of Ottawa
Ottawa
Canada

Editorial Office
MDPI AG
Grosspeteranlage 5
4052 Basel, Switzerland

This is a reprint of articles from the Special Issue published online in the open access journal *Buildings* (ISSN 2075-5309) (available at: https://www.mdpi.com/journal/buildings/special_issues/0HVHIP39MU).

For citation purposes, cite each article independently as indicated on the article page online and as indicated below:

Lastname, A.A.; Lastname, B.B. Article Title. *Journal Name* **Year**, *Volume Number*, Page Range.

ISBN 978-3-7258-1941-6 (Hbk)
ISBN 978-3-7258-1942-3 (PDF)
doi.org/10.3390/books978-3-7258-1942-3

Contents

buildings

Editorial

Advances in Road Engineering: Innovation in Road Pavements and Materials

Andrea Baliello [1,*] and Di Wang [2,*]

1 Department of Civil, Environmental and Architectural Engineering (ICEA), University of Padua, 35122 Padua, Italy
2 Department of Civil Engineering, University of Ottawa, Ottawa, ON K1N 6N5, Canada
* Correspondence: andrea.baliello@unipd.it (A.B.); dwang6@uottawa.ca (D.W.)

The road pavement sector is currently gaining more and more attractivity as a peculiar field of civil engineering. This takes effect considering the high extent of roads in the urban and non-urban texture, as well as the high importance of road systems as a distinctive part of an infrastructure's network that is improving in terms of competitiveness, adaptivity, functionality and global interconnection. Thus, higher and higher investments are going to be allocated towards the pavement engineering field by governments, public and private agencies, contractors, etc. Not least, the related sector became of actual interest in view of the present perspectives related to sustainability, the targets of which aspire towards innovative technical, societal, economic, and environmental frames for smart and integrated mobility and management, climate-neutral policies, and solutions. Within them, the topic of eco-friendly materials and structures promptly emerges, considering the well-known themes about recycling, since environmental and economic benefits due to low-impact materials, waste reuse, saving of finite natural resources, disposal reduction, and pollution mitigation are widely documented. In turn, innovation in pavement engineering is progressing towards a sustainable circular economy apprising certain energy and climate advantages. In this context, the significant research efforts that have been and are still spent by academies, authorities, agencies, etc. (see the huge investments often allocated to R&D departments of certain institutions), are promoting progressive advances and innovations in civil and pavement engineering. This regards improvements about a wide range of applications in road pavements, such as for instance the materials' design, optimization and testing, the structures' design, construction operations and practices, the utilization of innovative/recycled materials, the maintenance optimization and the increase in the durability, the management of investments and the intervention plans, the pavement management systems, and many more. Given this trend, the present Special Issue aims at collecting high-value scientific contributions while spreading advanced technical knowledge about all the mentioned areas by dealing with innovative, smart, eco-friendly road pavements addressed to the general sustainability of the road construction industry.

Based on the current state-of-the-art, countless articles about the presented arguments can be cited in the scientific productions of the last years and decades. However, given the current speed of technological progress (materials and technologies, laboratory and real-scale tests, analysis methods, investigation techniques, data science, etc.), further advancements for high-value scientific production must be actually promoted: this is the principal objective of the present Special Issue. As conceived, this Special Issue takes charge of digging out the majority of the above-discussed topics. This is developed through multiple investigation approaches that, from the strictly scientific point of view, couple more conventional applications (laboratory and field testing) with innovative smart assessment techniques (artificial intelligence, machine learning, digital image processing, etc.).

Innovation in materials for paving can be in regard to several technical improvements; e.g., related to the upgrade of binder performance, the enhancement of lithic skeleton

Citation: Baliello, A.; Wang, D. Advances in Road Engineering: Innovation in Road Pavements and Materials. *Buildings* **2024**, *14*, 2250. https://doi.org/10.3390/buildings14072250

Received: 11 July 2024
Accepted: 16 July 2024
Published: 22 July 2024

properties for the aggregate portion of mixture, the design and implementation of new additives and modifiers, the optimization of the mixture constituents to obtain satisfactory structural characteristics of the pavement layers, etc. A multi-scale analysis approach (evaluating the responses of binders and other constituents, mastics, mortars, and mixtures) is advised to comprehensively investigate the rheological properties and the mechanical behaviors of modified bituminous mixtures for the manufacturing of flexible pavements, studying the materials at different scales with overall consideration of performance under several environmental in-service conditions (temperature, weather, climate, etc.). In this sense, based on the SI content, the viscosity concerns of asphalt concretes are discussed thanks to advanced laboratory testing regarding high- and low-temperature performance of high-viscosity modified asphalt binders [1]. These are studied since the higher viscosity is beneficial to bonding force and adhesion to aggregates, allowing an efficient binder to be adopted in several innovative asphalt mixture types (porous asphalt, stress absorption layers, overlays, etc.), included recycled ones. Specific recommendations are given in terms of binder testing details (rheological indexes for performance evaluation) to identify suitable high-viscosity modified bitumen. Other modifiers for asphalt pavements are investigated in this SI, proposing the integration of epoxy resin and crumb rubber powder in polymer-modified asphalt binders. Here, a styrene–butadiene–styrene bitumen (inclusion of synthetic polymers to confer elastomeric and thermoplastic properties, produced through sequential polymerization operations) is modified to reduce the conventional binder and promote the recycling of waste materials. Epoxy resin is primarily used in the pavement industry to improve the flow and permanent deformation resistance thanks to the crosslinked structures formed during the manufacturing process. When using diluents, fillers, curing and toughening agents in combination with the polymeric structure of epoxy resin, great resistance to fatigue, moisture and oxidation aging can be conferred to the mixture also. The use of crumb rubber powder allows tires which have reached their end-of-life to be recycled for productive outcomes, rather than sent to landfill where they are often burned. Crumb rubber within asphalts enhances their durability, promotes noise absorption while improving the driving comfort, increases resistance to weather-related stress and confers flexibility and skid resistance to upper pavement layers. Based on research included in this SI, conventional asphalt concrete pavements are deteriorating rapidly due to the current increased traffic and extreme climate impacts. In addition to the upgrading in the construction quality, there is an urgent need to expand the utilization of modified asphalt binders to improve road capacity and traffic safety. This study envisions the incorporation of such modifiers to improve the road capacity with traffic safety purposes, being pushed by the current increase in traffic and the extreme climate impacts [2]. The analysis provides various laboratory rheological tests to measures the performance of modified mixtures, which are furtherly assessed via bearing capacity field trials. The strain accumulated by the control reference mix can be strongly reduced when adding modifiers thanks to improved stiffness and resistance to permanent deformations.

This SI also makes high-value contributions to the topic of reclaimed asphalt pavement recycling (RAP). This technique is used in asphalt pavement reconstruction and rehabilitation with several maintenance techniques (hot, warm, or cold recycling, in-site and in-plant recycling, etc.). RAP permits to achieve sustainable pavements thanks to cheaper alternatives to virgin bituminous binder and natural aggregates, while reutilizing end-of-life material that is not discharged to landfills. In general, since the asphalt industry is currently encouraging best practices to maximize the RAP percentage recycling, various technical challenges are arising to counterbalance the difficulties involved in the use of RAP as it is (large degree of aging and oxidizing, high stiffness, brittleness, etc.). The presence of these difficulties can compromise efficient production of good performance asphalt concretes. In this perspective, the refined separation method (set of plant procedures and equipment to integrate the RAP in the asphalt concrete) is deeply analyzed and discussed to evaluate the various effects of the mixing ratio of RAP that reflect on the recycled mixture performance. The related study examines the various technical and operational details of the treatment, assessing the variability of aggregate gradation, the asphalt content, the mix properties and the maximum mixing percentage of the coarse fraction [3]. The outcomes showed that the variability of several parameters is often considerable, but a refined separation process can definitely reduce the gradation variance. Another SI publication dealing with RAP proposes a novel approach to enhance the final performance of recycled asphalt concrete [4]. This consists of integrating an innovative kind of rejuvenator agent into the bituminous mixture, composed of constituents of multiple origins (wax, oil, chemical additive, and water). Based on the abovementioned multi-scale approach, various asphalt concretes are assessed at the small binder-scale and the laboratory mixture-scale (multiple stress creep recovery test, analysis of stiffness and visco-elasticity, study of water resistance and adhesion performance, resistance to cracking and rutting). Thanks to the designed rejuvenator, it is demonstrated how it is possible to obtain superior performance for the modified mixture with respect to common RAP-recycling mixture, while also obtaining a general cost convenience.

With respect to in-site testing, this SI also provides some experiences based on specific case studies in particular field conditions. This is the case of a research dealing with prefabricated prestressed concrete pavements (structural layer manufactured in the factory and transported to the construction site for assembly) [5]. Such a construction technique involves the installation of the prefabricated rigid layers and the following treatment of the joints, achieving fast times for construction or pavement maintenance, low material consumption and carbon emissions, high reusability, and long structural lifespan. It is often preferred for low-speed traffic roads in the urban context, as well as for sidewalk paving. Traditionally, the joint bars (dilatation and expansion) are recognized as a weakness domain of the rigid concrete pavements; these metallic bars must transmit the loads through their own shear and bending stiffness, but they are subjected to rust deterioration and can damage the surrounding concrete portions, leading to cracking, water seepage, and vertical misalignments threatening the driving comfort or the adequate bearing capacity of the whole rigid pavement. The proposed research regards the selection of an innovative joint material that is able to improve the bonding effects between the transfer bars and the prefabricated panels, as well as to prevent the water infiltration and the damage of the layers. The material consists of a steel fiber-reinforced self-stressing concrete. In this case, a preliminary laboratory testing is coupled with a real-time monitoring of the rigid pavement installed at a real intersection open to traffic (long-term outcomes are also provided). Results show that the innovative material can lead to long-term high stability at the joints, overcoming the brittle fracture characteristics of the ordinary concrete. The included steel fibers primarily limit the development of self-stress in the early stage, enabling higher self-stress to be retained in the concrete over the long period to improve the cracking resistance of pavement. Another SI research experience related to in-service conditions regards the ice-melting characteristics of a flexible asphalt pavement treated with slow-release deicing agents [6]. Ice and snow occurring during winter weather lay on carriage surfaces and form a thin film that is dangerous for driving safety, and can also be a primary cause of pavement distress. Various practices are usually employed to clean the surfaces (mechanical plowing, manual sweeping, salt spreading, etc.). However, these methods involve specialized machinery and labor costs and may cause traffic congestion. Thus, a slow-release deicing agent formula is designed as a filler replacement and is tested in the lab, assessing its effectiveness and durability on an asphalt pavement (high-temperature stability, cracking resistance and water sensitivity). The experimental findings indicate a slight reduction in such performance when increasing the deicing agent content (a 50% replacement of conventional filler with deicing agent is identified as the optimum concentration for the purpose). In turn, the modified asphalt mixture limits the interface adhesion between the surface and the ice film, with a maximum release concentration within two hours under rain and snow. The predicted durability of such an asphalt mixture is estimated to be around 5–8 years after paving. An additional in-field case study concerns a composite pavement constituted of a roller compacted cement base layer, over which a two-layered bituminous surface treatment is realized. The proposed structure is designed in Cambodia as a solution for the construction and rehabilitation of rural roads [7]. The research is aimed at gauging the bad status of such local roads, which are often unpaved or distressed, also considering the massive rainfalls that occur locally and, in general, the extraordinary growth in the national light and heavy transportation sector experienced by the country in a recent economical evolution. Several in-place trials (visual inspection of surface conditions, deflectometer for bearing capacity, sand patch test for roughness, noise level) were organized to evaluate the performance of the pavement. Effectively, the proposed composite structure showed superior performance to the existing surfaces, also in the presence of heavy truck vehicles and severe climatic conditions. Advanced analysis techniques in the field are presented in another SI article that deals with the crystallization mechanisms of asphalt mixture in a bridge deck pavement [8]. This research focuses on unknown crystal precipitates and is developed through innovative investigation methods employed in a bridge after paving, during ad hoc closures to traffic (fluorescence spectroscopy, infiltration, porosity and splitting tests, multi-wheel rutting, nuclear-free densitometry, ground-penetrating radar). The precipitation areas are analyzed to explain the influence of crystals on the performance of the asphalt pavement, studying the whole section to understand the crystal formation process. It was found that the main components of crystals are water molecules and machine-made sand, and no significant difference in pavement performance in the areas where crystals precipitated were detected.

A further set of SI publications also involve the adoption of innovative smart analysis techniques applied to several fields of study. This is typically referred to as the machine learning approach, i.e., a branch of artificial intelligence and computer science focused on the managing of data and algorithms to imitate the way that humans learn, ingesting a large amount of unstructured raw data to automatically determine a set of distinguishable features and categories. This method can be leveraged as a cutting-edge technology to address challenges in pavement infrastructure management, ultimately supporting the sustainability and longevity of transportation networks. More specifically, machine learning is used in combination with advanced image detection techniques to predict and detect the reflective cracking phenomenon in some pavement infrastructures [9]. The target of this study concerns the improvement in the predictive capabilities of pavement management systems, seeking to integrate advanced image processing to support real-time monitoring and assessment of pavement conditions. The developed machine learning model is characterized by a high predictive accuracy that can be further incremented through a mask region-based convolutional neural network for image detection. Thus, it can be implemented to reliably detect and monitor the reflective cracking, even across different pavement types and weather conditions. This SI also collects a contribution in which the machine learning technique is used to predict the bearing capacity of the soil (expressed in terms of CBR value) starting from known soil index properties. A model based on the natural gradient boosting algorithm is coupled with artificial intelligence to gap the current lack of reliable probabilistic estimations of soaked CBR values (conventional CBR predictions provide only deterministic results, without accounting for aleatoric uncertainty) [10]. A database of more than two thousand CBR results of different types of subgrade soils is used to instruct the algorithms, obtaining robust predictive indications of the soaked CBR performance with reliable confidence intervals. The method improvement is considered of extreme importance since the bearing properties of the subgrade layers are used as common input parameters for all the dimensioning methods of flexible or rigid road pavements. An algorithm for a finite element analysis is also discussed in a SI publication that investigates the stress–strain transmission and the cracking mechanisms in joints for rigid concrete pavement [11]. Transverse joints are constructed in plain concrete pavements to mitigate the cracking risk caused by shrinkage and temperature dilatations. However, the structural behavior of this kind of rigid pavement can be deeply influenced by the characteristics of the transverse joints that generate discontinuity through cement slabs. For these reasons, steel dowel bars can be installed in the traffic direction to improve the load transfer in the joint proximity (loaded vs. unloaded sides), relieving the slabs from common distresses. The research furnishes a three-dimensional pavement model able to simulate the pavement structural response and the flexural stress concentration in the slab by considering different dowel bar characteristics and configurations (positions and depth, size, bonding properties, etc.). Based on the results, it is possible to establish an optimum configuration for the joint design in the rigid pavement.

In conclusion, it is believed that the contributions given in this Special Issue could be able to move specific and narrow applications to the broader pavement engineering context by exploring many in-depth aspects of materials, structures, construction, and climate. Additionally, it is conceived that the published contents could drive technical future applications, further advancing the actual state of the art and knowledge, laying the foundations for future high-level scientific production regarding road engineering.

Conflicts of Interest: The authors declare no conflict of interest.

References

1. Huang, Z.; Ling, X.; Wang, D.; Li, P.; Li, H.; Wang, X.; Wang, Z.; Wei, R.; Zhu, W.; Cannone Falchetto, A. The research on high- and low-temperature rheological properties of high-viscosity modified asphalt binder. *Buildings* **2023**, *13*, 1077. [CrossRef]
2. Lee, S.Y.; Le, T.H.M. Laboratory and full-scale testbed study in the feasibility of styrene-butadiene-styrene asphalt pavement having epoxy resin and crumb rubber powder. *Buildings* **2023**, *13*, 652. [CrossRef]
3. Wang, Y.; Dou, G.; Wang, S.; Wang, J. The effect of refined separation on the properties of reclaimed asphalt pavement materials. *Buildings* **2024**, *14*, 1608. [CrossRef]
4. Kim, K.; Le, T.H.M. Feasibility and sustainable performance of rap mixtures with low-viscosity binder and castor wax–corn oil rejuvenators. *Buildings* **2023**, *13*, 1578. [CrossRef]
5. Xi, Y.F.; Ren, S.J.; Chen, B.L.; Yang, B.; Lee, J.; Zhu, G.H.; Zhou, T.C.; Xu, H. Application of steel-fiber-reinforced self-stressing concrete in prefabricated pavement joints. *Buildings* **2023**, *13*, 2129. [CrossRef]
6. Zhang, J.; Wang, W.; Liu, J.; Wang, S.; Qin, X.; Yu, B. Pavement performance and ice-melting characteristics of asphalt mixtures incorporating slow-release deicing agent. *Buildings* **2023**, *13*, 306. [CrossRef]
7. Elipse, C.; Kim, J.W.; Ohm, B.S. Performance evaluation on the application of MAST and RCC on a Cambodian rural road: A case study. *Buildings* **2023**, *13*, 2993. [CrossRef]
8. Luo, C.; Wang, D.; Li, J.; He, J. Study on crystallization mechanism of asphalt mixture in bridge deck. *Buildings* **2023**, *13*, 1527. [CrossRef]
9. Shin, S.P.; Kim, K.; Le, T.H.M. Feasibility of advanced reflective cracking prediction and detection for pavement management systems using machine learning and image detection. *Buildings* **2024**, *14*, 1808. [CrossRef]
10. Díaz, E.; Spagnoli, G. Natural gradient boosting for probabilistic prediction of soaked CBR values using an explainable artificial intelligence approach. *Buildings* **2024**, *14*, 352. [CrossRef]
11. Yaqoob, S.; Silfwerbrand, J.; Balieu, R.G.R. A parametric study investigating the dowel bar load transfer efficiency in jointed plain concrete pavement using a finite element model. *Buildings* **2024**, *14*, 1039. [CrossRef]

buildings

MDPI

Article

Feasibility of Advanced Reflective Cracking Prediction and Detection for Pavement Management Systems Using Machine Learning and Image Detection

Sung-Pil Shin [1], Kyungnam Kim [2,*] and Tri Ho Minh Le [3,*]

1 Department of Highway & Transportation Research, Korea Institute of Civil Engineering and Building Technology, 283 Goyandae-ro, Ilsanseo-gu, Goyang-si 10233, Republic of Korea; spshin@kict.re.kr
2 Pavement Research Division, Korea Expressway Corporation Research Institute, Dong-tansunhwan-daero 17-gil, Hwaseong-si 18489, Republic of Korea
3 Faculty of Civil Engineering, Nguyen Tat Thanh University, 300A Nguyen Tat Thanh Street, District 4, Ho Chi Minh City 70000, Vietnam
* Correspondence: kkn248@ex.co.kr (K.K.); lhmtri@ntt.edu.vn (T.H.M.L.)

Abstract: This research manuscript presents a comprehensive investigation into the prediction and detection of reflective cracking in pavement infrastructure through a combination of machine learning approaches and advanced image detection techniques. Leveraging machine learning algorithms, reflective cracking prediction models were developed and optimized for accuracy and efficiency. Additionally, the efficacy of image detection methods, particularly utilizing Mask R-CNN, was explored for robust and precise identification of reflective cracking on pavement surfaces. The study not only aims to enhance the predictive capabilities of pavement management systems (PMSs) through machine learning-based models but also seeks to integrate advanced image detection technologies to support real-time monitoring and assessment of pavement conditions. By providing accurate and timely detection of reflective cracking, these methodologies contribute to the optimization of pavement maintenance strategies and the overall improvement of pavement infrastructure management practices. Results indicate that the developed machine learning models achieve an average predictive accuracy of over 85%, with some models achieving accuracies exceeding 90%. Moreover, the utilization of a mask region-based convolutional neural network (Mask R-CNN) for image detection demonstrates exceptional precision, with a detection accuracy of over 95% on average across different pavement types and weather conditions. The results demonstrate the promising performance of the developed machine learning models in predicting reflective cracking, while the utilization of Mask R-CNN showcases exceptional accuracy in the detection of reflective cracking from images. This research underscores the importance of leveraging cutting-edge technologies to address challenges in pavement infrastructure management, ultimately supporting the sustainability and longevity of transportation networks.

Keywords: reflective cracking prediction; machine learning algorithms; image detection techniques; mask R-CNN; pavement management systems

Citation: Shin, S.-P.; Kim, K.; Le, T.H.M. Feasibility of Advanced Reflective Cracking Prediction and Detection for Pavement Management Systems Using Machine Learning and Image Detection. *Buildings* **2024**, *14*, 1808. https://doi.org/10.3390/buildings14061808

Academic Editor: Pengfei Liu

Received: 10 May 2024
Revised: 2 June 2024
Accepted: 13 June 2024
Published: 14 June 2024

1. Introduction

The construction and maintenance of road infrastructure are vital for facilitating economic development and ensuring safe transportation systems [1]. However, roads are susceptible to various environmental factors and climatic conditions that can lead to pavement distress [2], including reflective cracking. Reflective cracking, a prevalent form of pavement distress, poses significant challenges to the durability and performance of road infrastructure [3]. These cracks typically originate from underlying cracks or joints in the pavement structure and propagate upwards [4], often exacerbated by factors such as temperature fluctuations, traffic loading, and moisture infiltration [5]. The formation of reflective cracks is a complex phenomenon influenced by various factors, including pavement design, material properties, construction practices, and environmental conditions [6]. Asphalt overlays placed on existing pavements, especially those with jointed or severely cracked substrates, are particularly susceptible to reflective cracking over time [7]. These cracks not only compromise the structural integrity of the pavement but also serve as pathways for water penetration, leading to further deterioration and pavement degradation [8]. Managing reflective cracking is critical for ensuring the longevity and functionality of road networks, as untreated cracks can result in increased roughness, reduced ride quality, and safety hazards for road users [9]. Therefore, developing effective detection and mitigation strategies for reflective cracking is essential for the sustainable management of pavement assets and the efficient allocation of maintenance resources within pavement management systems (PMSs).

Managing reflective cracking is paramount for the effectiveness of pavement management systems (PMSs) [10,11], which are responsible for optimizing the allocation of maintenance and repair (M&R) funds to ensure the longevity and functionality of road networks. Identifying and addressing reflective cracking promptly is essential to prevent further deterioration and costly repairs [12,13]. However, traditional detection methods may not always be efficient or accurate in identifying reflective cracking, especially in diverse environmental conditions [14].

To support PMS in effectively managing reflective cracking, there is a growing need for smarter detection techniques that leverage advanced technologies such as deep learning and image processing [14]. By developing automated and intelligent systems capable of detecting reflective cracking with high accuracy and efficiency [15,16], government agencies can streamline maintenance and rehabilitation (M&R) efforts and allocate resources more effectively [17,18]. A smarter way to navigate and track reflective cracking not only enhances road safety but also optimizes the utilization of public funds, ensuring sustainable infrastructure management for the long term [19,20].

Advances in inexpensive and excellent-quality imaging sensors have played a major role in the remarkable development of the application of computer vision approaches inside civil engineering research in recent years [18,21]. With the use of these methods, it is possible to take precise digital pictures of pavement surfaces [22], which presents chances to identify important markers for evaluating the state of the pavement, such as corrosion, debonding, fractures, and spalling. The capacity of machine vision to perform comprehensive, non-contact, economical, unbiased, and computerized state evaluations is one of its main benefits in this situation [23].

Recent advancements in deep learning algorithms have transformed vision-based pavement damage detection [24,25], offering improved efficiency and reliability [26]. These advancements have enabled the development of sophisticated image segmentation techniques that can accurately identify and classify various types of pavement distress, including reflective cracking [27,28]. Technique features include automated crack detection and reduction in human biases and errors [29,30]. Object detection, crucial for civil engineering infrastructure maintenance and safety, has seen significant progress with algorithms like YOLOv6, v7, and v8, each introducing new features and improved performance [31–34]. Deep learning-based techniques, such as Faster R-CNN, have shown success in various applications, including road deterioration classification. Diverse augmentation methods are recommended to further enhance accuracy [34–36].

This research manuscript presents a comprehensive investigation into the prediction and detection of reflective cracking in pavement infrastructure through a combination of machine learning approaches and advanced image detection techniques. Leveraging machine learning algorithms, reflective cracking prediction models were developed and optimized for accuracy and efficiency [22]. Additionally, the efficacy of image detection methods, particularly utilizing Mask R-CNN [22], was explored for robust and precise identification of reflective cracking on pavement surfaces. The study not only aims to enhance the predictive capabilities of PMS through machine learning-based models but also seeks to integrate advanced image detection technologies to support real-time monitoring and assessment of pavement conditions [37–39]. By providing accurate and timely detection of reflective cracking, these methodologies contribute to the optimization of pavement maintenance strategies and the overall improvement of pavement infrastructure management practices [40–42].

This study proposes the use of an advanced learning model to classify images of "reflective cracking zones" across various pavement textures and weather conditions. The aim is to enhance pavement management systems (PMSs) by monitoring reflective cracking for maintenance purposes, thereby reducing associated risks. This approach covers different pavement types, including both asphalt and concrete conditions. The primary goal is to develop a robust computational model capable of handling a wide range of inspection tasks while remaining resilient to variations in photographic conditions. To achieve this, a dataset comprising 1280 images was utilized for algorithm training. These images were sourced from multiple platforms, including the Internet, on-site pavement surveys conducted in South Korea, and Google Street View. Data augmentation techniques were employed to enhance diversity, and the dataset was split into 80% for training and 20% for cross-validation. The research progressed through two pivotal stages: first through the development of a Convolutional Neural Network (CNN) architecture for the classification of reflective cracking, and second through the implementation of an image segmentation technique for reflective cracking detection, followed by an analysis of the training dataset.

2. Methodology

2.1. *Development of Deep Learning Prediction Models for Reflective Cracking in Highways*

2.1.1. Overview

This section explores the development of deep learning prediction models for reflective cracking in highways, focusing on both empirical and analytical approaches. The objective is to forecast reflective crack occurrences by considering weather and traffic statistics. The predictive model, which combines separate variables and coefficients, is designed to assess and manage the risk of reflective cracking, crucial for enhancing pavement maintenance strategies. Supervised learning methods for machine learning forecasting, including decision trees, multiple regression, support vector machines (SVMs), and Gaussian process analysis, are examined to provide a comprehensive analysis of the data and improve forecasting accuracy.

Previous research efforts have focused on enhancing forecasts of pavement deformation and fractures through weather-specific adjustments. However, in this study, situated within the uniform meteorological framework of the Seoul Metropolitan Government, we chose not to implement such modifications, capitalizing on the consistent environmental conditions present across the examined sites. Using both empirical and analytical approaches, the prior research aimed to develop a reliable technique for forecasting reflective cracking on specific expressways in Seoul. The empirical model concentrated on variables such as average temperature, precipitation, and traffic volume, while the analytical model included variables like maximum temperature, precipitation, and minimum relative humidity. Despite having a similar computational structure, these models examined different sets of independent factors [22].

The simulations were designed to help authorities plan maintenance and repair activities more efficiently by offering insights into the likelihood of reflective cracking development through regression modeling and variable normalization.

2.1.2. Predictive Model and Enhancements for Improved Accuracy

In order to anticipate the development of reflective cracking, this study presents a predictive model that combines separate variables and coefficients. The reflective cracking prediction amount (E) depends on a combination of independent factors and unknown coefficients ($\beta_0, \ldots, \beta_{45}$). The empirical forecasting approach employs factors like mean temperature, rainfall, maximum snowfall, maximum consecutive days with precipitation, and traffic volume to predict reflective cracking, whereas the analytical framework incorporates the highest temperature, precipitation, and the amount of traffic. These variables are inputs and are standardized between −1 and 1, where a value of −1 denotes a low probability of reflective cracking development and a value of 1 indicates a higher probability. This model provides a methodical way to assess and control reflective cracking risk, which is crucial for improving pavement management and upkeep techniques. Equation (1) illustrates how the design of experiments (DOEs) method was used to create the regression model [22].

$$\begin{aligned} E = \beta_0 + \beta_{10}x_1 + \beta_{20}x_2 + \beta_{30}x_3 + \beta_{40}x_4 + \beta_{50}x_5 + \beta_{11}x_1^2 + \beta_{22}x_2^2 \\ +\beta_{33}x_3^2 + \beta_{44}x_4^2 + \beta_{55}x_5^2 + \beta_{12}x_1x_2 + \beta_{13}x_1x_3 + \beta_{14}x_1x_4 + \beta_{15}x_1x_5 \\ +\beta_{23}x_2x_3 + \beta_{24}x_2x_4 + \beta_{25}x_2x_5 + \beta_{34}x_3x_4 + \beta_{35}x_3x_5 + \beta_{45}x_4x_5 \end{aligned} \quad (1)$$

Here is a detailed explanation of the factors and coefficients:

- E: This represents the reflective cracking prediction amount, which is the dependent variable we attempt to estimate.
- β_0 to β_{45}: These are the coefficients of the independent variables in the predictive model. Each coefficient represents the contribution of the corresponding independent variable to the prediction of reflective cracking.
- x_1 to x_5: These are the independent variables used in the predictive model. Each variable represents a different environmental or exploitation factor that may influence the occurrence of reflective cracking. Here is a breakdown of what each variable represents:
  - x_1: Mean temperature;
  - x_2: Relative humidity;
  - x_3: Largest amount of fresh snowfall;
  - x_4: Precipitation days;
  - x_5: Traffic volume.

In the empirical forecasting approach, x_1 to x_5 are standardized among −1 and 1, where a value of −1 denotes a low probability of reflective cracking development and a value of 1 indicates a higher probability. Each method provides unique insights into the prediction model, and these factors are selected based on their possible impact on the occurrence of reflective cracking [22].

Although helpful for broad forecasts in Seoul, the prior predictive model has many drawbacks. Its forecasts may not be accurate outside of its limited range because it mostly uses local data. Its dependability is also in doubt because it is purely based on past data and has not been verified by actual forecasts. The model was tested for accuracy by comparing it with real data for regions such as Buk-bu Expressway. The comparison showed that the model's success rate declined over time, suggesting that it might only be trustworthy for information up until 2022.

This shows that additional elements that might influence the occurrence of reflective cracking, such as changes in policies, were not taken into consideration by the model. The current study intends to improve the model's accuracy by taking into account further elements and upgrading its database to solve these problems.

In general, the focus is on forecasting reflective crack occurrences, considering weather and traffic statistics. Data from the National Meteorological Service and the Seoul Traffic Information Center revealed that high temperatures, rainfall, snowfall, humidity, and traffic volume significantly affect reflective crack predictions. Using these factors, a model was developed and tested, with data normalized to minimize errors. The analysis also accounted for traffic volume's impact on road deterioration, despite regulations on large vehicles. Multiple regression analysis was then used to assess the relationship between these variables and monthly reflective crack.

2.1.3. Supervised Learning Methods for Machine Learning Forecasting

This investigation employed machine learning methods using supervised learning techniques, including decision and regression trees, multiple regression, support vector machines (SVMs) [30], and Gaussian process analysis [43]. These methods, which segment data according to optimization levels, were used to analyze data and forecast results. For example, multiple regression explains the link between independent (X) and dependent (Y) variables, while SVMs use hyperplanes to group variables into clusters. Regression and decision trees make decisions based on reducing uncertainty and maximizing purity [44]. Gaussian process regression uses kernel functions to estimate continuous dependent variables, representing the relationship between input and output variables.

Each method offers a unique perspective to forecasting, contributing to a comprehensive examination of the data. The connection between the variables in the independent group (X) and the dependent variable (Y) is described by the analysis technique known as multiple regression [30]. This relationship can be determined by calculating the matrix's determinant for the estimation of parameters, as indicated in Equation (2):

$$min \sum_{i=1}^{N} \hat{\varepsilon}_i^2 = \sum_{i=1}^{N} (Y_i - \hat{Y}_i)^2 = \sum_{i=1}^{N} \left[[Y_i] - [\hat{\beta}_i][X_i]^T \right]^2 \\ \hat{\beta} = (X'X)^{-1} X'Y \tag{2}$$

where $\hat{\varepsilon}$ stands for the residual sum of squares and N is the number of independent variables. Regression analysis was used to obtain the $\hat{\beta}$ value, which led to the prediction model.

2.2. *Reflection Cracking Detection Model*

The second goal of the project was to develop effective techniques for creating a deep learning framework that is adaptable to different environments. Compiling the image collection required taking pictures of various pavement surfaces and ambient lighting. Preliminary data analysis highlighted these variables' potentially significant effects on reflective cracking detection performance. Moreover, anchor boxes are frequently produced by traditional bounding box methodologies [45], which are often used for reflective cracking surveillance but may not be precise in measuring and evaluating the amount of reflective cracking. To address this, our work used state-of-the-art Image Segmentation techniques to accurately delineate the area of reflective cracking areas [37].

2.2.1. Data Collecting and Pre-Processing

For this study, a training set of data was collected from various sources, including websites, in-person surveys, and Google Street View, to capture a wide range of pavement and air quality situations. For quality control reasons, the gathered images showed a range of resolutions, usually from 360p to 720p, with careful attention paid to obtaining a well-balanced depiction across various meteorological situations, lighting circumstances, and traffic settings [46]. After that, these pictures were manually labeled using the MATLAB Image Labeler application [47]. Eight hundred pictures were carefully classified as "Reflective cracking" based on professional advice and earlier studies. Comprehensive and exhaustive coverage of the data was ensured by the dataset's painstaking organization in COCO format, which included full annotations and captions for every image [47].

2.2.2. Data Labeling

Under the direction of a Korean pavement assessment specialist, reflection crack layouts were sketched and transcribed into JSON archives. The 1280 photos in the dataset were labeled using MATLAB 2023b Image Labeler, with the categories of "Reflective cracking" separated evenly [47]. The COCO-formatted dataset contains bounding polygon labels and comprehensive picture specs.

Table 1 presents a breakdown of the dataset used for training, validation, and testing in the context of pavement damage analysis. It categorizes the data by pavement types and weather conditions, indicating the number of samples allocated for each combination. For asphalt pavement, both clear and multiple weather scenarios were considered, with 256 samples each for training, 32 for validation, and 32 for testing, totaling 320 samples per condition. Similarly, concrete pavement data follows the same distribution, resulting in a total dataset size of 1280 samples [48]. This structured dataset ensures comprehensive coverage across various conditions, facilitating robust model development and evaluation for pavement damage detection.

Table 1. Overview of the composition of the dataset (number of images).

Pavement Types	Weather Types	Train Data	Validation Data	Test Data	Total
Asphalt pavement	Clear	256	32	32	320
	Multiple weather	256	32	32	320
Concrete pavement	Clear	256	32	32	320
	Multiple weather	256	32	32	320
Total		1024	128	128	1280

2.2.3. Data Augmentation

Image Data Generator enhanced photos to increase the diversity of the collection and make up for the lack of high-quality reflective cracking examples. Techniques included color upgrades (contrast, saturation) as well as positioning changes (flipping, scaling, and rotating) [16,49,50]. Processing was expedited by standardizing photos to 300 × 300 pixels and enlarging them from 200 to 600 pixels. To avoid over-augmentation for efficient model convergence, the dataset was divided into two halves: 80% for training and 20% for testing. This guided the creation of the model with the suggested architecture.

2.2.4. Deep Learning Driven Object Identification

As shown in Figure 1, the automated reflection cracking segment was made possible by the modified Mask R-CNN design. This version of Mask R-CNN was specifically designed to enhance the recognition of reflective cracking, and it was trained using labeled photos [47]. Modeling creation was made simpler by using Detectron2, a modern object identification technique, which removed the need for a new Mask R-CNN network. Quick image segmentation was made possible by utilizing the Detectron2's pre-trained model, which facilitated transfer learning for object recognition in a variety of domains. Furthermore, three convolutional blocks and max pooling layers from the Keras deep learning framework were smoothly included in the sequential network.

2.2.5. Mask R-CNN

By adding object masks to bounding boxes, Mask R-CNN—a Faster R-CNN extension—improves the identification and segmentation of objects efficiency as shown in Figure 2. Pixel-level segmentation is made possible by RoIAlign, which guarantees accurate spatial features in region-of-interest pooling [51]. The Mask head comes next. The already trained models from Detectron2 were mainly employed in their default setups, with some hyperparameter value alterations [36]. In Mask R-CNN, models such as R101-FPN perform better than others, despite longer training times and the occasional excessive fitting issue.

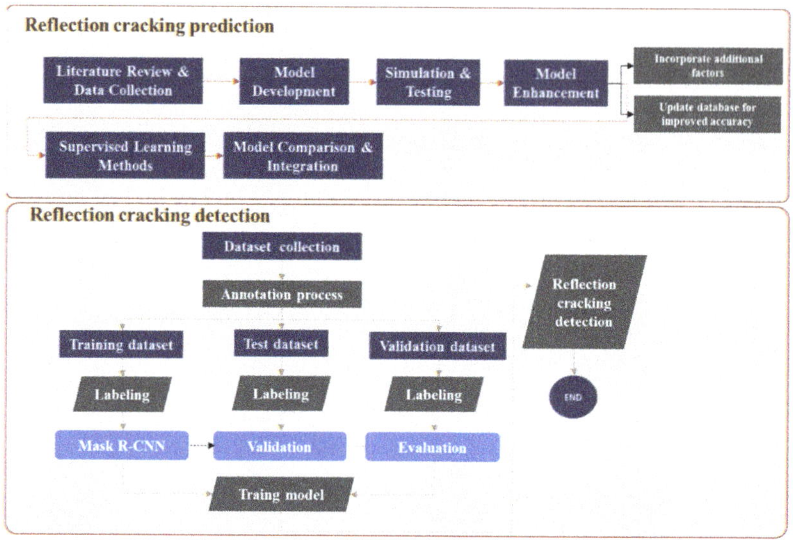

Figure 1. Framework for reflection cracking prediction and detection.

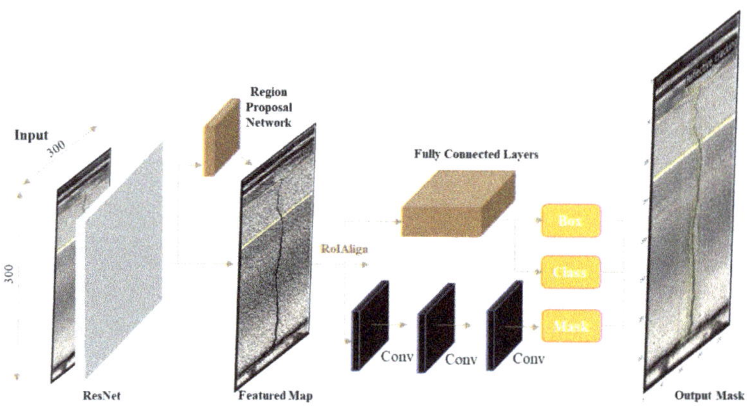

Figure 2. Mask R-CNN structure.

2.2.6. Applications

The main programming language used in this study was Python, and Google Colab provided the GPU environment—a Nvidia Tesla P100/K80/T4—for the development of all deep learning systems [52]. The popular data-flow computing and neural network creation tools TensorFlow and Keras were used to create the Mask R-CNN approach to reflective cracking identification. A model that was trained from the Common Objects in Context (COCO) database modified to fit the reflective cracking database was used for training.

2.2.7. Hyperparameters

After network topologies and datasets are finalized, it is imperative to set network hyperparameters before training. These parameters were found via heuristics as opposed to direct dataset estimates because they are not dependent on any particular dataset [48]. Although Mask R-CNN provides a wide range of hyperparameters to be adjusted during training, it can take a lot of time and resources to explore every possible configuration. Because of this, class numbers in the current study were designated as reflective cracking zones, and default settings were mostly employed. The default settings from Detectron2 were kept, while the parameters of the model, such as the learning algorithm, were adjusted using Optuna, a hyperparameter tuning tool [48].

Table 2 outlines the key hyperparameters for Mask R-CNN, crucial for its performance. These include a base learning rate of 0.00027, five images per batch, a gamma value of 0.06 for learning rate adjustment, and a maximum of 2000 iterations [47]. It also specifies 18 regions of interest per image and handles three distinct classes. Additionally, parameter cfg.MODEL.ROI_HEADS.SCORE_THRESH_TEST is set at 0.6 to balance recall and precision during evaluation. These settings collectively ensure the model's robustness and effectiveness across various computer vision tasks [47].

Table 2. Mask R-CNN hyperparameter setting.

Model Parameter	Value
cfg.SOLVER.BASE_LR	0.00027
cfg.SOLVER.IMS_PER_BATCH	5
cfg.SOLVER.GAMMA	0.06
cfg.SOLVER.MAX_ITER	2000
cfg.MODEL.ROI_HEADS.BATCH_SIZE_PER_IMAGE	18
cfg.MODEL.ROI_HEADS.NUM_CLASSES	3
cfg.MODEL.ROI_HEADS.SCORE_THRESH_TEST	0.6

2.2.8. Comparable Architectures

The most recent version of Ultralytics' YOLO V8 significantly increases object detection tasks' speed and accuracy [53]. Its integrated framework supports tasks like picture categorization and instance segmentation, while its rebuilt backbone network and an-chor-free head refine detecting powers. It provides models that have been trained with the recognition of object stages and is flexible in terms of export forms and CPU/GPU compatibility. YOLO V8, which comes in detection, segmentation, and classification variations, is a reliable choice that has gained recognition for its contributions to artificial recognition [53].

The authors of this study used a Kaggle dataset of 1280 reflective cracking photos, and they used the Pascal VOC annotation format to accurately pinpoint the reflective cracking spots in the photos. The dataset's separation into training and testing sets and intricacy were overcome by converting it into COCO format to satisfy the YOLOV8 model's needs. After preprocessing and splitting, the efficacy of the YOLOV8 framework for identifying reflective cracking forecasts was assessed during training. Improving the efficiency of models and guaranteeing correct object detection was made possible by data annotation with the help of tools such as RoboFlow. A crucial method that also became apparent was data augmentation, which improved the model's robustness versus changes in source data and enhanced the training dataset, increasing its precision and dependability in real-world projections.

Table 3 presents a comparative analysis of hyperparameter configurations across several notable architectures in the realm of object detection. The Yolov4 model [31], with an input size of 300 $\times$ 300, utilizes a momentum of 0.9 alongside a decay rate of 0.00005 and a learning rate of 0.0013, employing Leaky ReLU activation. In contrast, Yolov5 adopts a larger input dimension of 416 $\times$ 416, maintaining a momentum of 0.9 [38], a slightly higher decay of 0.0005, and a learning rate of 0.001, employing the ReLU activation function. Lastly, Yolov8 [53], also with a size of 416 $\times$ 416, employs a momentum of 0.85, a decay rate of 0.0003, and a learning rate of 0.002, utilizing the Mish activation function. These distinct configurations highlight the nuanced choices made in adjusting hyperparameters, catering to the specific requirements and architectural intricacies of each model, ultimately impacting their performance in object detection tasks.

Table 3. Hyperparameter configuration of comparable architectures.

Model	Width × Height	Momentum	Decay	Learning Rate	Activation
Yolov4	300 × 300	0.9	0.00005	0.0013	Leaky ReLU
Yolov5	416 × 416	0.9	0.0005	0.001	ReLU
Yolov8	416 × 416	0.85	0.0003	0.002	Mish

2.2.9. Comparative Analysis and Evaluation

Using a method that takes into account multiple parameters, an in-depth comparison was carried out to assess segmentation algorithms based on pixel recognition precision. Global accuracy (GA) is computed using an assessment method that includes true positives (TPs), true negatives (TNs), false positives (FPs), and false negatives (FNs). In order to additionally improve the segmentation framework, accuracy and Intersection-over-Union (IoU) scores were combined. IoU quantifies the overlap across predicted and reality segments [47]. Accuracy was ensured by establishing standards: a class was considered existent only if the final result possibility was more than 0.5, and true detection necessitated comparing the indicated bounding box with real data and requiring an IoU score of at least 50% [35]. This all-encompassing strategy made it possible to evaluate segmentation methods in depth, which helped with accurate choices on model improvement.

The following Equations (3) and (4) describe a comparison analysis conducted using the proportion of properly recognized pixels:

$$\mathrm{GA} = \frac{\mathrm{TP} + \mathrm{TN}}{\mathrm{TP} + \mathrm{TN} + \mathrm{FP} + \mathrm{FN}} \tag{3}$$

$$\mathrm{IoU} = \frac{\mathrm{TP}}{\mathrm{TP} + \mathrm{FP} + \mathrm{FN}} \tag{4}$$

2.2.10. Cross-Entropy Loss Function

The initial use of the traditional cross-entropy loss function (Equation (5)) in this study treats background and reflective cracking pixels equitably. However, this approach may not be the best choice for datasets including instances of uneven reflective cracking. To address this problem, a weighted binary cross-entropy loss function was devised, which is represented by Equation (6), where various weights are allocated to different types of pixels [54,55].

The traditional binary cross-entropy loss function is defined as follows:

$$\mathrm{L}(\mathrm{y}, \hat{y}) = -[\mathrm{y}\log(\hat{y}) + (1 - \mathrm{y})\log(1 - \hat{y})], \tag{5}$$

where

- y is the true label (0 for background, 1 for reflective cracking);
- $\hat{y}$ is the predicted probability;
- log denotes the natural logarithm.

To account for the uneven distribution of background and reflective cracking pixels, a weighted binary cross-entropy loss function is introduced. The weighted loss function is defined as

$$\mathrm{L}(\mathrm{y}, \hat{y}) = -[w_1 y \log(\hat{y}) + w_2(1 - y)\log(1 - \hat{y})], \tag{6}$$

where

- y is the true label (0 for background, 1 for reflective cracking);
- $\hat{y}$ is the predicted probability;
- w_1 is the weight assigned to the reflective cracking pixels;
- w_2 is the weight assigned to the background pixels;

- log denotes the natural logarithm.

By assigning different weights w_1 and w_2 to the different types of pixels, this weighted loss function better accommodates the uneven distribution in the dataset, improving the model's performance in identifying reflective cracking.

3. Results and Discussions

3.1. Reflection Cracking Prediction Results

3.1.1. Initial Multilinear Regression Analysis

Table 4 outlines the findings of multilinear regression analysis. It shows a strong correlation (R = 0.858) and a reasonable coefficient of determination (R^2 = 0.731). The Root Mean Squared Error (RMSE) is 1795.00, indicating the average deviation between observed and predicted values. The Mean Percentage Error (MPE) is 30.60%, suggesting overall accuracy.

Table 4. Result of multilinear regression analysis.

Model	R	R2	RMSE	MPE (%)	Std. Error of the Estimate
Regression Equation	0.858	0.731	1795.00	30.60	845.00

Table 5 uses multiple linear regression and correlation analysis to illustrate how different factors affect the prediction of reflective cracking. Both analyses highlight the strong positive relationships between reflective cracking frequency and traffic volume and precipitation, as well as the significant positive influence of the lowest temperature. The correlation and regression studies indicate that the maximum continuous precipitation day and average temperature are positively correlated with the occurrence of reflective cracking.

Table 5. Correlation Analysis Results for Predictive Factors in Pavement Damage.

Rank	Correlation Analysis	Multi-Linear Regression Analysis
	Correlation Factor	p-Value
1	Precipitation	0.612
2	Traffic Volume	0.550
3	Min. Temperature	0.510
4	Max. Continuous Precipitation Day	0.505
5	Avg. Temperature	0.502
6	Precipitation Day	0.446
7	Max. Snowfall	0.420
8	Max. New Snowfall	0.382
9	Relative Humidity	0.369
10	Min. Relative Humidity	0.317
11	Max. Temperature	0.192
12	Evaporation Loss	0.081

Additionally, there are further positive relationships between reflective cracking occurrence and variables such as maximum snowfall, maximum day of precipitation, and maximum amount of fresh snowfall. Regression analysis also supports the moderate positive associations between reflective cracking occurrence and minimum relative humidity, as well as the weaker positive relationships between evaporation loss and maximum temperature and reflective cracking occurrence. Overall, temperature-related factors, traffic volume, and precipitation appear to be the main determinants of the likelihood of reflective cracking.

According to the multilinear regression analysis, the key predictors for reflective cracking are monthly minimum temperature, traffic volume, total precipitation, highest consecutive days with precipitation, and average temperature. Each of these variables has a significant correlation of 50% or more. The monthly minimum and average temperatures, relative humidity, and precipitation appear to be independent in terms of collinearity among the independent variables. Additionally, a distinct group is formed by the lowest relative humidity, the greatest number of consecutive days with precipitation, and the maximum depth of newly fallen snow.

The lowest temperature has the highest correlation with the occurrence of reflective cracking, followed by relative humidity, precipitation, and traffic volume, as indicated by the standardization coefficient ranking. Despite its strong link with reflective cracking, minimum relative humidity is not selected because it overlaps with other variables. Instead, the ultimate independent factors chosen are the greatest number of consecutive days with rainfall and the maximum depth of newly fallen snow.

3.1.2. Comparison of Machine Learning Techniques for Model Optimization

Machine learning techniques were applied to optimize both empirical and analytical models, aiming to minimize errors using RMSE, MSE, and MAE (Table 6). Training velocity and time were also evaluated. The multilinear regression model emerged as the optimal choice due to its minimal error, while stepwise regression, despite its fast training speed, exhibited lower prediction efficiency and larger margins of error.

Table 6. Machine learning yielded optimization results.

Model	Metric	NH19 Empirical	NH19 Analytical	NH23 Empirical	NH23 Analytical
Linear Regression	RMSE	0.278	0.295	0.238	0.248
	MSE	0.077	0.087	0.057	0.062
	MAE	0.195	0.198	0.164	0.184
	T.V (n/s)	2000	2100	2100	1800
	T.T (s)	1.467	1.563	1.476	1.524
Stepwise Linear Reg.	RMSE	4.737	0.948	2.815	0.496
	MSE	22.439	0.898	7.923	0.246
	MAE	1.115	0.583	0.926	0.383
	T.V (n/s)	3100	2900	2800	3000
	T.T (s)	101.97	110.48	113.97	114.96
Decision Tree	RMSE	0.413	0.429	0.432	0.353
	MSE	0.171	0.184	0.186	0.125
	MAE	0.298	0.323	0.318	0.244
	T.V (n/s)	4300	3600	3200	3800
	T.T (s)	0.81	1.016	0.876	0.938
Support Vector Machine	RMSE	0.368	0.359	0.334	0.331
	MSE	0.136	0.129	0.112	0.109
	MAE	0.248	0.241	0.229	0.231
	T.V (n/s)	4400	3600	4700	3700
	T.T (s)	0.414	0.609	0.476	0.568

Table 6. *Cont.*

Model	Metric	NH19 Empirical	NH19 Analytical	NH23 Empirical	NH23 Analytical
Ensemble	RMSE	0.379	0.405	0.365	0.325
	MSE	0.144	0.164	0.133	0.106
	MAE	0.257	0.297	0.268	0.227
	T.V (n/s)	1400	1300	1400	1600
	T.T (s)	1.532	1.8	1.45	1.34
Gaussian Process Reg.	RMSE	0.406	0.383	0.343	0.317
	MSE	0.165	0.146	0.118	0.1
	MAE	0.28	0.263	0.252	0.214
	T.V (n/s)	3500	2100	3500	1500
	T.T (s)	0.797	1.676	0.441	0.465

Root Mean Squared Error (RMSE); Mean Squared Error (MSE); Mean Absolute Error (MAE); Traffic volume (T.V); Travel time (T.T).

The optimized multilinear regression model underwent coefficient adjustments based on previous studies, and the updated variables of Equation (1) are presented in Table 7. Using these refined coefficients, occurrences of reflective cracking were predicted. Figure 3 illustrates the comparison between actual and predicted values, showcasing the improved performance of the new predictive model, which integrates weather, traffic volume, and reflective cracking survey data from 2022. Standardization of variables was crucial to enhance accuracy, especially given the diverse ranges in traffic and temperature.

Table 7. Determination of regression coefficients.

	Empirical Approach		Analytical Approach	
	NH19 Roadway	**NH23 Expressway**	**NH19 Roadway**	**NH23 Expressway**
R^2	0.686	0.803	0.635	0.783
MSE	0.077	0.057	0.087	0.062
MAE	0.195	0.164	0.198	0.184
β_0	−1.363	−2.571	−1.043	1.365
β_{10}	2.328	2.102	0.744	−0.427
β_{20}	−3.154	−0.711	−0.813	0.608
β_{30}	0.001	−3.793	−0.491	−0.437
β_{40}	1.128	−0.271	−0.420	2.439
β_{50}	0.473	−1.635	−1.000	3.224
β_{11}	0.311	0.197	−0.212	0.079
β_{22}	−1.855	1.258	0.306	−2.471
β_{33}	0.862	−0.728	−0.247	0.327
β_{44}	−0.537	0.336	0.151	−0.667
β_{55}	0.285	0.546	0.162	−0.363
β_{12}	1.271	−0.815	0.389	−0.633
β_{13}	1.696	0.929	0.995	0.050
β_{14}	−0.499	−0.465	−0.426	0.235
β_{15}	−0.100	1.542	−0.007	0.143
β_{23}	−1.857	−0.627	−1.380	−0.696
β_{24}	1.878	0.856	0.128	3.515
β_{25}	0.571	−0.962	0.010	4.274
β_{34}	0.155	0.053	0.130	0.412
β_{35}	−0.015	−3.320	−1.132	−0.182
β_{45}	0.133	−0.521	−0.391	−0.218

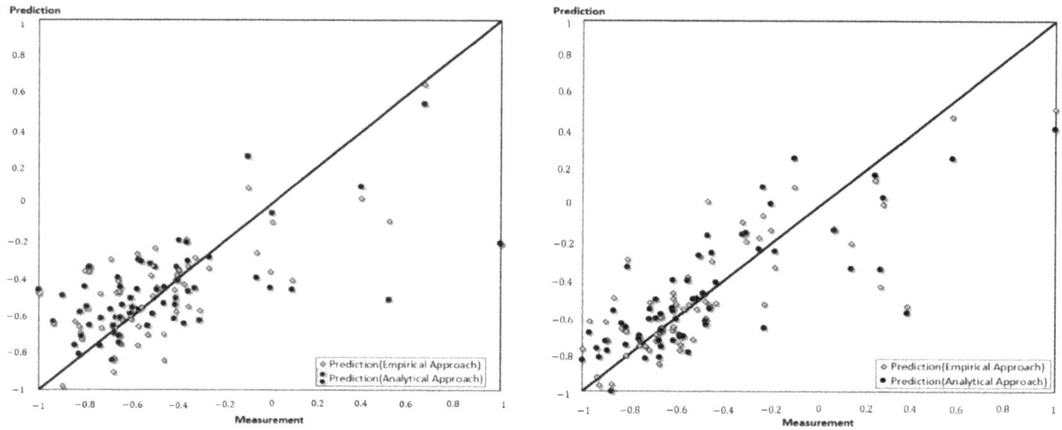

(**a**) (**b**)

Figure 3. The numbers of predicted reflective cracking: (**a**) NH19 Roadway and (**b**) NH23 Expressway.

Further evaluation involved a comparative analysis between analytical and empirical approaches, with the predictive model's efficacy confirmed by forecasting reflective cracking numbers for 2022, a period not included in the initial training or verification phase.

Figure 4 illustrates the performance of the prediction model based on actual reflective cracking occurrences, utilizing standardized values ranging from −1 to 1.

Figure 4a displays the prediction outcomes from the NH19 roadway area, whereas Figure 4b showcases the prediction outcomes from the NH23 roadway area. This finding presents the performance evaluation of the developed prediction models, with the empirical method showing proper performance, particularly on NH19 and NH23 expressways. Given its efficiency across locations, the empirical approach is recommended for standardized reflective cracking prediction in pavement management systems. Additionally, the recalibrated model can aid in cost analysis by accurately predicting reflective cracking occurrences, facilitating optimal budget allocation, and preventing budgetary discrepancies.

Furthermore, Table 8 provides a comprehensive assessment of the developed prediction models, considering both the standardization method and the methodologies used (empirical versus analytical). The performance metrics evaluated include Root Mean Squared Error (RMSE), Mean Absolute Error (MAE), Mean Squared Error (MSE), coefficient of determination (R^2), and the p-value.

Table 8. Assessment of the effectiveness of developed prediction models.

	Standardization									
	Empirical Approach					Analytical Approach				
	RMSE	MAE	MSE	R^2	p-Value	RMSE	MAE	MSE	R^2	p-Value
NH19	0.5355	0.4217	0.2867	0.5612	5.81×10^{-6}	0.5743	0.4182	0.3299	0.7151	0.0003
NH23	0.6138	0.4549	0.3767	0.4314	0.4548	0.5691	0.4619	0.3239	0.4302	9.68×10^{-10}
	Reflective cracking Prediction									
	Empirical Approach					Analytical Approach				
	RMSE	MAE	MSE	R^2	p-Value	RMSE	MAE	MSE	R^2	p-Value
NH19	344	272	118,598	0.5613	0.2495	369	269	136,390	0.7148	0.0869
NH23	490	363	239,995	0.4314	0.0231	454	369	206,142	0.4303	0.0040

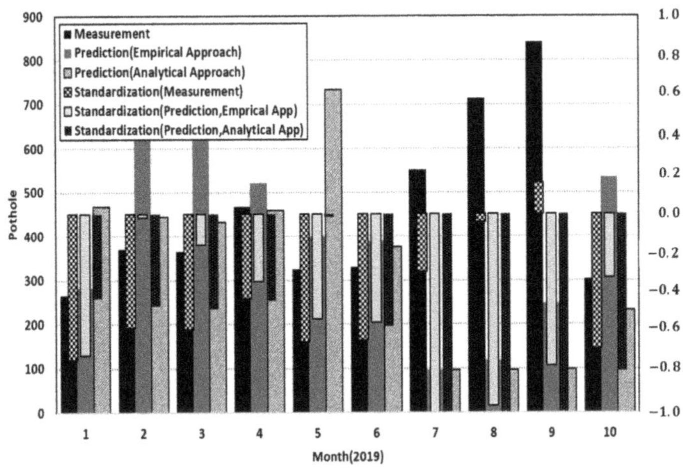

(a)

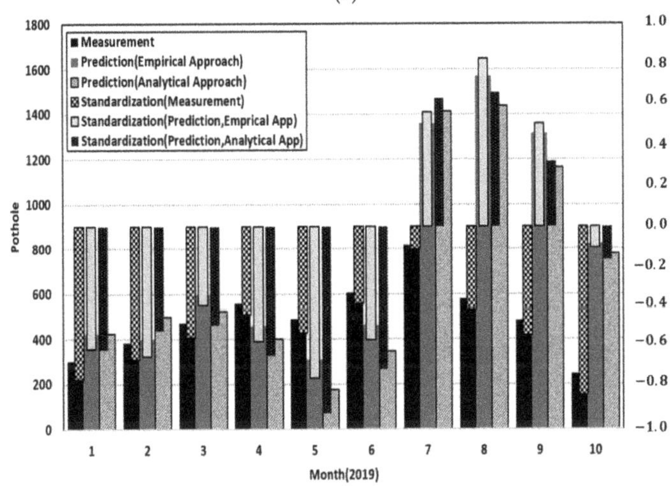

(b)

Figure 4. Assessment of predictive model performance for NH19 and NH23. (**a**) Prediction results from NH19 roadway zone. (**b**) Prediction results from NH23 roadway zone.

For the NH19 district, both empirical and analytical approaches demonstrate competitive results, with slight variations observed in RMSE, MAE, and MSE values. However, the R^2 values indicate better explanatory power for the analytical approach, suggesting a stronger fit to the data. The corresponding *p*-values underscore the statistical significance of the models, further validating their reliability.

Similarly, for the NH23 district, both approaches yield comparable results in terms of RMSE, MAE, and MSE. However, the analytical approach exhibits higher R^2 values, indicating superior predictive performance. Additionally, the significantly low *p*-values emphasize the statistical significance of the analytical model in capturing the underlying patterns of reflective cracking occurrence.

In summary, both empirical and analytical approaches demonstrate promising performance in predicting reflective cracking in the NH19 and NH23 districts. However, the analytical approach appears to offer slightly better predictive accuracy and statistical significance, particularly evident in the R^2 values and *p*-values. These findings highlight the importance of considering both methodological approaches and standardization techniques in developing robust predictive models for reflective cracking.

3.1.3. Limitations

Although our finding has provided valuable insights into predictive modeling for pavement reflective cracking using multilinear regression, it is important to acknowledge its limitations. One notable limitation is the exclusive reliance on polynomial functions within the regression model, neglecting the potential benefits of non-polynomial functions. Non-polynomial functions, particularly in the context of mechanics of cracked media, offer a more nuanced representation of complex phenomena, including possible singularities that may arise in pavement degradation processes. Therefore, in the next stage of the research, the aim is to address this limitation by considering non-polynomial functions and leveraging insights from this method [56]. This expansion of modeling techniques will enable a more comprehensive understanding of reflective cracking mechanisms and pave the way for enhanced predictive accuracy and robustness in pavement management practices.

3.2. Reflective Cracking Categorization

Table 9 presents the average precision scores obtained from different types of datasets for reflective cracking categorization. Two machine learning methods, including image classification, were employed across various pavement types to assess their precision rates under distinct conditions. For concrete pavement, image classification achieved an average precision of 95.9% under clear weather conditions and 91.4% under multiple weather scenarios. Similarly, for asphalt pavement, the average precision was recorded at 92.7% under clear weather conditions and 82.6% under multiple weather conditions. These precision scores provide insights into the effectiveness of image classification techniques in accurately categorizing reflective cracking across different pavement types and weather conditions.

Table 9. The accuracy results of reflective cracking categorization.

No.	Machine Learning Methods	Pavement Types	Precision Clear (%)	Multiple Weather (%)
1	Image classification	Concrete pavement	95.9	91.4
2	Image classification	Asphalt pavement	92.7	82.6

In concrete pavement, the detection effectiveness of reflective cracking is typically better due to several factors. One key reason is the stark contrast between the cracks and the concrete background. Cracks in concrete pavement often appear as dark lines against a light or white background, making them easier to detect using image processing techniques. The high contrast between the cracks and the background facilitates the segmentation of cracks from the surrounding pavement surface, leading to more accurate detection.

Additionally, concrete pavement tends to have a smoother surface texture compared to asphalt pavement, which can further aid in the detection process. The relatively uniform texture of concrete makes it easier to distinguish cracks from other surface irregularities, reducing the likelihood of false positives in the detection results. Moreover, concrete pavement is often characterized by its durability and resistance to deformation, which can result in more distinct and well-defined cracks compared to asphalt pavement. These well-defined cracks provide clearer visual cues for automated detection algorithms, improving the overall effectiveness of crack detection systems.

3.3. Reflective Cracking Segmentations

3.3.1. Overall Results

The results validate the effectiveness of detecting reflective cracking on roadways using the Image Segmentation technique, as illustrated in Figure 5. The method developed for detecting reflective cracking with Mask R-CNN performs satisfactorily, effectively differentiating between good pavement and reflective cracking pavement.

(**a**)

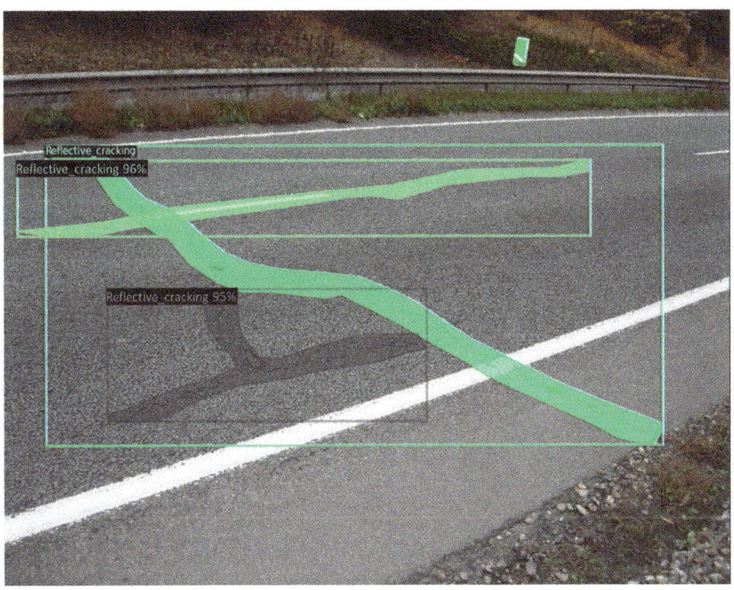

(**b**)

Figure 5. *Cont.*

(c)

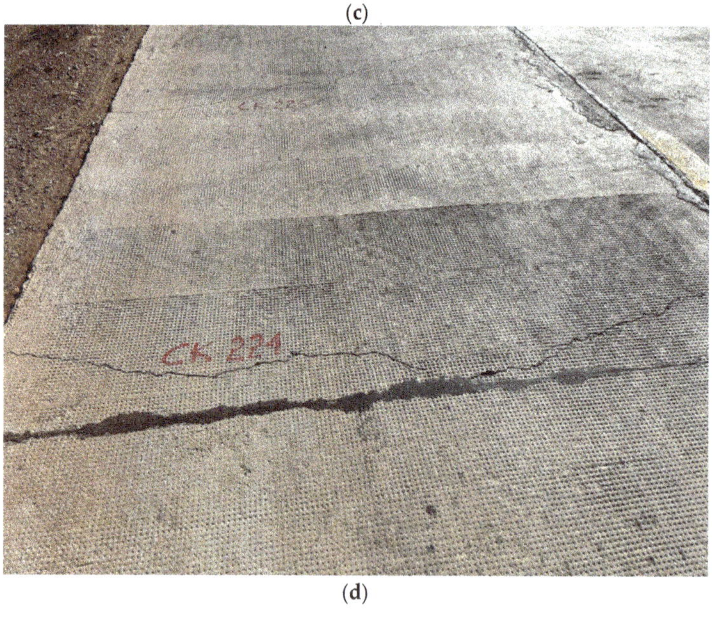

(d)

Figure 5. *Cont.*

(e)

Figure 5. Results from the trained model for detecting reflective cracking. (**a**) Original photo of reflective cracking on urban road; (**b**) Classified photo of reflective cracking on urban road; (**c**) Original vs. classified photo of reflective cracking on a rural road; (**d**) Original photo of reflective cracking on concrete road; (**e**) Classified photo of reflective cracking on concrete road.

In the examples provided, the detection performance for reflective cracking is impressive on both types of pavements, with accuracy exceeding 90% in most cases. However, there is a notable distinction in detection effectiveness between concrete and asphalt pavements. On concrete pavement, the detection score approaches nearly 99%, showcasing a superior capability to accurately identify reflective cracks. Conversely, on asphalt pavement, both in rural and urban settings, the detection rates hover around 96%, indicating a slightly lower but still commendable performance. This discrepancy suggests that while detection remains highly effective on both pavement types, the inherent characteristics of concrete pavement, such as its smoother texture and the high contrast between cracks and the surface, contribute to its exceptional detection capabilities.

3.3.2. Weather Impact on Training Efficiency

Training scenarios involving loss and precision in relation to various climate variables are illustrated in Figure 6. Overall, the proposed model for reflective cracking detection demonstrates satisfactory efficiency indicators. As a detection approach, Mask R-CNN performs well, consistently achieving a total loss of less than 0.3 and a precision greater than 0.9. Convergence typically occurs after 2000 iterations, except for databases containing diverse meteorological conditions, which require additional training up to 4000 iterations.

The impact of climate information on training effectiveness is significant; introducing multiple meteorological photographs is particularly impactful, with larger databases correlating with lower results. This outcome highlights the negative effect of varied weather conditions on accuracy, aligning with other findings in image classification. The performance variance between datasets containing only clear weather photographs and those with all meteorological types underscores the crucial role that climate parameters play in successful training. This can be attributed to the similar characteristics of reflective cracking and other surface irregularities, especially in cloudy and rainy weather.

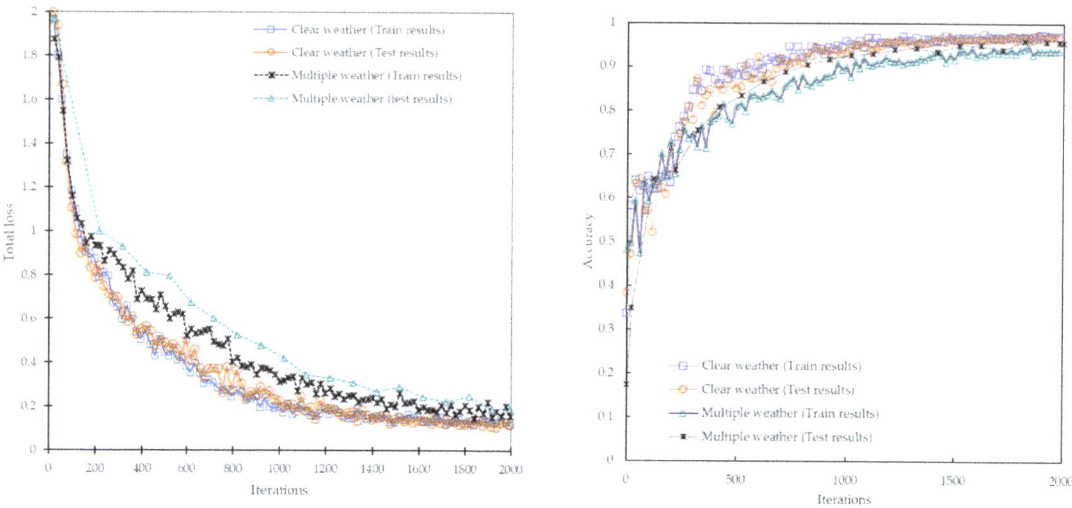

(**a**) (**b**)

Figure 6. Bounding box-based object identification method training outcomes are shown as (**a**) Total loss versus Iterations and (**b**) Mask-RCNN accuracy versus Iterations.

The model's efficacy is validated by the test dataset evaluation, with all testing groups achieving an accuracy of more than 80% using Mask R-CNN. Despite a slight discrepancy in performance between the training and test datasets—possibly due to fluctuations in surrounding conditions—the model performs consistently in all circumstances.

Even with a few small errors, the results support the applicability of this method for computerized reflective cracking assessment on roads. There were times when overestimations of reflective cracking severity occurred, especially in the identification of climatic conditions. Expanding the quantity of the training dataset may improve the classifier's accuracy and flexibility for future uses. In addition, a comprehensive analysis of the model's parameters was carried out to maximize performance. Using a mini-batch size of four, the loss function was minimized iteratively throughout learning in order to guarantee thorough convergence and avoid excessive fitting. The learning rate of 0.00025, the momentum of 0.8, the regularization of 0.0001, and the mini-batch size of 4 are the optimum model parameters.

3.3.3. Results of Average Precision at 50%

Table 10 provides an overview of the segmentation accuracy used to identify reflective cracking. The findings show that, out of the two categories, the second circumstance has the least accuracy, while the "Clear settings" have the most reliability. For example, the mean accuracy for the first and second circumstances is 92.5% and 83.7%, respectively, for IOU = 0.5 (AP50). These results imply that different contexts may have an effect on reflective cracking identification by picture segmentation. Furthermore, the incorporation of concrete pavement brings uncertainty to the accuracy of the model.

Since the segmentation may mistakenly see reflected cracking as a pattern on the pavement surface, the "black" color of asphalt pavement could lead to errors throughout training. Once reflective cracking occurs throughout the training period, it could be mistaken for the black pattern on the road surface. These findings demonstrate how difficult it can be to recognize reflective cracking in practical environments.

Table 10. AP50 results.

No.	Pavement Types	AP50: the Average Precision at IOU = 0.5	
		Combined Weather Data	
		1st Cond. Clear	2nd Cond. Multiple Weathers
1	Concrete pavement	92.5%	83.7%
2	Asphalt pavement and Concrete pavement	89.1%	80.3%

3.4. Architecture Comparison

Table 11 provides a comprehensive comparison of performance metrics across multiple architectures utilized for the pavement damage detection task. The evaluated architectures include Mask R-CNN, Yolov8, Yolov5, and Yolov4. Notably, the table showcases the time required per iteration for each architecture, with Mask R-CNN having a time of 0.92 s, while Yolov8 demonstrates the most efficient performance with a reduced time requirement of 0.75 s. Furthermore, the table presents the average precision at 50% Intersection over Union (AP50) on clear weather conditions, specifically on asphalt pavement. Mask R-CNN achieves an impressive AP50 score of 92.5%, closely followed by Yolov8 with 91.3%. These findings underscore the efficiency and accuracy of different architectures in the context of pavement damage detection, offering valuable insights for practitioners and researchers in the field.

Table 11. Comparisons of performance across several architectures.

	Mask R-CNN	Yolov8	Yolov5	Yolov4
Time needed per iteration (in seconds)	0.92	0.75	0.81	0.87
AP50 on clear weather	92.5%	91.3%	87.8%	83.4%

3.5. Challenges in Implementation

During the experimental process of implementing the proposed method, several challenges surfaced, reflecting the complex nature of pavement management and reflective cracking detection. One significant obstacle pertained to the variability of environmental conditions and pavement types across different regions. The diverse weather patterns and road surface characteristics posed challenges in training and validating the models effectively. To address this, a rigorous approach to data collection was adopted, incorporating a wide range of weather data and pavement images from various sources. Additionally, fine-tuning the deep learning algorithms required iterative experimentation and parameter optimization to ensure robust performance across different scenarios. Moreover, the interpretability of the models and the potential biases in the training data presented ongoing challenges in achieving generalizability and reliability. To mitigate these issues, comprehensive sensitivity analyses and model validations were conducted, leveraging cross-validation techniques and external validation datasets where possible. Despite these challenges, the study underscores the importance of continuous refinement and validation of predictive models in real-world applications, paving the way for more accurate and reliable pavement management strategies in the future.

4. Conclusions

This study investigates predictive and detection methods for reflective cracking in pavement infrastructure, combining machine learning and advanced image detection techniques. By employing algorithms such as linear regression and Mask R-CNN, predictive models and precise detection methods were developed. These approaches aim to enhance pavement management systems by enabling automatic monitoring and assessment of pavement conditions, ultimately improving maintenance strategies.

- The refined multilinear regression model exhibited improved predictive performance for reflective cracking occurrences. By integrating data from weather, traffic volume, and reflective cracking surveys spanning 2014 to 2018, the model achieved heightened accuracy. Standardization of variables was crucial for accuracy enhancement, particularly given the diverse ranges in traffic and temperature. Comparative analysis between analytical and empirical approaches further validated the model's efficacy, as it successfully forecasted reflective cracking numbers for 2019, a period not included in its initial training.
- The comprehensive evaluation of prediction models highlights the competitive performance of both empirical and analytical approaches across the NH19 and NH23 districts. For the NH19 district, the empirical approach yielded RMSE, MAE, and MSE values of 0.5355, 0.4217, and 0.2867, respectively, while the analytical approach demonstrated slightly improved values of 0.5743, 0.4182, and 0.3299, indicating a stronger fit to the data.
- The findings highlight the effectiveness of image classification techniques in categorizing reflective cracking across various pavement types and weather conditions, as evidenced by precision rates derived from extensive datasets. Notably, for concrete pavement, image classification achieved remarkable precision rates of 95.9% under clear weather and 91.4% under various weather scenarios. Conversely, for asphalt pavement, slightly lower but still impressive average precision scores of 92.7% under clear weather and 82.6% under multiple weather conditions were attained. Concrete pavement's superior detection effectiveness can be attributed to several factors, including its high contrast with cracks against the background, smoother surface texture aiding segmentation, and durability leading to well-defined cracks.
- The Mask R-CNN model showed strong performance in detecting reflective cracking, maintaining a total loss below 0.3 and a precision above 0.9. While convergence typically happens after 2000 iterations, datasets with varied weather conditions may require up to 4000 iterations for optimal training. The impact of climate data on training is significant, with diverse weather conditions correlating with lower results. The model's effectiveness was validated by achieving over 80% accuracy in all testing scenarios, despite slight performance variations, indicating its reliability across different conditions.
- The AP50 results illustrate segmentation accuracy for reflective cracking identification, revealing varied reliability across contexts. "Clear settings" exhibited the highest reliability, while the "multiple weather" scenario displayed the least reliability. Mean accuracies for these conditions were 94.7% and 82.4%, respectively, for IOU = 0.5 (AP50). The integration of the "black" color of asphalt pavement poses challenges, potentially leading to mistaken identification. These findings underscore the practical difficulty in recognizing reflective cracking.
- Mask R-CNN and Yolov8 exhibited top performance in pavement damage detection, with AP50 scores of 92.5% and 91.3%, respectively, under clear weather conditions for asphalt pavement.

Author Contributions: S.-P.S. and T.H.M.L.: conceptualization, methodology, writing—original draft. K.K., S.-P.S. and T.H.M.L.: visualization, investigation, writing—review, and editing. S.-P.S. and T.H.M.L.: data curation, software. All authors have read and agreed to the published version of the manuscript.

Funding: This research received no external funding.

Data Availability Statement: The original contributions presented in the study are included in the article, further inquiries can be directed to the corresponding author.

Conflicts of Interest: The authors declare no conflicts of interest, financial or otherwise.

References

1. Dhakal, N.; Elseifi, M.A.; Zhang, Z. Mitigation Strategies for Reflection Cracking in Rehabilitated Pavements—A Synthesis. *Int. J. Pavement Res. Technol.* **2016**, *9*, 228–239. [CrossRef]
2. Tam, A.B.; Park, D.W.; Le, T.H.M.; Kim, J.S. Evaluation on Fatigue Cracking Resistance of Fiber Grid Reinforced Asphalt Concrete with Reflection Cracking Rate Computation. *Constr. Build. Mater.* **2020**, *239*, 117873. [CrossRef]
3. Fallah, S.; Khodaii, A. Reinforcing Overlay to Reduce Reflection Cracking; An Experimental Investigation. *Geotext. Geomembr.* **2015**, *43*, 216–227. [CrossRef]
4. Ji, R.Y.; Mandal, T.; Yin, H. Laboratory Characterization of Temperature Induced Reflection Cracks. *J. Traffic Transp. Eng.* **2020**, *7*, 668–677. [CrossRef]
5. Doh, Y.S.; Baek, S.H.; Kim, K.W. Estimation of Relative Performance of Reinforced Overlaid Asphalt Concretes against Reflection Cracking Due to Bending More Fracture. *Constr. Build. Mater.* **2009**, *23*, 1803–1807. [CrossRef]
6. Khodaii, A.; Fallah, S.; Moghadas Nejad, F. Effects of Geosynthetics on Reduction of Reflection Cracking in Asphalt Overlays. *Geotext. Geomembr.* **2009**, *27*, 1–8. [CrossRef]
7. Chen, Y.; Zhu, Z.; Lin, Z.; Zhou, Y. Building Surface Crack Detection Using Deep Learning Technology. *Buildings* **2023**, *13*, 1814. [CrossRef]
8. Xi, Y.F.; Ren, S.J.; Chen, B.L.; Yang, B.; Lee, J.; Zhu, G.H.; Zhou, T.C.; Xu, H. Application of Steel-Fiber-Reinforced Self-Stressing Concrete in Prefabricated Pavement Joints. *Buildings* **2023**, *13*, 2129. [CrossRef]
9. Lu, X.; Yan, G. A Quasi-2D Exploration of Mixed-Mode Fracture Propagation in Concrete Semi-Circular Chevron-Notched Disks. *Buildings* **2023**, *13*, 2633. [CrossRef]
10. Di Mascio, P.; Moretti, L. Implementation of a Pavement Management System for Maintenance and Rehabilitation of Airport Surfaces. *Case Stud. Constr. Mater.* **2019**, *11*, e00251. [CrossRef]
11. Zhao, Y.; Goulias, D.; Peterson, D. Recycled Asphalt Pavement Materials in Transport Pavement Infrastructure: Sustainability Analysis & Metrics. *Sustainability* **2021**, *13*, 8071. [CrossRef]
12. Moradi, M.; Assaf, G.J. Building an Augmented Reality Experience on Top of a Smart Pavement Management System. *Buildings* **2022**, *12*, 1915. [CrossRef]
13. Shu, X.; Wang, Z.; Basheer, I.A. Large-Scale Evaluation of Pavement Performance Models Utilizing Automated Pavement Condition Survey Data. *Int. J. Transp. Sci. Technol.* **2022**, *11*, 678–689. [CrossRef]
14. Justo-Silva, R.; Ferreira, A.; Flintsch, G. Review on Machine Learning Techniques for Developing Pavement Performance Prediction Models. *Sustainability* **2021**, *13*, 5248. [CrossRef]
15. Pei, L.; Sun, Z.; Xiao, L.; Li, W.; Sun, J.; Zhang, H. Virtual Generation of Pavement Crack Images Based on Improved Deep Convolutional Generative Adversarial Network. *Eng. Appl. Artif. Intell.* **2021**, *104*, 104376. [CrossRef]
16. Que, Y.; Dai, Y.; Ji, X.; Kwan Leung, A.; Chen, Z.; Tang, Y.; Jiang, Z. Automatic Classification of Asphalt Pavement Cracks Using a Novel Integrated Generative Adversarial Networks and Improved VGG Model. *Eng. Struct.* **2023**, *277*, 115406. [CrossRef]
17. Patel, T.; Guo, B.H.W.; van der Walt, J.D.; Zou, Y. Effective Motion Sensors and Deep Learning Techniques for Unmanned Ground Vehicle (UGV)-Based Automated Pavement Layer Change Detection in Road Construction. *Buildings* **2023**, *13*, 5. [CrossRef]
18. Tabatabai, H.; Aljuboori, M. A Novel Concrete-Based Sensor for Detection of Ice and Water on Roads and Bridges. *Sensors* **2017**, *17*, 2912. [CrossRef]
19. Rhee, J.Y.; Park, K.T.; Cho, J.W.; Lee, S.Y. A Study of the Application and the Limitations of Gpr Investigation on Underground Survey of the Korean Expressways. *Remote Sens.* **2021**, *13*, 1805. [CrossRef]
20. Ramanna, S.; Sengoz, C.; Kehler, S.; Pham, D. Near Real-Time Map Building with Multi-Class Image Set Labeling and Classification of Road Conditions Using Convolutional Neural Networks. *Appl. Artif. Intell.* **2021**, *35*, 803–833. [CrossRef]
21. Zhang, L.; Yang, F.; Daniel Zhang, Y.; Zhu, Y.J. Road Crack Detection Using Deep Convolutional Neural Network. In Proceedings of the 2016 IEEE International Conference on Image Processing—ICIP, Phoenix, AZ, USA, 25–28 September 2016; pp. 3708–3712. [CrossRef]
22. Lee, S.Y.; Le, T.H.M.; Kim, Y.M. Prediction and Detection of Potholes in Urban Roads: Machine Learning and Deep Learning Based Image Segmentation Approaches. *Dev. Built Environ.* **2023**, *13*, 100109. [CrossRef]
23. Dong, C.Z.; Catbas, F.N. A Review of Computer Vision–Based Structural Health Monitoring at Local and Global Levels. *Struct. Health Monit.* **2021**, *20*, 692–743. [CrossRef]
24. Harrou, F.; Zeroual, A.; Hittawe, M.M.; Sun, Y. Recurrent and Convolutional Neural Networks for Traffic Management. *Road Traffic Model. Manag.* **2022**, 197–246. [CrossRef]
25. Harrou, F.; Zeroual, A.; Hittawe, M.M.; Sun, Y. *Road Traffic Modeling and Management*; Elsevier: Amsterdam, The Netherlands, 2022. [CrossRef]
26. He, K.; Zhang, X.; Ren, S.; Sun, J. Deep Residual Learning for Image Recognition. In Proceedings of the IEEE Conference on Computer Vision and Pattern Recognition, Las Vegas, NV, USA, 27–30 June 2016; pp. 770–778. [CrossRef]

27. Hittawe, M.M.; Langodan, S.; Beya, O.; Hoteit, I.; Knio, O. Efficient SST Prediction in the Red Sea Using Hybrid Deep Learning-Based Approach. In Proceedings of the 2022 IEEE 20th International Conference on Industrial Informatics (INDIN), Perth, Australia, 25–28 July 2022; pp. 107–114. [CrossRef]
28. Hittawe, M.M.; Sidibé, D.; Beya, O.; Mériaudeau, F. Machine Vision for Timber Grading Singularities Detection and Applications. *J. Electron. Imaging* **2017**, *26*, 063015. [CrossRef]
29. Zhang, J.; Qian, S.; Tan, C. Automated Bridge Surface Crack Detection and Segmentation Using Computer Vision-Based Deep Learning Model. *Eng. Appl. Artif. Intell.* **2022**, *115*, 105225. [CrossRef]
30. Xu, Y.; Zhou, Y.; Sekula, P.; Ding, L. Machine Learning in Construction: From Shallow to Deep Learning. *Dev. Built Environ.* **2021**, *6*, 100045. [CrossRef]
31. Bochkovskiy, A.; Wang, C.-Y.; Liao, H.-Y.M. YOLOv4: Optimal Speed and Accuracy of Object Detection. *arXiv* **2020**, arXiv:2004.10934.
32. Li, C.; Li, L.; Jiang, H.; Weng, K.; Geng, Y.; Li, L.; Ke, Z.; Li, Q.; Cheng, M.; Nie, W.; et al. YOLOv6: A Single-Stage Object Detection Framework for Industrial Applications. *arXiv* **2022**, arXiv:2209.02976.
33. Li, C.; Li, L.; Geng, Y.; Jiang, H.; Cheng, M.; Zhang, B.; Ke, Z.; Xu, X.; Chu, X. YOLOv6 v3.0: A Full-Scale Reloading. *arXiv* **2023**, arXiv:2301.05586.
34. Wang, C.-Y.; Bochkovskiy, A.; Liao, H.-Y.M. YOLOv7: Trainable Bag-of-Freebies Sets New State-of-the-Art for Real-Time Object Detectors. In Proceedings of the IEEE/CVF Conference on Computer Vision and Pattern Recognition, New Orleans, LA, USA, 18–24 June 2022.
35. He, K.; Gkioxari, G.; Dollár, P.; Girshick, R. Mask R-CNN. In Proceedings of the IEEE International Conference on Computer Vision, Venice, Italy, 22–29 October 2017.
36. Ansari, S.; Rennie, C.D.; Clark, S.P.; Seidou, O. IceMaskNet: River Ice Detection and Characterization Using Deep Learning Algorithms Applied to Aerial Photography. *Cold Reg. Sci. Technol.* **2021**, *189*, 103324. [CrossRef]
37. Zhang, H.; Qian, Z.; Tan, Y.; Xie, Y.; Li, M. Investigation of Pavement Crack Detection Based on Deep Learning Method Using Weakly Supervised Instance Segmentation Framework. *Constr. Build. Mater.* **2022**, *358*, 129117. [CrossRef]
38. Liu, F.; Liu, J.; Wang, L.; Al-Qadi, I.L. Multiple-Type Distress Detection in Asphalt Concrete Pavement Using Infrared Thermography and Deep Learning. *Autom. Constr.* **2024**, *161*, 105355. [CrossRef]
39. Baduge, S.K.; Thilakarathna, S.; Perera, J.S.; Ruwanpathirana, G.P.; Doyle, L.; Duckett, M.; Lee, J.; Saenda, J.; Mendis, P. Assessment of Crack Severity of Asphalt Pavements Using Deep Learning Algorithms and Geospatial System. *Constr. Build. Mater.* **2023**, *401*, 132684. [CrossRef]
40. Xiong, X.; Meng, A.; Lu, J.; Tan, Y.; Chen, B.; Tang, J.; Zhang, C.; Xiao, S.; Hu, J. Automatic Detection and Location of Pavement Internal Distresses from Ground Penetrating Radar Images Based on Deep Learning. *Constr. Build. Mater.* **2024**, *411*, 134483. [CrossRef]
41. Al-Huda, Z.; Peng, B.; Algburi, R.N.A.; Al-antari, M.A.; AL-Jarazi, R.; Zhai, D. A Hybrid Deep Learning Pavement Crack Semantic Segmentation. *Eng. Appl. Artif. Intell.* **2023**, *122*, 106142. [CrossRef]
42. Song, Q.; Liu, L.; Lu, N.; Zhang, Y.; Muniyandi, R.C.; An, Y. A Three-Stage Pavement Image Crack Detection Framework with Positive Sample Augmentation. *Eng. Appl. Artif. Intell.* **2024**, *129*, 107624. [CrossRef]
43. Ounpraseuth, S.T. Gaussian Processes for Machine Learning. *J. Am. Stat. Assoc.* **2008**, *103*, 429. [CrossRef]
44. Shai, S.; Shai, B. *Understanding Machine Learning from Theory to Algorithm*; Cambridge University: Cambridge, UK, 2014.
45. Ali, R.; Chuah, J.H.; Talip, M.S.A.; Mokhtar, N.; Shoaib, M.A. Crack Segmentation Network Using Additive Attention Gate—CSN-II. *Eng. Appl. Artif. Intell.* **2022**, *114*, 105130. [CrossRef]
46. Vishwakarma, R.; Vennelakanti, R. CNN Model Tuning for Global Road Damage Detection. In Proceedings of the 2020 IEEE International Conference on Big Data, Atlanta, GA, USA, 10–13 December 2020; pp. 5609–5615. [CrossRef]
47. Pham, V.; Pham, C.; Dang, T. Road Damage Detection and Classification with Detectron2 and Faster R-CNN. In Proceedings of the 2020 IEEE International Conference on Big Data, Atlanta, GA, USA, 10–13 December 2020; pp. 5592–5601. [CrossRef]
48. Wu, Y.; Kirillov, A.; Massa, F.; Lo, W.; Girshick, R. Detectron2. Available online: https://github.com/facebookresearch/detectron2 (accessed on 15 May 2024).
49. Mokhtar, M.M.; Morsy, M.; Taha, N.A.; Ahmed, E.M. Investigating the Mechanical Performance of Nano Additives Reinforced High-Performance Concrete. *Constr. Build. Mater.* **2022**, *320*, 125537. [CrossRef]
50. Buslaev, A.; Iglovikov, V.I.; Khvedchenya, E.; Parinov, A.; Druzhinin, M.; Kalinin, A.A. Albumentations: Fast and Flexible Image Augmentations. *Information* **2020**, *11*, 125. [CrossRef]
51. Lee, S.Y.; Jeon, J.S.; Le, T.H.M. Feasibility of Automated Black Ice Segmentation in Various Climate Conditions Using Deep Learning. *Buildings* **2023**, *13*, 767. [CrossRef]
52. Singh, J.; Shekhar, S. Road Damage Detection and Classification in Smartphone Captured Images Using Mask R-CNN. *arXiv* **2018**, arXiv:1811.04535.
53. Ultralytics Ultralytics YOLOv8. Available online: https://docs.ultralytics.com/vi (accessed on 12 June 2024).
54. Zhang, Z.; Sabuncu, M.R. Generalized Cross Entropy Loss for Noisy Labels. Available online: https://neurips.cc/media/nips-2018/Slides/12761.pdf (accessed on 12 June 2024).

55. Nar, K.; Ocal, O.; Sastry, S.S.; Ramchandran, K. Cross-Entropy Loss and Low-Rank Features Have Responsibility for Adversarial Examples. *arXiv* **2019**, arXiv:1901.08360.
56. Figiel, Ł.; Kamiński, M. Numerical Probabilistic Approach to Sensitivity Analysis in a Fatigue Delamination Problem of a Two Layer Composite. *Appl. Math. Comput.* **2009**, *209*, 75–90. [CrossRef]

buildings

MDPI

Article

The Effect of Refined Separation on the Properties of Reclaimed Asphalt Pavement Materials

Yuanfu Wang [1], Guangwu Dou [2], Sudi Wang [2,*] and Jie Wang [2,*]

1 Enforcement Detachment of Project Quality Supervision and Law Enforcement of Transportation Comprehensive Administrative Law Enforcement Bureau of Xinjiang Uygur Autonomous Region, Urumqi 830000, China
2 Research Institute of Highway Ministry of Transport, Beijing 100088, China
* Correspondence: sd.wang@rioh.cn (S.W.); j.wang@rioh.cn (J.W.)

Abstract: Refined separation not only controls the variability of reclaimed asphalt pavement (RAP), but also improves the mixing ratio of RAP and the quality of recycled asphalt mixtures. This study examines RAP treated with various refined separation frequency parameters, analyzes the variation rules and the variability of RAP aggregate gradation, asphalt content, asphalt properties, and aggregate properties, and calculates the maximum mixing percentage of coarse RAP material by using the gradation variability control method and the asphalt content variability control method. The results show that the variability of gradation and asphalt content of coarsely separated RAP is considerable, and a refined separation process significantly reduces the variability of gradation and asphalt content of RAP; the agglomeration of RAP decreases with an increase in the refined separation frequency; and the RAP agglomeration of three kinds of RAPs (E1, E2, and E3) under a refined separation frequency of 55 Hz reduces by 6.40%, 4.30%, and 4.30%, respectively, as compared with that of coarsely separated RAPs. The asphalt content of the refined separation RAP gradually decreases with an increase in frequency, and the asphalt content of E1 and E2 (55 Hz) was only 0.95% and 1.10%, respectively. The maximum percentage of RAP in recycled asphalt mixtures was calculated using the gradation variability control method and the asphalt content variability control method, respectively. The maximum proportions of RAP were 45% and 33% for A1 (0 Hz), respectively, and the maximum proportions of RAP for E1 (55 Hz) were all 100%. The results of the two methods show that the process of refined separation can increase the maximum proportion of blended RAP materials. They also demonstrate that the refined separation process can increase the maximum blending ratio of coarse RAP materials, thereby improving the quality of the RAP, increasing the proportion of RAP blending, and ensuring the quality of the recycled asphalt mixture. In conclusion, the refined separation process holds promise for maximizing the potential value of RAP and optimizing its recycling, environmental, and economic benefits.

Keywords: reclaimed asphalt pavement; refined separation; gradation variability; aggregate property; asphalt property

Citation: Wang, Y.; Dou, G.; Wang, S.; Wang, J. The Effect of Refined Separation on the Properties of Reclaimed Asphalt Pavement Materials. *Buildings* **2024**, *14*, 1608. https://doi.org/10.3390/buildings14061608

Academic Editors: Bjorn Birgisson, Andrea Baliello and Di Wang

Received: 27 March 2024
Revised: 23 May 2024
Accepted: 27 May 2024
Published: 1 June 2024

1. Introduction

Reclaimed asphalt pavement (RAP) recycling has become an important development direction for highway transportation [1–3]. However, due to the large number of "false-particles" in RAP, there is a high degree of variability in its gradation and asphalt content. The variability of the properties of RAP limits how much of it can be added to recycled materials and poses a challenge for its full utilization [4–7]. In order to ensure the quality of recycled asphalt mixtures, the general plant-mixed thermal regeneration requires a ratio of RAP addition of 15% to 30% [8–10]. For this reason, researchers have proposed asphalt and mineral separation techniques, including physical refined separation methods [11–13], chemical solvent separation methods [13,14], and microbial separation methods [15]. Refined separation methods have been emphasized by the industry due to their advantages of high separation efficiency, low environmental impact, and low separation cost [13,16–20].

The variability of RAP materials seriously affects the stability of the performance of hot recycled asphalt mixtures [21], which is one of the main reasons for limiting the maximum addition ratio of RAP. Montañez [22] found that there is significant variability in the mechanical response and performance of the evaluated RAPs, suggesting the necessity of taking into account the variability of RAP when fabricating asphalt mixtures with RAP obtained from different sources. Liu et al. [23] proposed the 12.5 mm particle size as the key sieve size for coarse and fine particles of RAP materials, which can significantly reduce the old asphalt content of RAP materials and the variability of aggregate gradation. Guo [24] found that by increasing the percentage of material passing through a 4.75 mm sieve in the gradation of hot recycled asphalt mixtures, the effect of internal friction angle on the shear strength of the mixture can be reduced, thus effectively improving the negative impact of the variability of RAP gradation on the performance of hot recycled asphalt mixtures. Some researchers focus on binder activation within RAP [25–28]. Preheating temperature and RAP source were generally considered in previous studies, and both of them were found to have a significant impact on the activation of RAP binders [25,29,30]. Mechanical separation only changes the asphalt content of RAP without changing the binder properties of RAP. Therefore, the effect of mechanical separation on the binder activation of RAP materials can be disregarded in related studies.

Numerous studies have shown that RAP pretreatment techniques can effectively reduce its variability [31–33], while refined separation techniques can further improve its utilization in recycled asphalt mixtures [34,35]. Zou et al. [36] used the refined separation process to deal with RAP, and it was found that the asphalt content of the coarse RAP after refined separation was less than 1%. Through refined separation, a large number of false particles are eliminated, and the gradation stability of the recycled asphalt mixture is Improved. Qiu et al. [37] utilized the refined separation process on RAP obtained from a highway expansion project. The results showed that the performances of a recycled asphalt mixture with 80% RAP and a mixture of new asphalt are comparable. Wang et al. [38] also found that the performance of recycled asphalt mixtures is basically the same as new asphalt mixtures. Road performance research based on recycled asphalt mixtures shows that the refined separation process is conducive to enhancing the proportion of RAP blending, improving the quality and stability of recycled asphalt mixtures.

Existing research focuses more on the performance of recycled asphalt mixtures with refined separation RAP, but less on the evaluation of the quality of refined separation RAP. The quality of RAP after refined separation is the key process of recycled asphalt mixtures, which directly affects the final road performance and the durability of recycled asphalt mixtures. Therefore, in order to ensure the quality of recycled asphalt mixtures, it is very necessary to carry out a systematic evaluation study of RAP produced by on-site refined separation equipment.

This paper focuses on the effect of refined separation on the properties of reclaimed asphalt pavement materials. The variability of RAP aggregate gradation, aggregate properties, asphalt content, and asphalt properties after refined separation were analyzed. Meanwhile, the effects of different refined separation frequencies on the properties of RAP materials were quantitatively evaluated. The method of calculating the maximum proportion of coarse RAP that can be added to a recycled asphalt mixture is provided in this study, and the maximum proportion of RAP that can be added after refined separation was calculated by the method of controlling the variability of gradation and the method of controlling the variability of asphalt content, respectively. Its findings will significantly contribute to maximizing the utilization of RAP materials and optimizing recycled asphalt mixture design.

2. Refined Separation Principle

The refined separation equipment comprises several sub-systems designed for RAP refined separation functionality. It includes the feeding system, screening system, centrifugal impact crusher, roller crusher, dust removal system, discharging system, etc. The refined separation equipment and principle are shown in Figure 1. The core of the equipment is the centrifugal impact crusher (Figure 1); RAP enters the centrifugal impact crusher, it obtains kinetic energy from the high-speed rotor, and it is thrown out. The thrown RAP collides and rubs violently against the crusher's baffle and falling RAPs. The RAP "false-particles" along the asphalt interface between the aggregates are crushed, so that the coarse RAP (≥5 mm) and the fine RAP (0–5 mm) are broken. The coarse RAP (≥5 mm) is separated from the fine RAP (0–5 mm). The surface of coarse RAP contains little asphalt due to collision and friction, which can realize the separation of asphalt and coarse RAP aggregate. The effect of refined separation of RAP is related to the rotational speed of the crusher, the nature of the RAP and its temperature, and the position of the crusher plate, of which the rotational speed of the crusher is the key influence factor. This test is based on the engineering site; it is difficult to change the position of the crusher plate and the nature and temperature of RAP. Therefore, the rotor speed was adjusted by changing the crusher frequency, which can change the refined separation effect for RAP. The ambient temperature during this test ranges from 23 to 28 °C, and the RAP used is the modified asphalt mixture from the middle surface layer on the highway.

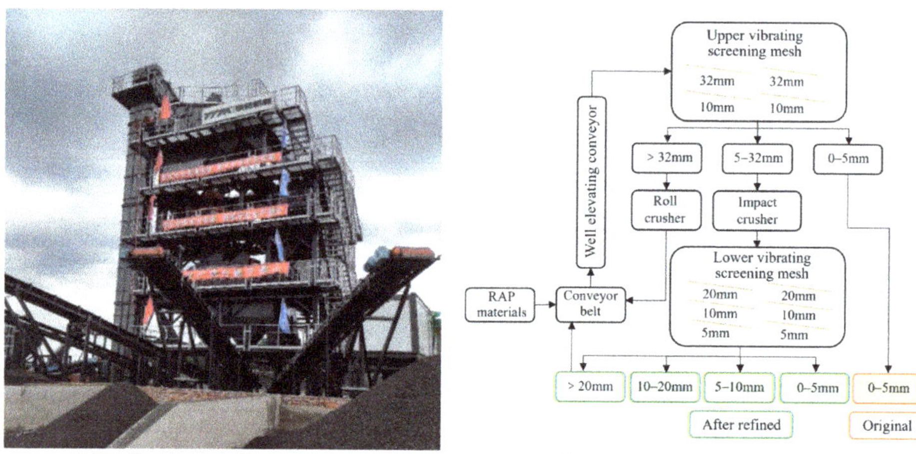

(**a**) Refined separation equipment (**b**) Refined separation principle

Figure 1. Refined separation principle of RAP materials.

3. Materials and Methods

3.1. RAP Materials

A 120 t/h processing capacity refined separation equipment was used to finely separate RAP into 3 grades (0–5 mm, 5–10 mm, 10–20 mm) at different frequencies (0 Hz (original material), 30 Hz, 40 Hz, 50 Hz, 55 Hz). The selected RAP sample is asphalt pavement surface material, which is milled on-site in Beijing, and the old asphalt in the RAP material is a mixture of modified bitumen and plain bitumen. The RAP sample numbers used for the test are shown in Table 1. The materials were taken from four positions of the same stockpile, which are indicated by ① ② ③ ④, respectively. The RAP (10–20 mm) samples under different refined separation frequencies are given in this paper, as shown in Figure 2.

Table 1. Abbreviations of RAP samples with different refined separation frequency.

Size (mm)	Refined Separation Frequency (Hz)				
	0	30	40	50	55
10–20	A1	B1	C1	D1	E1
5–10	A2	B2	C2	D2	E2
0–5	A3	B3	C3	D3	E3

Figure 2. RAP materials with different refined separation frequency parameters. (RAP particles in the red square: false particles; (**A1**), (**B1**), (**C1**), (**D1**) and (**E1**) represent 10–20 mm RAP particles after separation at 0 Hz, 30 Hz, 40 Hz, 50 Hz, and 55 Hz, respectively.)

3.2. Test Methods

In this paper, after refined separation, the aggregate and old asphalt in RAP of each grain size were separated by extraction and sieving, and the aggregate gradation and asphalt content were tested and the variability was calculated after RAP extraction. The properties of the aggregate (angularity, elongated and flat particle content, crushing value) and the properties of the old asphalt (penetration, softening point, and ductility) were tested. Furthermore, based on the evaluation of the test results, the maximum addition ratio of the coarse RAP material was calculated by using the variability control method of the gradation and the variability control method of the asphalt content, respectively.

The RAP Is subjected to different collision Intensities at different refined separation frequencies, which have different effects on the morphology of the aggregates. Therefore, aggregate angularity was tested with a coarse aggregate analyzer. The RAP with different refined separation frequency was extracted and sieved, and the elongated and flat content of the coarse aggregate was tested according to the JTG E42-2005 (T0312) [39], and the crushing value was tested according to the JTG E42-2005 (T0316) [39]. Asphalt penetration, softening point and ductility indexes were tested according to the JTG E20-2011 [40].

The refined separation evaluation indexes selected for this study are summarized in Figure 3, and the calculation of the agglomeration degree is described in Section 3.2 of this paper.

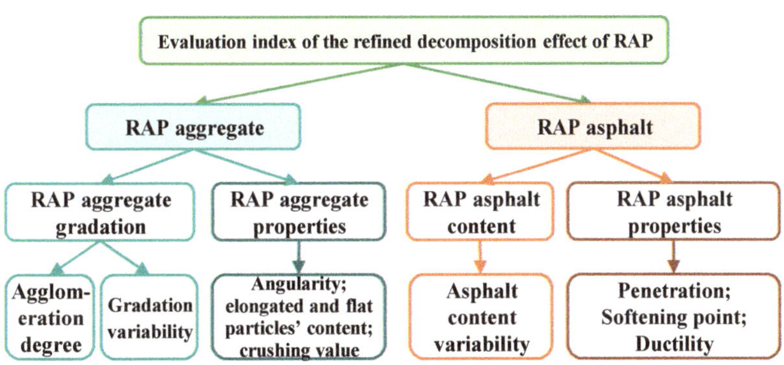

Figure 3. RAP material property evaluation index.

4. The Effect Evaluation of Refined Separation on RAP

4.1. Gradation Variability of RAP

Controlling the variability of RAP gradation is an important objective of the refined separation process. Because the test needs to analyze the old asphalt, each gradation of RAP aggregate is obtained by the centrifugal separation method. The specific method is as follows. Firstly, the RAP samples were soaked with trichloroethylene, and then the mixture and solvent were poured into the centrifugal extractor to separate the asphalt from the minerals. After refined separation, the results of extraction and sieving of RAP aggregate gradation are shown in Figure 4, and the coefficients of variation (CVs) of the gradation of the aggregates are shown in Figure 5.

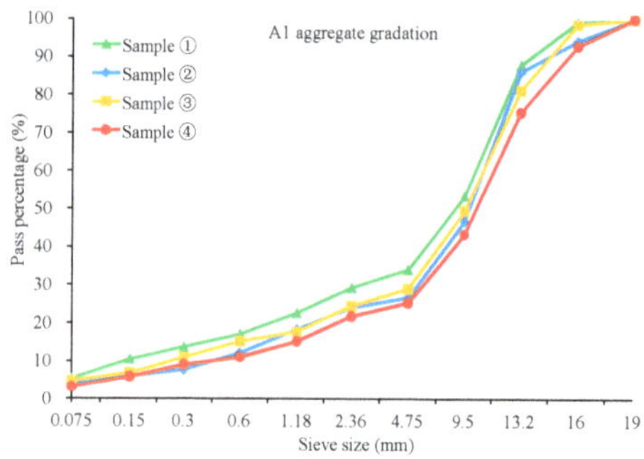

(**a**) A1 aggregate gradation

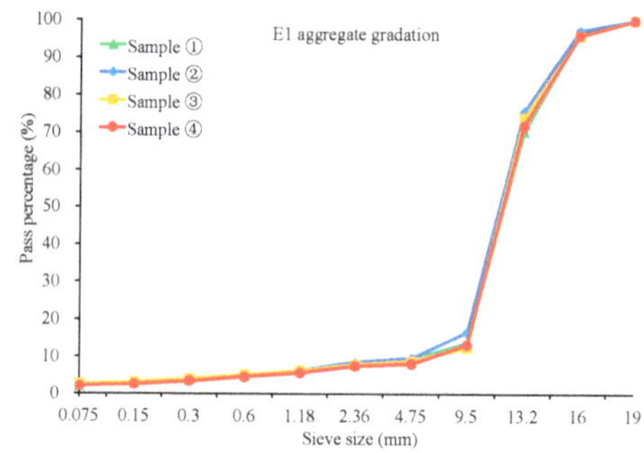

(**b**) E1 aggregate gradation

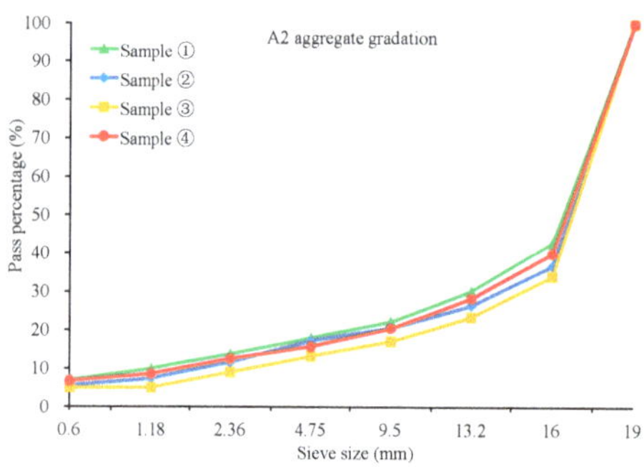

(**c**) A2 aggregate gradation

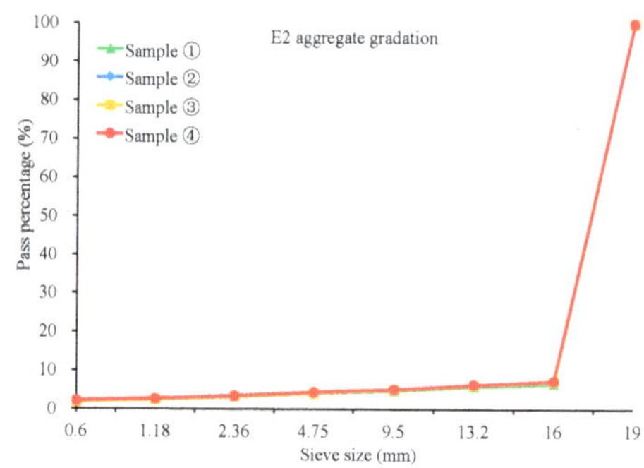

(**d**) E2 aggregate gradation

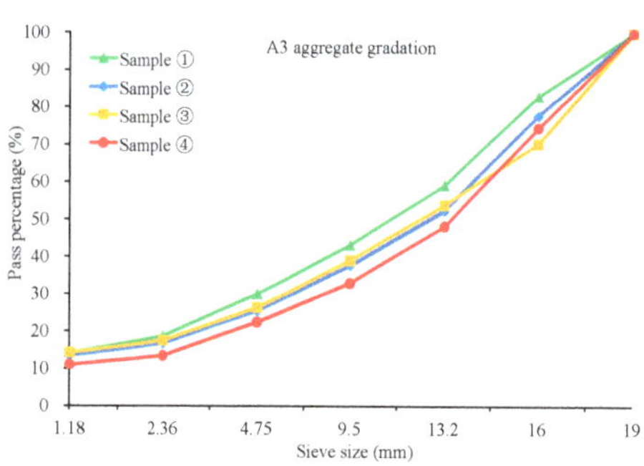

(**e**) A3 aggregate gradation

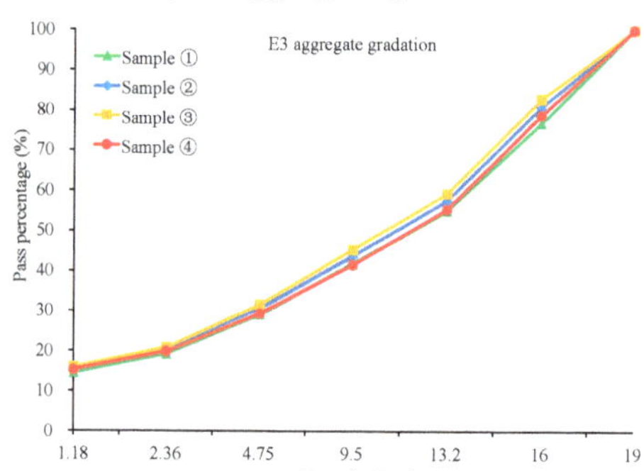

(**f**) E3 aggregate gradation

Figure 4. Gradation of RAP (extracted) with different sizes.

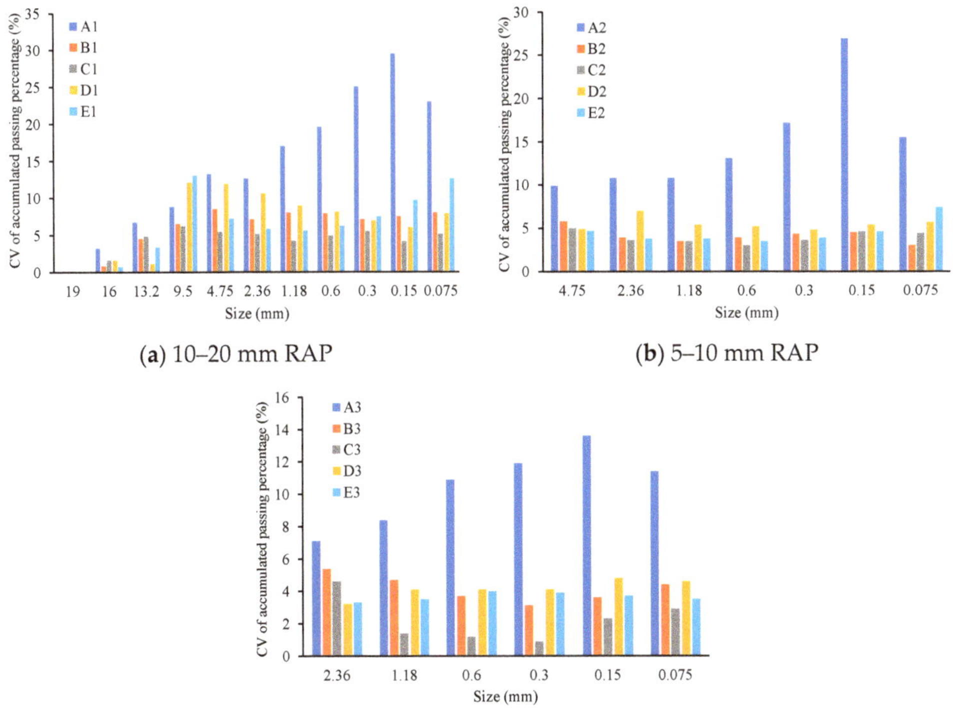

(**a**) 10–20 mm RAP

(**b**) 5–10 mm RAP

(**c**) 0–5 mm RAP

Figure 5. Coefficient of variation of RAP with different sizes.

As seen from Figure 4, by comparing the gradation curves, there is a notable difference in the sieve passing percentage between the sieve gradation curves of A1 and A2, while the consistency of the sieve gradation curves of E1 and E2 is better. From Figure 5, for the three sizes of RAP, the coefficients of variation of A1, A2, and A3 (0 Hz) are larger, and the coefficients of variation tend to increase while the particle size decreases. In addition, the coefficient of variation of 4.75 mm in A1 (0 Hz) is 13.2%, while the coefficient of variation of 4.75 mm in E1 (55 Hz) is 7.2%, and the coefficients of variation for RAPs treated with refined separation are significantly lower compared with those of the unfine separated RAPs (0 Hz). For the 5–10 mm RAP, the difference in E2 (after refined separation) was significantly smaller than that of A2. For the 0–5 mm RAP, the difference in E3 was slightly smaller than A3. Compared with the coarse RAP (>5 mm), the effect of refined separation on the RAP of 0–5 mm particle size was significantly smaller. By analyzing the results of the three sizes of RAP, the refined separation process reduces the variability of aggregate gradation of the RAP and makes RAP gradation have better stability.

4.2. *Agglomeration Degree of RAP*

The agglomeration degree characteristic of RAP, also known as a false particle, is an important reason for the large variability of RAP gradation. In order to evaluate the agglomeration of RAP, this paper establishes the agglomeration degree evaluation index of RAP based on the difference of the sieve residue before and after the extraction of old material, and the calculation model is shown as follows:

$$I_c = \frac{\sum_{i=1}^{n} \left| P_i^b - P_i^a \right|}{n} \quad (1)$$

where I_c is RAP agglomeration degree (%); P_i^b is the passing percentage of each sieve size before RAP is extracted; P_i^a is the passing percentage of each sieve size after RAP is extracted.

The results of agglomeration degree (I_c) value of RAP materials with different refined separation frequencies are presented in Figure 6.

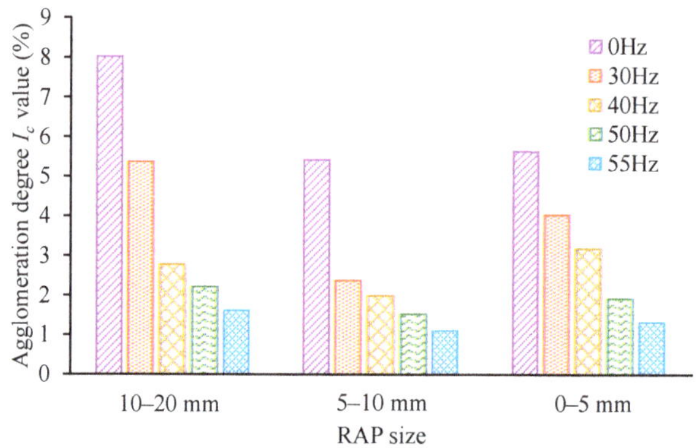

Figure 6. Agglomeration degree (I_c) value of RAP materials with different refined separation.

As shown in Figure 6, the RAP agglomeration gradually decreased with the increase in the frequency of refined separation, and the trends of the three particle sizes of RAP agglomeration were basically the same. For A1, A2, and A3, the I_c values of the agglomeration indexes were 8.02%, 5.42%, and 5.64%, respectively. And I_c values for E1, E2, and E3 are 1.62%, 1.12%, and 1.34%, respectively. Compared with I_c of A1, A2, and A3, the I_c values of E1, E2, and E3 decreased by 6.40%, 4.30%, and 4.30%, respectively. The agglomeration of RAP after refined separation decreased significantly, i.e., the content of false particles was reduced, which made RAP closer to the real mineral material. This is mainly due to the fact that the RAP entering the centrifugal impact crusher obtains kinetic energy, and the kinetic energy increases with the increase in frequency. It causes the RAP clusters to break after collision and friction. And after the friction and impact effect, the old asphalt binder is stripped from the surface of the coarse RAP particles. This effectively reduces the content of false particles in the RAP and improves the quality and stability of the RAP.

4.3. Aggregate Properties of RAP

The angularity, elongated and flat particle content, and crushing values of the aggregates after RAP extraction at different refined separation frequencies are presented in Table 2.

Table 2. Aggregate properties of RAP after extraction.

Index	Size (mm)	Refined Separation Frequency (Hz)				
		0	30	40	50	55
Angularity	5–10	17.90	17.53	17.29	17.18	17.16
	10–20	17.63	17.25	17.19	16.89	16.82
Elongated and flat particle content (%)	5–10	6.39	5.29	4.79	3.99	3.19
	10–20	6.68	5.39	4.80	4.20	3.20
Crushed value (%)	10–20	6.19	5.60	5.30	5.10	5.00

As can be seen from Table 2, the coarse aggregate angularity gradually decreases with the increase in refined separation frequency. This is mainly due to the fact that the collision and friction of RAP in the crusher become stronger with the increase in frequency, and the shape of aggregate tends to be "spherical", resulting in the decrease in aggregate angularity. Aggregate angularity has a great influence on the performance of asphalt mixtures, especially on the high-temperature performance of asphalt mixtures. Therefore, it is crucial not to increase the frequency of refined separation too much to improve the separation effect, as it may significantly impact the angularity of the coarse aggregate, thereby affecting the high-temperature performance of recycled asphalt mixtures.

From Table 2, the elongated and flat particle content of aggregate after refined separation is lower than that of the original aggregate, and the elongated and flat particle content of the aggregates decreases gradually with the increase in the refined separation frequency. When the frequency of refined separation is 55 Hz, the elongated and flat particle contents of 5–10 mm and 10–20 mm aggregates are 3.19% and 3.20%, respectively, which is 3.20% and 3.48% lower than that of non-refined separated aggregates (0 Hz). The elongated and flat particles are easily broken and crushed during the refined separation process, which results in a decrease in the elongated and flat content of the aggregate.

From the crushing value of aggregates in Table 2, it can be seen that the crushing value of aggregates after refined separation is lower than that of aggregates without fine separation, and with the increase in the frequency of fine separation, the crushing value of aggregates gradually decreases. This is due to the fact that the refined separation process reduces the aggregate elongated and flat particle content, and the aggregate surface angle is also reduced, which in turn reduces the aggregate crushed, and with the increase in the frequency of fine separation, the aggregate tends to be more "rounded", and the aggregate is crushed less.

4.4. Asphalt Content Variability of RAP

The asphalt contents of RAP with different refined separation frequencies are shown in Figure 7.

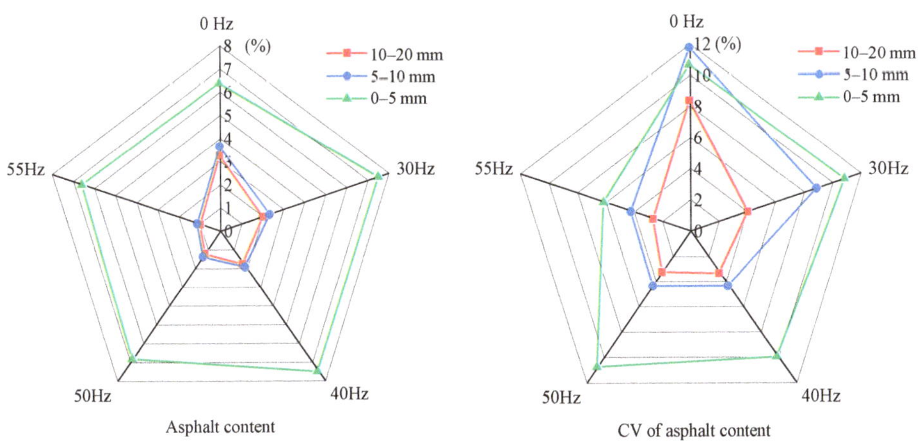

(**a**) Asphalt content of RAP (**b**) CV of asphalt content of RAP

Figure 7. Asphalt content of RAP and asphalt content variability of RAP.

As shown in Figure 7, for the 0–5 mm RAP, the asphalt content of A3 (0 Hz) has a small change from the asphalt content of the refined separated RAP. The asphalt content of the 5–10 mm and 10–20 mm RAP decreases with the increase in the frequency. The asphalt content of A1 and A2 (0 Hz) is 3.28% and 3.68%, and that of E1 and E2 (55 Hz) is 0.95% and 1.10%, respectively. Compared with the same specifications, the asphalt content of RAP after refined separation (55 Hz) decreased by 2.33% and 2.58%, respectively. In coarser RAP after the refined separation process, the surface old asphalt binder is stripped after the repeated collision friction effect, resulting in a significant decrease in the RAP asphalt content.

The asphalt content and its coefficient of variation after extraction of unfine separation RAPs (A1, A2, and A3) and 55 Hz refined separation RAPs (E1, E2, and E3) are shown in Figure 7.

As can be seen in Figure 7, the coefficients of variation for asphalt content of the 55 Hz refined separation RAPs (E1, E2 and E3) are significantly lower than those of the unfine separation RAPs (A1, A2, and A3). The coefficients of variation are 10.7%, 11.8%, and 8.4% for A1, A2, and A3, respectively, and 6.1%, 7.4%, and 2.6% for the fine separated E1, E2, and E3, respectively. Compared with A1, A2, and A3 (0 Hz), the coefficients of variation of asphalt content of E1, E2, and E3 (55 Hz) are reduced by 4.6%, 4.4%, and 5.8%, respectively. The refined separation process significantly reduces the variability of asphalt content in RAP.

4.5. Asphalt Properties of RAP

The asphalt properties after extraction of RAP are shown in Figure 8.

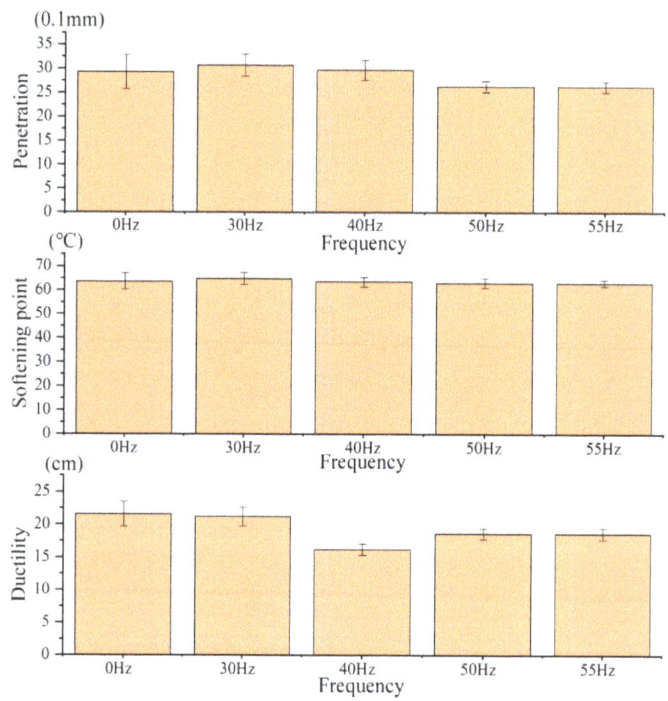

Figure 8. Asphalt properties of RAP.

As can be seen from Figure 8, the average values of penetration, softening point, and ductility of asphalt obtained by extraction before and after RAP refined separation change less. Moreover, with the increase in separation frequency, the standard error of the three performance indicators of asphalt gradually decreases. This is due to the RAP in the centrifugal impact crusher collision process undergoing further mixing, so that the asphalt performance variability is reduced. However, during the refined separation process, the RAPs collide and rub against each other, and this process is a physical process, which has less influence on the asphalt performance index.

5. The Calculation Method of Maximum Addition Proportion of RAP for Recycled Asphalt Mixture

5.1. Maximum Addition Proportion of RAP under the Control of Gradation Variability

The Technical Specification for construction of Highway Asphalt Pavement (JTG F40-2004) specifies that the sieve passing percentage of each mineral aggregate material should meet the corresponding quality requirements [41]. The coefficient of variation of RAP is larger than that of the new mineral material, and when the proportion of RAP added to the recycled mixture is too large, it will result in the recycled asphalt mixture not meeting the quality requirements. Therefore, in order to ensure the quality of the recycled asphalt mixtures, it is necessary to calculate the maximum proportion of RAP based on different coefficients of variation. Assuming the RAP blending ratio X, the gradation variability should meet the requirements of Equation (2) [5]:

$$VX \leq V' \tag{2}$$

where V is the standard deviation of RAP passing percentage of each sieve size, %; X is the RAP addition ratio in asphalt mixture; V' is the specification value of hot mix recycled asphalt gradation passing percentage of each sieve size, %.

Due to the new aggregate gradation sieve, throughput variability is not considered, so the recycled asphalt mixture variability requirements should be slightly more stringent than the ordinary hot mix asphalt mixture. It is recommended that the hot mix recycled asphalt mixture variability meet the requirements of Table 3.

Table 3. Grading quality requirements of hot mixed recycled asphalt mixtures.

Sieve Size (mm)	Number of Samples	Quality Requirements
		Highway
0.075 mm		$XV_{0.075} \leq 1\%$
≤2.36 mm	RAP material sample number n: $n \geq 4$	$XV_{\leq 2.36} \leq 3\%$
≥4.75 mm		$XV_{\geq 4.75} \leq 4\%$

As can be seen from Table 3, when the RAP variability is large, the quality requirements can be satisfied by reducing the RAP addition ratio, and when the RAP variability is small, it is favorable to increase the RAP addition ratio.

Relevant studies [18,42] show that the passing percentage of each sieve size of RAP aggregate material can be characterized by being distributed, and the overall standard deviation can be estimated from the samples according to the mathematical and statistical methods. According to the theory of confidence interval, the overall standard deviation is calculated by Equation (3).

$$S = \frac{\sigma}{\sqrt{n}} t_{\alpha/2} \tag{3}$$

where S is the RAP overall standard deviation, %; σ is the RAP sample standard deviation, %; n is the number of samples; α is the confidence level; for $t_{\alpha/2}$, check the t-distribution table.

When the confidence level $\alpha = 0.05$ and the degree of freedom $(n - 1)$ is 3, check the table to obtain $t_{\alpha/2} = 3.182$. The overall standard deviation of the four specifications of RAP (A1, A2, E1, E2), i.e., V value, can be obtained, which is brought into the formula in Table 3 to obtain the quality requirement under the control of gradation variability, as shown in Table 4.

In order to make the recycled asphalt mixture meet the quality requirements, there exists a maximum addition ratio of RAP additions, and the maximum additive ratio is calculated according to Equation (4) [5]:

$$X_{max} = \min\{0.04/S_{\geq 4.75}, 0.03/S_{\leq 2.36}, 0.01/S_{0.075}\} \tag{4}$$

where X_{max} is the maximum addition proportion of RAP; $S_{\geq 4.75}$ is the overall standard deviation of the passing percentage with sieve size greater than 4.75, %; $S_{\leq 2.36}$ is the overall standard deviation of the passing percentage with sieve size less than 2.36, %; $S_{0.075}$ is the overall standard deviation of the passing percentage with sieve size 0.075, %. It should be noted that when recycled asphalt mixtures are designed, the addition ratio of RAP should not exceed X_{max}. The specific RAP addition ratio should be determined based on the mixture design requirements and the properties of the recycled asphalt mixture.

Table 4. Quality requirements for aggregate gradation.

Sieve Size (mm)	Quality Requirements			
	A1	A2	E1	E2
19	0.0%X ≤ 4%	0.0%X ≤ 4%	-	-
16	4.9%X ≤ 4%	1.1%X ≤ 4%	-	-
13.2	8.8%X ≤ 4%	3.8%X ≤ 4%	-	-
9.5	6.8%X ≤ 4%	2.9%X ≤ 4%	0.0%X ≤ 4%	0.0%X ≤ 4%
4.75	6.1%X ≤ 4%	1.0%X ≤ 4%	6.1%X ≤ 4%	0.5%X ≤ 4%
2.36	5.0%X ≤ 3%	0.7%X ≤ 3%	4.7%X ≤ 3%	0.4%X ≤ 3%
1.18	5.0%X ≤ 3%	0.6%X ≤ 3%	3.5%X ≤ 3%	0.3%X ≤ 3%
0.6	4.3%X ≤ 3%	0.5%X ≤ 3%	3.4%X ≤ 3%	0.2%X ≤ 3%
0.3	4.2%X ≤ 3%	0.5%X ≤ 3%	3.2%X ≤ 3%	0.2%X ≤ 3%
0.15	3.5%X ≤ 3%	0.4%X ≤ 3%	3.4%X ≤ 3%	0.2%X ≤ 3%
0.075	1.6%X ≤ 1%	0.5%X ≤ 1%	1.5%X ≤ 1%	0.2%X ≤ 1%

The maximum addition proportion of RAP was calculated according to Equation (4) and the results are shown in Figure 9.

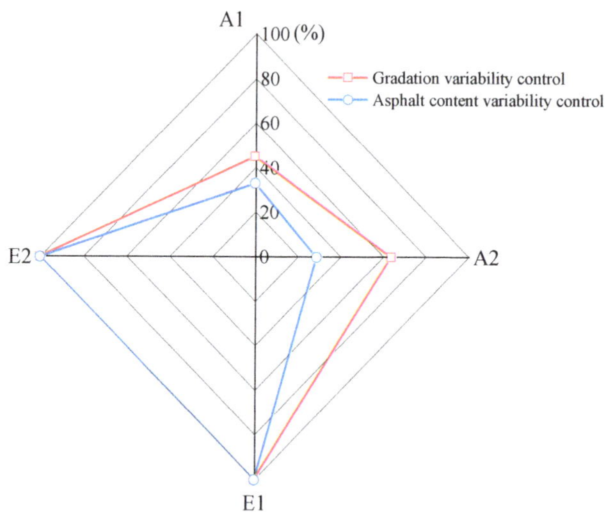

Figure 9. Maximum addition ratio of RAP under different control methods.

From Figure 9, it can be seen that the calculated values for the maximum addition proportion without refined separation of A1 and A2 are 45% and 64%, respectively. After refined separation, the maximum addition proportion for both E1 and E2 was 100%. The above results demonstrate that the maximum addition proportion of RAP is significantly increased after refined separation. The calculated value of the maximum addition proportion of RAP can be determined by the control method. In order to ensure the stability of the quality of the recycled pavement, in the actual recycled asphalt mixture design and application, it can effectively control the variability of the recycled asphalt mixture when the addition proportion of RAP is less than the calculated maximum proportion value.

5.2. Maximum Addition Proportion of RAP under the Control of Asphalt Content Variability

In order to investigate the effect of asphalt content variability on the maximum addition ratio of RAP, the asphalt content variability control method was used to calculate the maximum addition ratio of RAP. The Technical Specification JTG F40-2004 requires that the asphalt–aggregate ratio not exceed ±0.3%, and considering the variability of new asphalt and construction processes, the variability of RAP asphalt content is controlled at ±0.2%. Referring to the calculation method of RAP gradation variability control (i.e., Table 4), the maximum addition ratio of RAP under asphalt content variability control was calculated, and the results are shown in Figure 9.

As can be seen from Figure 9, the maximum addition ratios of A1 and A2 (0 Hz) are 33% and 29%, respectively. The maximum addition ratios of E1 and E2 (55 Hz) are all 100%. Compared to the unfine separation RAP, the addition ratio of RAP after refined separation increased by 67% and 71%, respectively. The results obtained from the calculations of the asphalt content variability control method and the gradation variability control method are consistent with the fact that the refined separation process increases the maximum addition percentage of coarse RAP material in the recycled asphalt mixtures. It should be noted that in order to ensure the stability of the quality of the recycled asphalt mixture, the addition proportion of RAP should be less than the minimum of the values calculated by the two control methods, so as to effectively control the variability of recycled asphalt mixtures.

6. Conclusions

This study systematically evaluated and analyzed the effect of refined separation on aggregate gradation, aggregate properties, asphalt content, and asphalt properties of RAP. The calculation method of the maximum addition proportion of RAP material using gradation variability control and asphalt content control is also given, respectively. The following conclusions can be drawn:

(1) The refined separation process can effectively reduce the variability of RAP and enhance the property stability of RAP.
(2) The RAP false particles decreased significantly after refined separation and were closer to the natural mineral material. RAP agglomeration degree decreases with increasing frequency parameters of refined separation. Comparing with the I_c values of A1, A2, and A3, those of E1, E2, and E3 decreased by 6.40%, 4.30%, and 4.30%, respectively.
(3) The asphalt content of the coarse RAP materials after refined separation decreases gradually with increasing frequency parameter. For 0–5 mm grain size RAP, the asphalt content of A3 changed less compared to that of the refined separated RAP.
(4) Based on the quality control requirements of hot mix asphalt mixtures, the maximum blending percentage of RAP coarse material was calculated using the gradation variability control method and asphalt content variability control method, respectively. The refined separation process can significantly increase the maximum addition ratio of RAP material.
(5) Through refined separation, it is conducive to improving the addition proportion of RAP in the recycled asphalt mixture, ensuring the quality of the recycled asphalt mixture, and further maximizing the potential value of RAP. This enhances the environmental benefits of recycling RAP and economic benefits.

Future research should focus on the application of fine RAP after refined decomposition to further maximize the utilization of RAP materials. The refined separation method of fine RAP should be improved to realize efficient asphalt–aggregate separation to reduce the variability of RAP materials. It is an important way to realize the high-value recycling of RAP. Currently, the research predominantly centers on the refined separation process of RAP materials, with limited consideration for practical applications. It is imperative to conduct in-depth research on the application of refined separation RAP.

Author Contributions: Conceptualization, J.W. and S.W.; methodology, J.W.; software, S.W.; validation, Y.W. and G.D.; formal analysis, Y.W. and G.D.; investigation, Y.W., G.D., S.W. and J.W.; resources, J.W.; data curation, Y.W., G.D., S.W. and J.W.; writing—original draft preparation, Y.W., G.D. and S.W.; writing—review and editing, J.W.; supervision, J.W.; funding acquisition, J.W. All authors have read and agreed to the published version of the manuscript.

Funding: This research was funded by the Fundamental Research Funds (2021-9038b) and Research and Development of Key Technologies and Equipment for RAP Refined Separation and Its Application in Micro Surfacing (2021-JKKJ-15).

Data Availability Statement: Data will be provided on request.

Conflicts of Interest: The authors declare no conflicts of interest.

References

1. Miao, Y.; Wang, S.; Guo, L.; Zheng, X.; Huang, Y.; Wang, L. Effect of temperature on deformation properties of unbound granular materials containing fine RAP. *Constr. Build. Mater.* **2018**, *169*, 443–451. [CrossRef]
2. He, Y.; Wang, Q.; Zhang, J.; Li, Y.; He, H.; Huang, G. Foamed crumb rubber asphalt binder: Preparation, rheological properties and adhesion characteristics. *J. Clean. Prod.* **2023**, *396*, 136516. [CrossRef]
3. Qu, L.; Wang, Y.; Wang, L.; Li, X.; Xiao, Y. Experimental evaluation of fatigue performance of recycled asphalt mixture using refined separation recycled aggregates. *Constr. Build. Mater.* **2024**, *411*, 134786. [CrossRef]
4. Gao, X.; He, W.; Zhou, X. Dynamic Mechanical Properties of Hot-in-Place Recycled Asphalt Mixtures. *Sci. Technol. Eng.* **2022**, *22*, 15742–15748.
5. Wang, J.; Qin, Y.; Huang, S.; Bo, Z.; Guo, S. Variability of Reclaimed Asphalt Pavement. *J. Beijing Univ. Technol.* **2018**, *44*, 244–250.
6. Gonçalves Luz, P.M.S.; Silva, I.M.; de Medeiros Melo Neto, O.; Dias, T.L.; de Figueiredo Lopes Lucena, L.C.; Sampaio, R.M.B. Analysis of the Degree of Blending (DoB) of recycled asphalt mixtures with variation in mixing temperature, type, and RAP content. *Constr. Build. Mater.* **2024**, *431*, 136577. [CrossRef]
7. Xu, J.-Z.; Hao, P.-W.; Guo, X.-G.; Li, H.-X.; Zhang, B.-J.; Le, C. Review of Mix Design Method of Hot In-plant Recycled Asphalt Mixture. *China J. Highw. Transp.* **2021**, *34*, 72–88.
8. Zou, F.; Ye, F.; Song, Q. Effect of RAP percentage on pavement performance of recycled asphalt mixtures. *J. Jilin Univ. (Sci. Ed.)* **2020**, *50*, 1403–1410.
9. Zou, X.; Li, N.; Wu, W.; Hu, Y.; Zhang, Y.; Tang, W. Study on Effect of Fine Separation on RAP Material Properties. *J. Dalian Jiaotong Univ.* **2023**, *44*, 71–76.
10. Ai, X.; Cao, J.; Feng, D.; Gao, L.; Hu, W.; Yi, J. Performance evaluation of recycled asphalt mixtures with various percentages of RAP from the rotary separation process. *Constr. Build. Mater.* **2022**, *321*, 126406. [CrossRef]
11. Yu, X.; Tang, W.; Li, N.; Jiang, M.; Huang, J.; Wang, D. Refined separation: A new separation method for RAP materials and its effect on aggregate properties. *Constr. Build. Mater.* **2022**, *358*, 129452. [CrossRef]
12. Akatsu, K.; Kanou, Y.; Aiba, S. Separation Recycling Technology for Restoring Reclaimed Asphalt Pavement. *J. JSCE* **2018**, *6*, 110–117. [CrossRef]
13. Feng, D.; Cao, J.; Gao, L.; Yi, J. Recent developments in asphalt-aggregate separation technology for reclaimed asphalt pavement. *J. Road Eng.* **2022**, *2*, 332–347. [CrossRef]
14. Wang, T.; Zhang, C.; Zhao, R.; Zhu, C.; Yang, C.; Liu, C. Solvent Extraction of Bitumen from Oil Sands. *Energy Fuels* **2014**, *28*, 2297–2304. [CrossRef]
15. Xue, J.; Yu, Y.; Bai, Y.; Wang, L.; Wu, Y. Marine Oil-Degrading Microorganisms and Biodegradation Process of Petroleum Hydrocarbon in Marine Environments: A Review. *Curr. Microbiol.* **2015**, *71*, 220–228. [CrossRef]
16. Zhang, Y.; Chen, H.; Wang, K.; Huang, G.; Shen, Z.; Sun, L. Effect of recycled aggregate gradation on the degree of blending and performance of recycled hot-mix asphalt (HMA). *J. Clean. Prod.* **2023**, *398*, 136550. [CrossRef]
17. Katla, B.; Raju, S.; Waim, A.R.; Danam, V.A. Utilization of Higher Percentages of RAP for Improved Mixture Performance by Adopting the Process of Fractionation. *Int. J. Pavement Res. Technol.* **2022**, *15*, 349–366. [CrossRef]
18. Yang, J.; Zhang, W.; Gao, J.; Yao, Y.; Sun, C. Optimizing RAP sieving efficiency of linear vibrating sieve using DEM simulation. *Constr. Build. Mater.* **2022**, *333*, 127442. [CrossRef]
19. Yang, J.; Tao, W.; Gao, J.; Yu, D.; Zhou, J.; He, L.; Yao, Y. Measurement of particle agglomeration and aggregate breakdown of reclaimed asphalt pavement. *Constr. Build. Mater.* **2021**, *296*, 123681. [CrossRef]
20. Bressi, S.; Dumont, A.G.; Partl, M.N. A new laboratory methodology for optimization of mixture design of asphalt concrete containing reclaimed asphalt pavement material. *Mater. Struct.* **2016**, *12*, 4975–4990. [CrossRef]
21. Xiao, F.; Su, N.; Yao, S.; Amirkhanian, S.; Wang, J. Performance grades, environmental and economic investigations of reclaimed asphalt pavement materials. *J. Clean. Prod.* **2019**, *211*, 1299–1312. [CrossRef]
22. Montañez, J.; Caro, S.; Carrizosa, D.; Calvo, A.; Sánchez, X. Variability of the mechanical properties of Reclaimed Asphalt Pavement (RAP) obtained from different sources. *Constr. Build. Mater.* **2020**, *230*, 116968. [CrossRef]

23. Liu, Y.; Qiu, Q.; Ji, W.; Pang, L.; Li, N.; Tang, W.; Zhan, H. Effect of RAP Classification on Road Performance Variability of Hot Recycled Asphalt Mixture. *Highw. Eng.* **2021**, *46*, 68–72.
24. Guo, P.; Wei, W.; Tang, B.; Yang, F. Evaluation of High-temperature Performance of Asphalt Mixture based on Angularity of Recycled Coarse Aggregate. *Highway* **2017**, *69*, 220–225.
25. Orešković, M.; Pires, G.M.; Bressi, S.; Vasconcelos, K.; Presti, D.L. Quantitative assessment of the parameters linked to the blending between reclaimed asphalt binder and recycling agent: A literature review. *Constr. Build. Mater.* **2020**, *234*, 117323. [CrossRef]
26. Liu, J.; Liu, Q.; Wang, S.; Zhang, X.; Xiao, C.; Yu, B. Molecular dynamics evaluation of activation mechanism of rejuvenator in reclaimed asphalt pavement (RAP) binder. *Constr. Build. Mater.* **2021**, *298*, 123898. [CrossRef]
27. Meng, Y.; Liu, L.; Huang, W.; Li, M. Effect of increasing preheating temperature on the activation and aging of asphalt binder in reclaimed asphalt pavement (RAP). *J. Clean. Prod.* **2023**, *402*, 136780. [CrossRef]
28. Xing, C.; Li, M.; Liu, L.; Lu, R.; Liu, N.; Wu, W.; Yuan, D. A comprehensive review on the blending condition between virgin and RAP asphalt binders in hot recycled asphalt mixtures: Mechanisms, evaluation methods, and influencing factors. *J. Clean. Prod.* **2023**, *398*, 136515. [CrossRef]
29. Pires, G.M.; Presti, D.L.; Airey, G.D. A practical approach to estimate the degree of binder activity of reclaimed asphalt materials. *Road Mater. Pavement Des.* **2021**, *22*, 1093–1116. [CrossRef]
30. Meng, Y.; Liu, L. Impact of Preheating Temperatures and RAP Characteristics on the Activation of RAP Binder. *Appl. Sci.* **2020**, *10*, 8378. [CrossRef]
31. Gao, J.; Yang, J.; Yu, D.; Jiang, Y.; Ruan, K.; Tao, W.; Sun, C.; Luo, L. Reducing the variability of multi-source reclaimed asphalt pavement materials: A practice in China. *Constr. Build. Mater.* **2021**, *278*, 122389. [CrossRef]
32. Zaumanis, M.; Oga, J.; Haritonovs, V. How to reduce reclaimed asphalt variability: A full-scale study. *Constr. Build. Mater.* **2018**, *188*, 546–554. [CrossRef]
33. Xue, Y.; Liu, C.; Qu, J.; Lv, S.; Ju, Z.; Ding, S.; An, H.; Ren, K. Research on pavement performance of recycled asphalt mixture based on separation technology of asphalt and aggregate in RAP. *Constr. Build. Mater.* **2023**, *393*, 132103. [CrossRef]
34. Pan, Y.; Li, J.; Yang, T.; Liu, G.; Zhou, J.; Guo, P.; Zhao, Y. Optimization of gradation design of recycled asphalt mixtures based on fractal and Mohr-Coulomb theories. *Constr. Build. Mater.* **2020**, *248*, 118649. [CrossRef]
35. Yao, Y.; Yang, J.; Gao, J.; Zheng, M.; Xu, J.; Zhang, W.; Song, L. Strategy for improving the effect of hot in-place recycling of asphalt pavement. *Constr. Build. Mater.* **2023**, *366*, 130054. [CrossRef]
36. Zou, G.; Zha, Z.; Dong, S.; An, Q. Effect of plant-mixed recycled RAP pretreatment process on separation efficiency. *Constr. Mach. Equip.* **2022**, *53*, 56–62.
37. Qiu, J.; Huurman, M.; de Bruin, B.; Demmink, E.; Frunt, M. Towards 90% warm re-use of porous asphalt using foaming technology. *J. Clean. Prod.* **2018**, *190*, 251–260. [CrossRef]
38. Peng, W. Research on recycling technology of waste asphalt mixture based on dry oil-aggregate separation. *Eng. Technol. Res.* **2021**, *6*, 9–11.
39. *JTG E42-2005*; Test Methods of Aggregate for Highway Engineering. Ministry-of-Transport-of-the-People's-Republic-of-China, China Communications Press: Beijing, China, 2005.
40. *JTG E20-2011*; Standards Test Methods of Bitumen and Bituminous Mixtures for Highway Engineering. Ministry-of-Transport-of-the-People's-Republic-of-China, China Communications Press: Beijing, China, 2011.
41. *JTG F40-2004*; Technical Specifications for Construction of Highway Asphalt Pavements. Ministry-of-Transport-of-the-People's-Republic-of-China, China Communications Press: Beijing, China, 2004.
42. Huang, X.; Zhao, Y. *Theory and Practice of Asphalt Pavement Recycling*; Science and Technology of China Press: Beijing, China, 2014.

buildings

MDPI

Article

A Parametric Study Investigating the Dowel Bar Load Transfer Efficiency in Jointed Plain Concrete Pavement Using a Finite Element Model

Saima Yaqoob [1,*], Johan Silfwerbrand [1] and Romain Gabriel Roger Balieu [2]

[1] Division of Concrete Structures, Department of Civil & Architectural Engineering, KTH Royal Institute of Technology, SE-10044 Stockholm, Sweden
[2] Division of Structural Engineering and Bridges, Department of Civil & Architectural Engineering, KTH Royal Institute of Technology, SE-10044 Stockholm, Sweden
* Correspondence: saimay@kth.se

Abstract: Transverse joints are introduced in jointed plain concrete pavement systems to mitigate the risk of cracks that can develop due to shrinkage and temperature variations. However, the structural behaviour of jointed plain concrete pavement (JPCP) is significantly affected by the transverse joint, as it creates a discontinuity between adjacent slabs. The performance of JPCP at the transverse joints is enhanced by providing steel dowel bars in the traffic direction. The dowel bar provides reliable transfer of traffic loads from the loaded side of the joint to the unloaded side, known as load transfer efficiency (LTE) or joint efficiency (JE). Furthermore, dowel bars contribute to the slab's alignment in the JPCP. Joints are the critical component of concrete pavements that can lead to various distresses, necessitating rehabilitation. The Swedish Transport Administration (Trafikverket) is concerned with the repair of concrete pavement. Precast concrete slabs are efficient for repairing concrete pavement, but their performance relies on well-functioning dowel bars. In this study, a three-dimensional finite element model (3D-FEM) was developed using the ABAQUS software to evaluate the structural response of JPCP and analyse the flexural stress concentration in the concrete slab by considering the dowel bar at three different locations (i.e., at the concrete slabs' top, bottom, and mid-height). Furthermore, the structural response of JPCP was also investigated for several important parameters, such as the joint opening between adjacent slabs, mispositioning of dowel bars (horizontal, vertical, and longitudinal translations), size (diameter) of the dowel bar, and bond between the slab and the dowel bar. The study found that the maximum LTE occurred when the dowel bar was positioned at the mid-depth of the concrete slab. An increase in the dowel bar diameter yielded a 3% increase in LTE. Conversely, the increase in the joint opening between slabs led to a 2.1% decrease in LTE. Additionally, the mispositioning of dowel bars in the horizontal and longitudinal directions showed a 2.1% difference in the LTE. However, a 0.5% reduction in the LTE was observed for a vertical translation. Moreover, an approximately 0.5% increase in LTE was observed when there was improved bonding between the concrete slab and dowel bar. These findings can be valuable in designing and evaluating dowel-jointed plain concrete pavements.

Keywords: jointed plain concrete pavement; finite element modelling; load transfer efficiency; steel dowel bar

Citation: Yaqoob, S.; Silfwerbrand, J.; Balieu, R.G.R. A Parametric Study Investigating the Dowel Bar Load Transfer Efficiency in Jointed Plain Concrete Pavement Using a Finite Element Model. *Buildings* **2024**, *14*, 1039. https://doi.org/10.3390/buildings14041039

Academic Editors: Andrea Baliello and Di Wang

Received: 15 January 2024
Revised: 28 March 2024
Accepted: 5 April 2024
Published: 8 April 2024

1. Introduction

The transverse joints are the key element in the jointed plain concrete pavement system. These joints are introduced to jointed plain concrete pavement systems to mitigate the risk of cracks that can develop due to shrinkage and temperature variations. However, the joints create a gap between adjacent slabs, thus reducing the pavement's performance. To maintain the structural integrity of the pavement system, adequate load transfer at the joints is essential. This can be achieved either through aggregate interlock (interlocking of the irregular faces of adjacent slabs) or using a mechanical load transfer device in the traffic direction. The standard mechanical device that is widely used in pavement construction is rounded, smooth steel dowel bars [1,2]. The dowel bars transmit traffic load through both shear and moment mechanisms. However, Guo et al. (1995) found that load transfer predominantly occurs via shear, particularly in joints with a width of less than 6 mm [0.25 in.] [3,4]. An efficient way of repairing JPCP is replacing damaged slabs with precast concrete slabs [5–7]. Since the precast concrete pavement incorporates the smooth faces of slabs, there is a lack of load transfer through aggregate interaction, and load transfer primarily relies on the dowel bars.

In jointed cast-in-place concrete pavement, the dowels are placed in the concrete slabs using dowel bar inserters [DBIs] or prefabricated dowel basket assemblies, for details, see [8]; meanwhile, in the jointed precast pavement systems, the dowel bars are placed in the dowel slots either in the factory or on the project site, and then slots are filled with the patching material. In Sweden, dowel bars are automatically installed by the slip-form pavers. To protect the dowel bars against possible corrosion and minimise their adherence to the concrete, they must be epoxy-coated and greased [9,10]. Improper greasing of dowel bars can restrict the horizontal movement of concrete slabs due to shrinkage and temperature variations, leading to cracks typically appearing around the mid-slab location [11].

When the concrete slab with the dowelled joint is loaded at the joint, it experiences greater flexural tensile stresses and deflection in the vicinity of joints. The dowel bars transmit the portion of the traffic load from the loaded side of the joint to the unloaded side, which minimises the critical stresses and deflection in the loaded slab and helps to prevent future deterioration such as spalling, faulting, and corner breaks, thereby improving the pavement's performance and service life. The reduced magnitude of stresses and deflection in the loaded slab is known as load transfer efficiency (LTE) or joint efficiency (JE) [10,12,13]. The load transfer mechanism between adjacent concrete slabs with dowelled and undowelled joints is illustrated in Figure 1. Various equations can be used to determine the LTE. The most common two equations are described in [14]. In this paper, the first equation is used to compute the LTE of a dowelled joint.

$$LTE_u = \frac{\triangle_{u_L}}{\triangle_L} \tag{1}$$

$$LTE_\sigma = \frac{\sigma_{u_L}}{\sigma_L} \tag{2}$$

where,

LTE_u = load transfer efficiency based on deflection;
$\triangle_{u_L}$ = deflection of the unloaded slab at the joint;
$\triangle_L$ = deflection of the loaded slab at the joint;
LTE_σ = load transfer efficiency based on stress;
σ_{u_L} = flexural stress of the unloaded slab at the joint;
σ_L = flexural stress of the loaded slab at the joint.

The proper placement of the dowel bar is critical. Ideally, the dowel bar must be placed parallel to the concrete pavement's surface and the centreline of the pavement's longitudinal axis. However, if the dowel bar deviates from its planned position, this is referred to as mispositioned or misaligned [15]. Different types of dowel bar mispositioning, i.e., horizontal, vertical, and longitudinal translations and misalignment, i.e., horizontal and vertical tilt, are discussed below and illustrated in Figure 2.

- **Lateral or horizontal translation**: When the dowel bar is located far enough from its specified position, this is defined as horizontal translation [16].

- **Depth deviation or vertical translation**: Depth deviation indicates the elevation difference between the dowel bar's planned position and its actual placement within the concrete slab [16,17].
- **Longitudinal translation**: If the dowel bar centroid does not match with the transverse joint, this is referred to as longitudinal translation [18].
- **Horizontal or vertical skew**: The rotation of the dowel bar in a horizontal plane relative to the pavement centreline is known as horizontal skew. Conversely, vertical tilt is the rotation of the dowel bar in a vertical plane relative to the pavement surface [17].

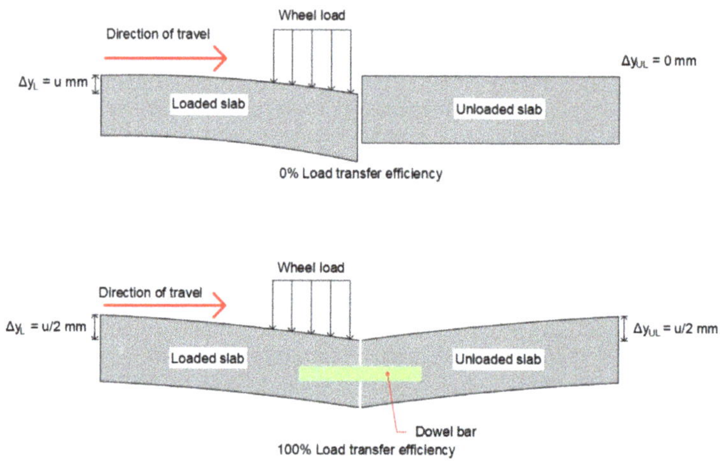

Figure 1. Load transfer mechanism in JPCP including joints with and without the dowel bar (Reproduced with permission of Transportation Research Record from [19]).

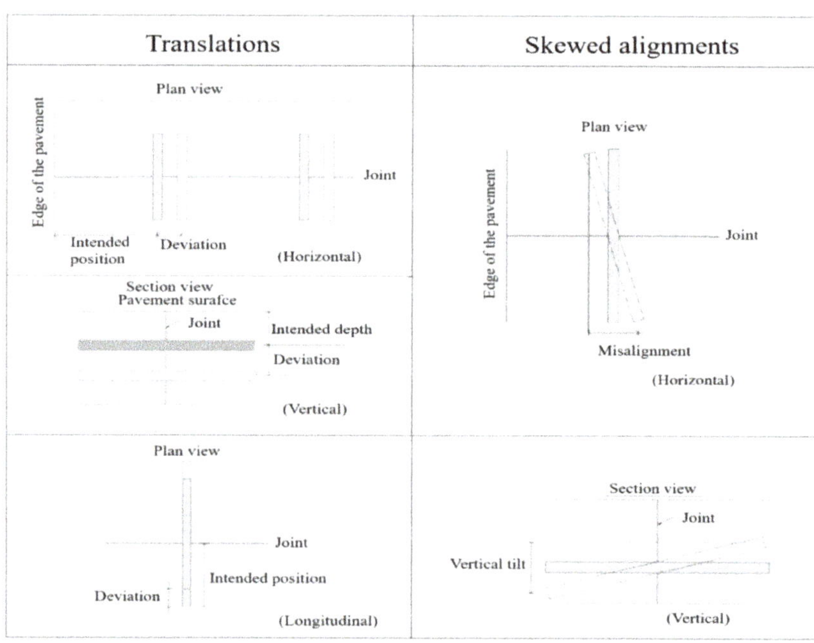

Figure 2. Schematic view of different types of dowel bar mislocation (Reproduced with permission of the Transportation Research Board from [20]).

The performance of the JPCP depends on several factors including the thickness of the concrete slab; the strength of the concrete slab, base, subbase and subgrade; the surrounding temperature; the number and size of the vehicle; the transverse joint spacing; the joint width; and the dowel bars' diameter and spacing [21,22]. Numerous studies have been conducted to enhance the performance of the JPCP system. Mackiewicz (2015) studied the stress concentration around the dowel bar and determined the efficiency of interaction between the loaded and unloaded slab depending on the dowel bar parameters, i.e., diameter, spacing, and length. The findings indicated that dowel parameters substantially influenced the vertical compressive stresses in the concrete slab [23,24]. Khazanovic et al. (2009) performed a laboratory test considering various embedment lengths, ranging from 51 mm to 229 mm [2 in. to 9 in.]. The results indicated that increasing the embedment length enhanced the shear capacity of the dowel bar [25].

Odden et al. (2003) performed a full-scale test to examine the influence of concrete cover resulting from vertical deviation of the dowel bar. Two different dowel bar configurations were considered. In the first arrangement, the dowel bars were placed at the mid-depth of 190 mm [7.5 in.] of a thick concrete slab with a 76 mm [3 in.] concrete cover, while in the second arrangement, the dowel bars were set with a 51 mm [2 in.] cover. It was found that reducing the concrete cover slightly lowered the LTE and increased the dowel looseness [26]. Furthermore, Khazanovic et al. (2009) proposed that a concrete cover greater than 3.5× diameter of the dowel bar does not contribute significantly to the shear capacity [25].

Shoukry (2002) conducted a numerical study to compute contact stress around the dowel bar by comparing the modified dowel design with the conventional dowel bar. The modified dowel design included adding two steel sleeves at the mid-length of the dowel bar while the sleeves were spaced 25–30 mm apart. The outer surface of the sleeves was directly in contact with the concrete slab, while the dowel bar could freely slide inside the sleeves. The results revealed that contact stress could be reduced by 52% through utilizing the modified dowel design. Encouraged by the successful demonstration of the new dowel design, highway engineers in West Virginia conducted field tests on Robert C. Byrd's Highway; for details, see [27]. Table 1 further summarises the important research concerning JPCP, providing a basis for this article.

Table 1. Research overview.

Theme(s)	Method/Methods	Year and References
Numerical study of various parameters, i.e, slab—base interaction and dowel bar locking due to thermal gradient and shrinkage, mislocation of transverse joints, and dowel looseness.	Computational analysis	2003 [28]
Effect of dowel bar deviation (horizontal skew and vertical tilt in rigid pavement), vertical displacement on load transfer capacity.	Computational analysis	2016 [29]
Evaluation of stress field around dowel bar, effect of the type of debonding agent and dowel bar diameter on the pull-out force magnitude and dowel-concrete friction.	Computational analysis and experimental testing	2003 [30]
Investigation of dowel bar installation in New Jersey based on various kinds of coatings (red paint, graphite oil, tar paint, transmission oil, and asphaltic oil) and protective treatments (hot rolling and galvanization on the dowel bar).	Experimental testing	1955 [31]

Table 1. *Cont.*

Theme(s)	Method/Methods	Year and References
Investigation of stress distribution in dowelled jointed concrete slabs due to temperature variations, considering different diameters of dowel bars.	Computational analysis	2014 [32]
Analysis of dowel bar group action using different pavement configurations (slab thickness, concrete elastic modulus, and modulus of subgrade reaction), dowel bar system and wheel loading.	Computational analysis	2009 [33]
Analysis of the structural response of JPCP due to dowel bar looseness utilizing the embedded formulation of a beam element technique for the dowel modelling.	Computational analysis	2000 [34], 2014 [35]
Evaluation of JPCP response based on dowel bar horizontal skew and vertical tilt, interface bond between concrete slab and base and base type (asphalt- and cement0treated base).	Computational analysis	2021 [36]
A study of the structural performance of JPCP considering different dowel bar configurations (standard and special), the influence of base layer on dowel joint behaviour and the effect of dowel bar arrangement on the stresses in concrete and dowel bar.	Computational analysis and experimental testing	2018 [37]
Investigation of dowel bar geometry, spacing and subbase stiffness on the stress in the concrete slab and dowel bar.	Computational analysis	2001 [38]
Investigation of LTE based on material properties (concrete and base layer), load magnitude and bond between the concrete slab and base.	Computational analysis	2018 [39]
Investigation of concrete pavement response subjected to traffic loading and different environmental conditions.	Experimental testing and computational analysis	1997 [40]
Evaluation of improperly aligned dowel bars (horizontal skew and vertical tilt) on joint performance by examining joint lockup and slab cracking.	Computational analysis	2006 [41]
Effect of different parameters related to misaligned dowel bars on the performance of joints and stress states in concrete pavement.	Experimental testing and computational analysis	2007 [42]

Although JPCPs are recognized for their longevity, they constitute less than one percent of the total highway network in Sweden [43,44]. This limited adoption is attributed to the experience with existing concrete pavements, which is mixed. The most recent JPCP was constructed 18 years ago in Uppsala and showed extensive rutting earlier than was anticipated. Another contributing factor to the scarcity is the potential for major deterioration, which is a concern for the Swedish Transport Administration (Trafikverket) due to the time-consuming repair process, including lengthy lane closures, necessitating circuitous routes and disruption for road users [45]. Precast concrete slabs can be installed overnight [46], but their performance relies on the well-functioning dowel bar [47,48].

Furthermore, the literature on precast concrete technology demonstrates variations in the placement of dowel bars, i.e., at the concrete slab's top, bottom, and mid-height [49,50].

However, the dowel bar position is crucial for ensuring the adequate performance of the pavement system. Despite several research studies being conducted over the past decades, the JPCP still requires further investigation to enhance the structural performance of the pavement system and thereby increase the service life. This is particularly important in Sweden, where research on concrete roads is relatively limited. This paper examined the structural response of JPCP through a parametric study of the dowel bar using numerical simulation with finite element software (ABAQUS) based on Swedish concrete pavement design.

2. Aims and Objectives

The aim of the paper is to investigate the structural behaviour of JPCP based on the dowel-related parameters using a three-dimensional finite element model (3D-FEM). The 3D model was developed using Abaqus software and was compared with the analytical solution, i.e., the Westergaard method. Subsequently, dowel bars were introduced in the model to analyse the load transfer capacity. The objectives of the study were as follows:

- Evaluate the effect of different positions of the dowel bar (at the top, bottom, and mid-height of the concrete slab) on LTE and flexural stress in the concrete slab;
- Identify the influence of the joint opening on the LTE;
- Identify the effect of mislocation of the dowel bar (horizontal, vertical, and longitudinal translations) on the LTE;
- Examine the influence of the dowel bar diameter on the LTE;
- Analyse the impact of the bond between the concrete slab and the dowel bar on the LTE.

3. Westergaard Method

In 1926, the Danish-American H.M. Westergaard developed a technique to determine the stresses and deflections due to single wheel load in concrete pavements. This method assumed that an infinite, thin concrete slab rests on an elastic foundation and that the concrete slab and subgrade are always in full contact. Furthermore, a circular contact area is assumed for wheel load [51,52]. Later, these equations were modified by Teller, Sutherland, Kelly, and Eisenmann [53–57]. The derived equations are available for three critical loading conditions, i.e., interior, edge, and corner loading; see Figure 3. The following stress and deflection equations are cited in [57–59], respectively.

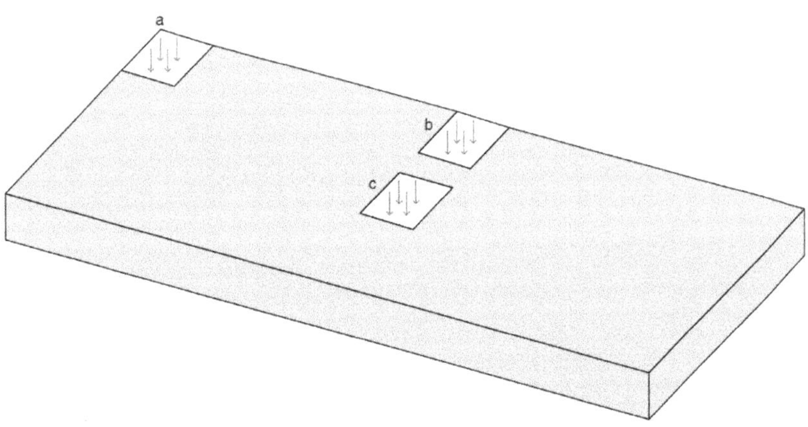

Figure 3. Different loading conditions. (**a**) Corner loading. (**b**) Edge loading. (**c**) Interior loading.

Stress and deflection equations for interior loading

$$\sigma_{\text{centre}} = 0.275(1+\nu)\frac{F}{h_c^2}\left\{\log\frac{E_c h_c^3}{kb^4} - 0.436\right\}$$

$$w_{0_m} = \frac{F}{8kl^2}$$

$$w_{\text{centre}} = w_{0_m}\left[1 + \left(0.3665\log\left(\frac{a}{l}\right) - 0.2174\right)\left(\frac{a}{l}\right)^2\right]$$

Stress and deflection equations for edge loading

$$\sigma_{\text{edge}} = 0.529(1+0.54\nu)\frac{F}{h_c^2}\left\{\log\left(\frac{E_c h_c^3}{kb^4}\right) + \log\left(\frac{b}{1-\nu^2}\right) - 2.484\right\}$$

$$w_{\text{edge}} = \frac{1}{\sqrt{6}}(1+0.4\nu)\frac{F}{kl^2}$$

Stress and deflection equations for corner loading

$$\sigma_{\text{corner}} = \frac{3F}{h_c^2}\left\{1 - 12(1-\nu^2)\frac{k}{E_c h_c^3}\right\}^{0.3}(\sqrt{2}a)^{1.2}$$

$$w_{\text{corner}} = \left(1.1 - 0.88\frac{\alpha}{l}\right)\frac{F}{kl^2}$$

where,

$$l = \left(\frac{E_c h_c^3}{12(1-\nu^2)k}\right)^{0.25}$$

$$k = \frac{E_u}{(h_1^*)^2\left(\frac{1}{h^*} - \frac{h_b}{h_2^*}\right)\left(\frac{1}{h^*} - \frac{1}{h_1^*}\right)}$$

$$h_1^* = 0.83h_c\sqrt[3]{\left(\frac{E_c}{E_u}\right)}$$

$$h_2^* = 0.9h_b\sqrt[3]{\left(\frac{E_b}{E_u}\right)}$$

$$h^* = h_1^* + h_2^*$$

$$b = \begin{cases} \sqrt{1.6a^2 + h_c^2} - 0.675h_c & \text{for } a < 1.724h_c \\ a & \text{for } a > 1.724h_c \end{cases}$$

F = point load (N)
h_c = slab thickness (mm)
h_b = base layer thickness (mm)
a = radius of the loaded circular surface (mm)
b = derived length parameter (mm)
ν = Poisson's ratio (−)
l = radius of relative stiffness (mm)
α = distance of the load from the corner (mm)
E_c = modulus of elasticity of concrete (MPa)
E_b = modulus of elasticity of the base (MPa)

E_u = modulus of elasticity of subgrade (MPa)

k = modulus of subgrade reaction (N/mm^3)

4. Development of a 3D Finite Element Model

ABAQUS is the product of two main analyses, i.e., ABAQUS/Standard and ABAQUS/Explicit. ABAQUS/Standard is a general-purpose program that can be used for both linear and nonlinear problems, while ABAQUS/Explicit is a special-purpose analysis that is suitable for dynamic events such as impact and blast problems and is also efficient for highly nonlinear problems [60]. In this study, ABAQUS/Standard was used to develop a finite element model due to the static nature of the applied load.

4.1. Geometry and Loads

The developed FE model consists of concrete slabs supported by the base, subbase, and subgrade. The slabs of Swedish jointed plain concrete pavements are 5 m long with a lane width of 3.5 m [61]. In this study, concrete slabs were modelled with 2 m × 1.75 m dimensions. This size reduction was intended to enhance computational efficiency and simplify the model, as the study primarily focused on analysing local stresses and deformations. The size reduction would hardly influence local behaviours close to the dowel bar. Although a finer mesh would provide more accurate results, it would also require more computational time than would a coarser mesh due to the greater number of mesh elements. Therefore, reducing the model's size is a practical approach to balancing computational efficiency with accuracy in the FE solution.

The slab, base, and subbase thicknesses were 200 mm, 160 mm, and 300 mm (including 80 mm of the unbounded road base), respectively [45,62]. The subgrade layer was extended to 2500 mm to approximate it as an infinite foundation. The joint width between adjacent slabs was considered 2 mm. The abutting slabs were connected by a steel dowel bar. The steel dowel bars were 32 mm in diameter and 358 mm in length and were placed at the mid-depth of the concrete slab with 300 mm centre-to-centre spacing. The standard axle (a single axle with dual wheels on each side) with 10 metric tons was considered for vehicle load [59]. The single wheel load in the standard axle was 25 kN with a tire pressure of 800 kPa. The dual-wheel loads were modelled on a small rectangular plate, i.e., 150 mm × 208 mm and were 125 mm away from the edge of the loaded slab. A three-dimensional view of the JPCP is shown in Figure 4.

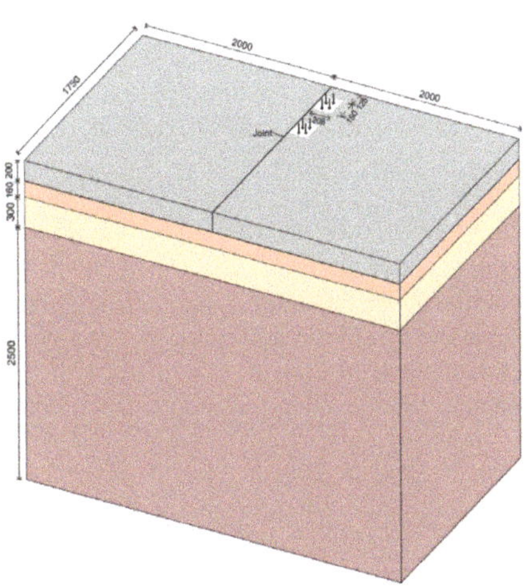

Figure 4. Three-dimensional view of the model (all dimensions are in mm).

4.2. Material Properties

Isotropic elastic materials are considered for concrete slabs, base, subbase, subgrade, and steel dowel bars. The concrete strength class was assumed to be C40/50, and the base property was selected from [59], while the properties of subbase and subgrade were chosen from [63]. The materials' properties are presented in Table 2.

Table 2. Materials properties.

Materials	Parameters	Values	Unit
Concrete	Modulus of elasticity Poisson's ratio	35,000 0.2	MPa -
Base (Cement-bound gravel)	Modulus of elasticity Poisson's ratio	8000 0.2	MPa -
Subbase (gravel)	Modulus of elasticity Poisson's ratio	160 0.35	MPa -
Subgrade (sand)	Modulus of elasticity Poisson's ratio	100 0.35	MPa -
Steel dowel bar	Modulus of elasticity Poisson's ratio	200,000 0.35	MPa -

4.3. Interactions

The tangential behaviour between the concrete slab and base, base and subbase, and subbase and subgrade was modelled using the Coulomb friction model, which considers frictional behaviour in terms of the coefficient of friction acting to oppose relative motion between contacting surfaces. In addition, normal hard contact was modelled to allow for the separation between the different surfaces. The interaction between the surfaces was modelled by considering the surface-to-surface contact [64]. The coefficients of friction between the concrete slab and base, base and subbase, and subbase and subgrade were all considered 1.0. The interaction between the dowel bar and the concrete slab was also developed using the Columb friction model. The interface between half of the dowel bar and the concrete slab was modelled as a perfect bond (i.e., ungreased dowel bar), while the other half interface between the dowel bar and the adjacent concrete slab was modelled in a way that allowed for the longitudinal movement of the dowel bar (i.e., greased dowel bar). Therefore, the coefficient of friction for the perfect bonded side was assumed to be 1.0, and for the greased dowel bar, was assumed to be 0.05 (no data are available in the literature). The separation between the interfaces was facilitated using hard contact model behaviour that considered surface-to-surface contact.

4.4. Boundary Conditions

The connection between the steel dowel bar and the concrete slab solely depends on the interaction properties. However, all degrees of freedom are constrained at the bottom of the subgrade, whereas the sides of the concrete slab, base, subbase, and subgrade are restrained in the vertical direction (i.e., Y-direction) and the respective degree-of-freedom direction. These boundary conditions were implemented by comparing the FE model with the analytical solution. The boundary condition of the 3D model is presented in Figure 5.

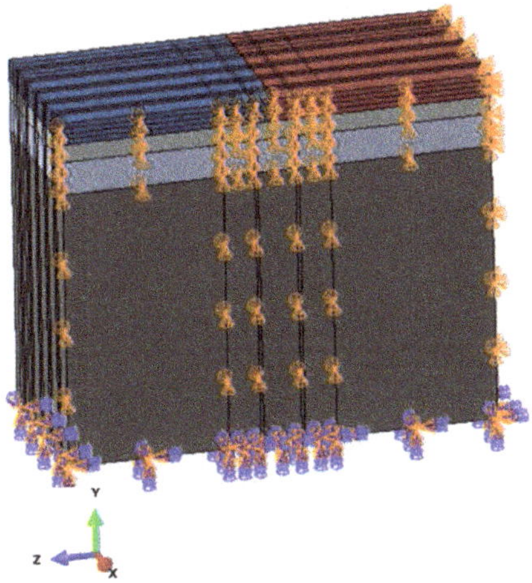

Figure 5. Boundary conditions in the 3D FE model.

4.5. Meshing

All model parts were meshed using eight-noded continuum three-dimensional brick elements (C3D8Rs) with reduced integration and hourglass control. The FE model uses more nodes in full integration, enhancing the element's stiffness. Therefore, reduced integration was chosen, as it accounts for fewer nodes, which, on the other hand, might exaggerate the deflection. However, to mitigate this potential issue, hourglass control was implemented. Additionally, the area around the dowel bar and the proximity of the applied load are critical stress zones, and thus the model was divided into different parts using the partition tool, and a finer mesh was applied in these regions. The model contains a total of 207,800 elements. Figure 6 illustrates the mesh density around the dowel bar.

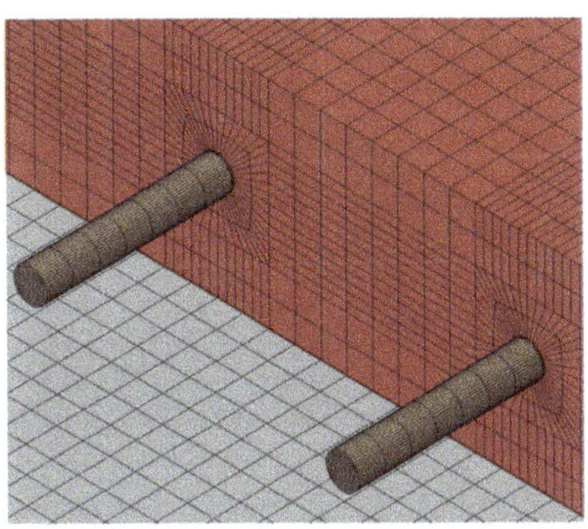

Figure 6. Mesh details around the dowel bar.

5. Results and Discussion

5.1. Comparison of the FE Model with the Westergaard Method

The FE model was checked by comparing the results of computational solutions with the Westergaard method for interior, edge, and corner loadings. The stress and deflection results of different loading conditions are presented in Table 3. It was observed that the difference in the stress values between the computational and analytical solutions ranged between 8 and 60%, while the values of deflection ranged between 28 and 35%. Figure 7 illustrates the locations where stress and deflection were analysed for both interior and edge loading, as acquired from the finite element (FE) analysis.

Table 3. Stress and deflection values for the different loading conditions.

Load Cases	Interior Loading		Edge Loading		Corner Loading	
Analysis Method	**W.M [a]**	**FE Model**	**W.M [a]**	**FE Model**	**W.M [a]**	**FE Model**
Stress (MPa)	0.835	0.659	1.47	1.35	1.613	0.651
Deflection (mm)	0.0675	0.0442	0.24	0.159	0.60	0.427
Difference in stress (%)	21.0		8.5		59.6	
Difference in deflection (%)	34.4		33.7		28.7	

[a] Westergaard's method.

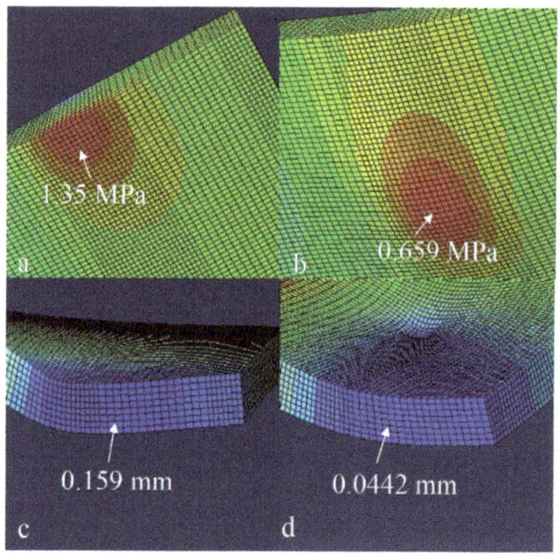

Figure 7. FE stress and deflection locations for different loading conditions. (**a**,**c**) Edge loading. (**b**,**d**) Interior loading.

The difference in the values between the ABAQUS and analytical solution could have been due to the Westergaard method accounting the circular wheel loading and infinite extension of the concrete slab. Furthermore, the Westergaard method considers the deflection beneath the applied load. In reality, the area outside the load within the boundary of the concrete slab also contributes to the deflection [64,65]. However, it was also noted that the Westergaard solution provided fairly good results for interior and edge loading while overestimating the stress for corner loading.

5.2. The Effect of Dowel Bar Position on the Load Transfer Efficiency

The appropriate position of the dowel bar is a critical factor in preventing transverse joint faulting, as it negatively affects ride smoothness and significantly impacts the durability. This section analysed the structural performance of concrete pavement by considering the dowel bar at the concrete slab's top, bottom, and mid-height. In addition, the flexural stress in the concrete slab was also evaluated for these positions of the dowel bar. The deflection values of the loaded and unloaded slabs and the flexural stress in the loaded concrete slab are shown in Table 4, while the LTE for each dowel position is presented in Figure 8.

Table 4. Stress and deflection values based on the position of the dowel bar.

Slab	Stress/Deflection	Units	Dowel Bar Positions		
			Dowel Bar at the Top (h′ * = 50 mm)	Dowel Bar at the Centre (h′ * = 100 mm)	Dowel Bar a at the Bottom (h′ * = 150 mm)
Loaded slab	Stress (σ_L)	MPa	0.59	0.61	0.71
	Deflection ($\triangle_L$)	mm	0.3462	0.3459	0.3028
Unloaded slab	Deflection ($\triangle_{uL}$)	mm	0.3234	0.3241	0.2815

* Vertical depth of dowel bar from the concrete slab surface.

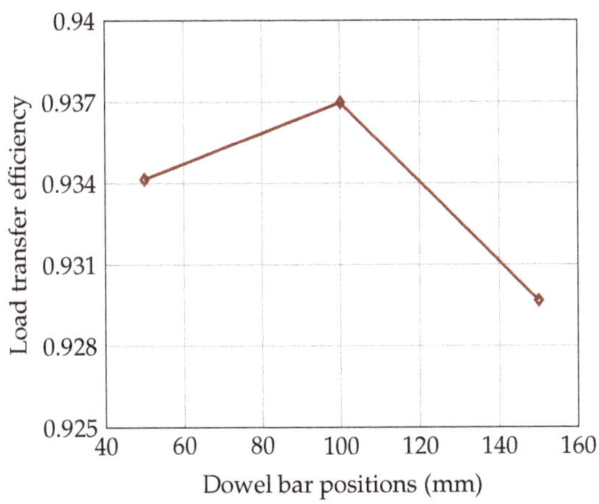

Figure 8. Impact of dowel bar's position on the load transfer efficiency.

Figure 8 shows that the maximum LTE was achieved when the dowel bar was at the mid-depth of the concrete slab. Furthermore, noticeable variations in LTE existed for the dowel bar at different locations, although these differences were small. It is also important to note that dowel bar slots located at the top of the slab increased the exposure of the grout, potentially causing roughness and loose materials on the pavement surface. It is worth mentioning that good slot materials that are properly installed should not deteriorate. However, construction problems can occur, and placing the slots at the centre or bottom of the concrete slab reduces the load and environmental effects on the dowel grouting material. A lower value of flexural stress was observed when the dowel bar was at the top height of the concrete slab. In contrast, higher flexural stress developed when the dowel bar was placed at the bottom height of the concrete slab. This disparity can be attributed to the dowel bar at the bottom height being farther away from the applied load. Consequently, the dowel bar was less effective at this position in supporting flexural stress than when placed at top height of the concrete slab.

5.3. The Effect of Joint Opening on the Load Transfer Efficiency

The concrete slabs contract and joints open when subjected to low temperatures. This section presents the evaluation of the effect of joint opening on dowel bar LTE. The deflection values of loaded and unloaded slabs for different joint widths, i.e., 2 mm, 4 mm, 6 mm, 8 mm, 10 mm, and 12 mm, are presented in Table 5, and the LTE is shown in Figure 9. It can be seen in Figure 9 that the joint width had a significant impact on the LTE; with increasing joint width, the LTE decreases. In reality, aggregate interlock also contributes to the LTE for small joint openings. In precast concrete pavement, joint openings are typically variable and wider than are those in cast-in-place pavement, which maintains uniform narrow widths. The potential variation in the joints of precast concrete pavement may be affected by the precise cut of the damaged slab and the accurate installation of the precast slab. Moreover, wider joint openings demand more sealant materials to maintain the integrity of the pavement structure, and wider sealant materials may face challenges in terms of durability due to increased exposure to traffic and environmental loads. Therefore, recognizing the role of the joint opening is paramount to ensuring the sustained performance and longevity of the pavement system.

Table 5. Computed deflection values based on joint width.

Joint Width mm	Deflection at the Loaded Slab Loaded Slab ($\triangle_L$) mm	Deflection at the Unloaded Slab ($\triangle_{u_L}$) mm
2	0.3459	0.3241
4	0.346	0.3231
6	0.3461	0.3219
8	0.3462	0.3207
10	0.3463	0.3195
12	0.3465	0.3182

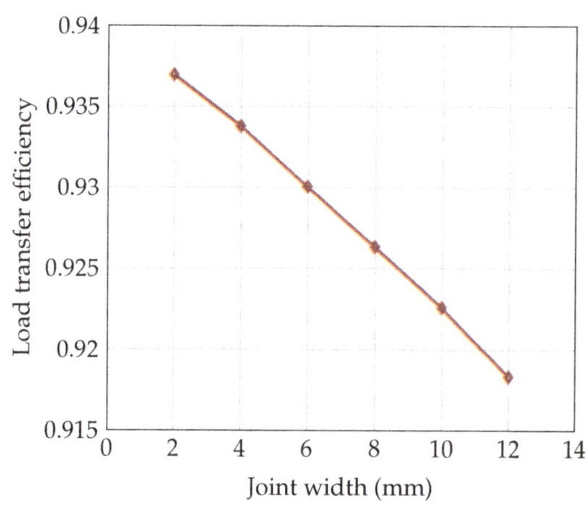

Figure 9. Impact of joint width on the load transfer efficiency.

5.4. The Effect of the Mislocation of Dowel Bar on the Load Transfer Efficiency

This section outlines the analyses of the LTE, which included consideration of the dowel bar's horizontal, vertical, and longitudinal translations; see Figure 10. The horizontal translation of the dowel bar was examined when the dowel bar was located close (i.e., 250 mm from the corner) and far (i.e., 350 mm from the corner) from its intended position (i.e., 275 mm from the corner). In the case of horizontal translation, it can be seen in Figure 10a that the applied load was further moved at 25 mm from the original position (i.e., 125 from the edge) to avoid complexity in the model. The vertical translation was evaluated when the dowel bar's depth deviated 25 mm from its intended position (vertical depth 100 mm); see Figure 10b. The longitudinal translation was analysed when the dowel bar centroid was 50 mm away from the transverse joint, see Figure 10c. The deflection values of loaded and unloaded slabs for each translation are presented in Tables 6–8, and their corresponding LTE is illustrated in Figure 11.

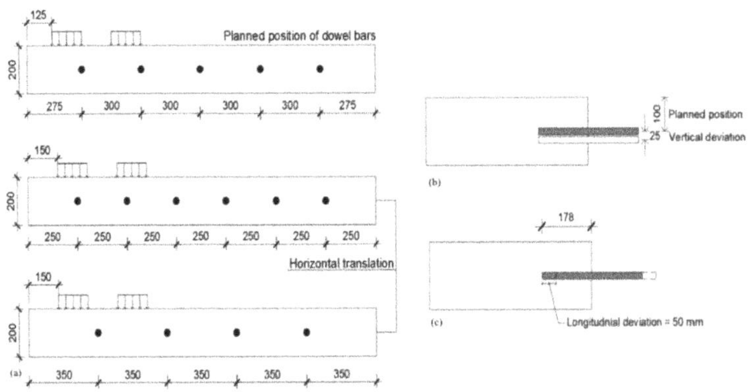

Figure 10. Dowel bars location in the FE model for considering various translations. (**a**) Horizontal translation. (**b**) Vertical translation. (**c**) Longitudinal translation.

Table 6. Computed deflection values based on horizontal translation.

Distance from Edge to the First Dowel Bar mm	Deflection at the Loaded Slab ($\triangle_L$) mm	Deflection at the Unloaded Slab ($\triangle_{u_L}$) mm
250	0.3331	0.3184
275	0.3459	0.3241
350	0.3375	0.3091

Table 7. Computed deflection values based on vertical translation.

Vertical Depth of Dowel Bar mm	Deflection at the Loaded Slab ($\triangle_L$) mm	Deflection at the Unloaded Slab ($\triangle_{u_L}$) mm
100	0.3459	0.3241
125	0.3025	0.2829

Table 8. Computed deflection values based on longitudinal translation.

Embedment Length of Dowel Bar at the Loaded Side of Slab mm	Deflection at the Loaded Slab ($\triangle_L$) mm	Deflection at the Unloaded Slab ($\triangle_{u_L}$) mm
178	0.3459	0.3241
128	0.3465	0.3186

Figure 11 demonstrates the influence of deviation of the dowel bar from its planned position on LTE. Figure 11a reveals that the proximity of the dowel bar to the corner enhances LTE, while the distant placement of the dowel bar from its planned position decreases LTE. Figure 11b,c shows that the deviation of the vertical depth and longitudinal translation of the dowel bar noticeably decreases the LTE. Additionally, improper vertical positioning may compromise the concrete cover, leading to potential corrosion of the dowel bar and resulting in concrete spalling or cracking. Moreover, deviation of the dowel bar centroid from the joint may result in higher bearing stresses, potentially causing concrete cracking and faulting. These findings underscore the importance of precise dowel bar placement for optimal load transfer and structural integrity.

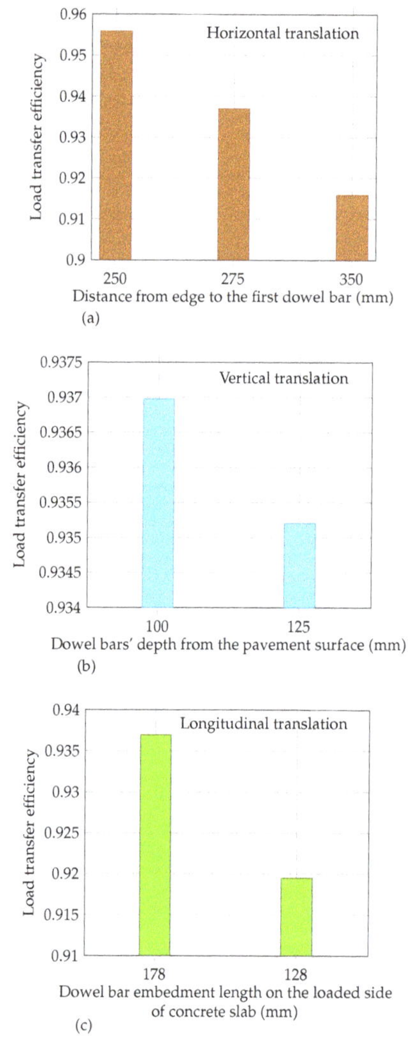

Figure 11. Impact of mislocation of the dowel bar on the load transfer efficiency. (**a**) Horizontal translation. (**b**) Vertical translation. (**c**) Longitudinal translation.

5.5. The Effect of Dowel Bar Diameter on the Load Transfer Efficiency

This section describes the assessment of the structural performance of concrete pavement with various diameters of dowel bars, including 20 mm, 25 mm, 32 mm, and 38 mm. The deflection values of the loaded and unloaded concrete slabs are presented in Table 9, and the corresponding LTE is illustrated in Figure 12. It can be seen in Figure 12 that the dowel's diameter significantly contributes to enhancing the LTE. This is because the cross-sectional area of the dowel bar increases with the diameter, ultimately reducing bearing stress and increasing joint stiffness. Therefore, the adequate size of the dowel bar is a critical factor in the overall concrete pavement design to avoid anticipated distresses such as corner cracking, faulting, and pumping.

Table 9. Computed deflection values based on dowel bar diameter.

Dowel Bar Diameter mm	Deflection at the Loaded Slab ($\triangle_L$) mm	Deflection at the Unloaded Slab ($\triangle_{u_L}$) mm
20	0.3484	0.3191
25	0.3472	0.3218
32	0.3459	0.3241
38	0.3448	0.3251

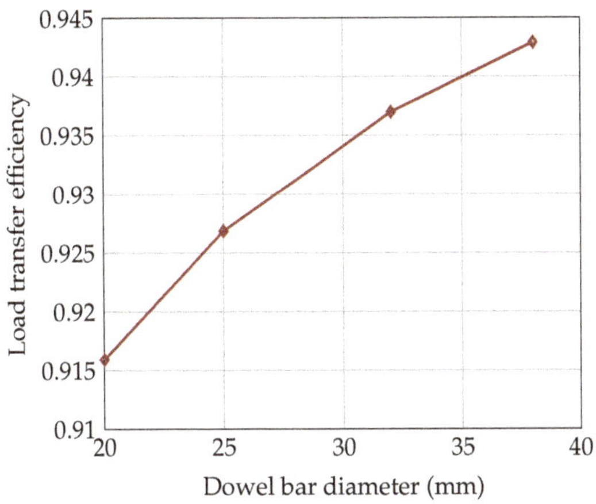

Figure 12. Impact of dowel bar diameter on the load transfer efficiency.

5.6. The Effect of Bonding between the Concrete Slab and Dowel Bar on Load Transfer Efficiency

This section described the examination of the influence of the bond between the concrete slab and dowel bar on LTE based on the coefficient of friction. Half of the interface was assumed to be perfectly bonded, simulating an ungreased dowel bar with a corresponding coefficient of friction of 1.0. Conversely, the other half of the interface exhibited varying bond strengths, resulting in a range of friction coefficients, i.e., 0.05, 0.5, and 1.0. The deflection values of the loaded and unloaded slabs are shown in Table 10, and the corresponding LTE is presented in Figure 13.

Table 10. Computed deflection values based on the coefficient of friction.

Coef. of Friction ν -	Deflection at the Loaded Slab ($\triangle_L$) mm	Deflection at the Unloaded Slab ($\triangle_{u_L}$) mm
0.05	0.3459	0.3241
0.5	0.3458	0.3243
1	0.3458	0.3244

Figure 13 illustrates that the bond at the interface of the concrete slab and dowel bar had a minor effect on LTE. However, Khazanovic et al. (2009) conducted research on the 152 mm embedded dowel bar, and the results demonstrated that greasing the dowel bar reduces the force required for pull-out failure, while an ungreased dowel bar requires a greater force. It is important to note that a higher pull-out force for an ungreased dowel bar may lead to increased resistance against the movement, which could potentially restrict the dowel bar's ability to accommodate the expansion and contraction of the slab due to temperature variations and concrete shrinkage in new construction. Consequently, such restriction in movement may affect the overall performance and durability of the pavement system.

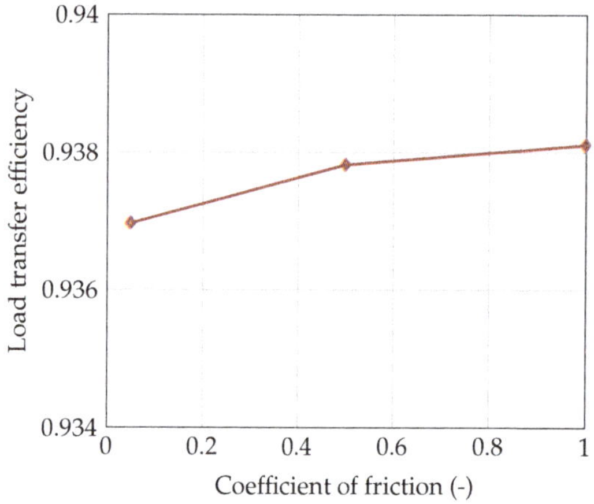

Figure 13. Impact of bond at the interface of concrete slab and dowel bar on the load transfer efficiency.

6. Conclusions

The dowel bar is an essential parameter of the jointed plain concrete pavement system as it provides structural support to the pavement at the joints without restricting the joints opening and closing due to shrinkage and temperature variations. The key function of the dowel bar is to transfer the applied load across the joint. In this study, a finite element model was developed to investigate the interaction between adjacent slabs in terms of LTE due to dowel-related parameters. The FE model was compared with the analytical solution. The following conclusions could be drawn from the numerical simulations:

- The results demonstrate that the maximum LTE is accomplished when the dowel bar is placed at the mid-height of the concrete slab. In addition, higher flexural stress develops in the concrete slab when the dowel bar is located at the bottom height of the concrete slab.
- During winter, the joints open as the concrete slab contracts, potentially causing issues in transferring wheel loads. This study observed a 2.1% reduction in LTE as the joint width increased from 2 mm to 12 mm. Conversely, the LTE improved by 3% with an increase in the diameter of the dowel bar. The dowel bar diameters considered in this study were 20 mm, 25 mm, 32 mm, and 38 mm.
- The study shows that the mislocation of the dowel bar (i.e., horizontal, vertical, and longitudinal translations) has a minimal impact on LTE. An approximate 2.1% difference in LTE was noted for horizontal and longitudinal translations of the dowel bar, while for vertical translation, the reduction in load transfer was around 0.5%. However, a marginal increase in LTE of approximately 0.5% was observed with an increased bond at the interface of the concrete slab and the dowel bar.

The findings of this study provide new knowledge that can be used for ensuring the effectiveness of dowel load transfer systems in jointed plain concrete pavement. Given the substantial anticipated loads during the pavement's service life, the impact of the investigated parameters is significant. Therefore, careful attention to factors such as joint width, dowel bar placement, diameter, and dowel bar deviation from its intended position is essential for optimizing pavement's structural and functional performance, as well as its durability. Prioritizing durability not only leads to lower maintenance costs but also allows transportation agencies to effectively manage infrastructure budgets and improve overall road network efficiency.

7. Future Research

The structural performance of jointed plain concrete pavement can be improved by reducing the bearing stresses at the interface between the concrete and dowel bar. This can be done using non-rounded dowel bars (such as elliptical and plate dowels) and corrosion-resistant materials (such as stainless steel and glass fibre-reinforced polymer dowels). For this, experimental tests can be conducted to assess the structural capacity of the non-rounded dowel bars and the dowel bars fabricated from different materials, with the results being compared with the those of a rounded steel dowel bar of a similar cross-sectional area. Furthermore, this study did not include the effect of skewed dowels on pavement performance, which can be investigated in future research.

Author Contributions: S.Y.: conceptualization, methodology, formal analysis, and writing—original draft preparation, review, and editing. J.S. and R.G.R.B.: validation and writing—review and editing. All authors have read and agreed to the published version of the manuscript.

Funding: This work is financially supported by the Swedish Transport Agency (Trafikverket).

Data Availability Statement: All datasets presented in this study are included in the article.

Conflicts of Interest: The authors declare no conflicts of interest.

References

1. Maitra, S.R.; Reddy, K.S.; Ramachandra, L.S. Load transfer characteristics of aggregate interlocking in concrete pavement. *J. Transp. Eng.* **2010**, *3*, 190–195. [CrossRef]
2. Delatte, N. *Concrete Pavement Design, Construction, and Performance*; Taylor & Francis: Abingdon, UK; New York, NY, USA, 2008; pp. 25–27.
3. Guo, H.; Sherwood, J.A.; Snyder, M.B. Component dowel-bar model for load-transfer systems in PCC pavement. *J. Transp. Eng.* **1995**, *121*, 289–298. [CrossRef]
4. Snyder, M. *Guide to Dowel Load Transfer Systems for Jointed Concrete Roadway Pavements*; Institute for Transportation: Ames, IA, USA, 2011.
5. Yaqoob, S.; Silfwerbrand, J.; Strömberg, L. Evaluation of rapid repair of concrete pavements using precast concrete technology: A sustainable and cost-effective solution. *Nord. Concr. Res.* **2021**, *65*, 107–128. [CrossRef]
6. Tayabji, S. *FHWA Project R05 IAP Funded Project Case Study: Honolulu Interstate H1 Precast Concrete Pavement Demonstration Project*; Federal Highway Administration (FHWA): Washington, DC, USA, 2016.
7. Tayabji, S. *FHWA Project R05 IAP Funded Project Case Study: Mobile Ramp Precast Concrete Pavement Demonstration Project*; Federal Highway Administration (FHWA): Washington, DC, USA, 2017.
8. Federal Highway Administration. *Concrete Pavement Joints (Technical Advisory, T 5040.30)*; Federal Highway Administration: Washington, DC, USA, 2019.
9. Tayabji, S. *Guide Specification for Jointed Precast Concrete Pavement: [techbrief]*; Federal Highway Administration: Washington, DC, USA, 2019.
10. Smith, K.; Harrington, D.; Pierce, L.; Ram, P.; Smith, K. *Concrete Pavement Preservation Guide*, 2nd ed.; National Concrete Pavement Technology Center: Ames, IA, USA, 2014.
11. ACPA. *Dowel Bar Alignment and Location for Placement by Mechanical Dowel Bar Insertion*; American Concrete Pavement Association Guide Specification: Skokie, IL, USA, 2013.
12. Sturges, T.; Frankhouser, A.; Abbas, A.R. Evaluation of dowel bar alignment from a two-step dowel bar inserter. *Int. J. Pavement Eng.* **2014**, *15*, 438–448. [CrossRef]
13. Khaki, A.M.; Azadravesh, E. Generating a 3D model for evaluating the joint opening effects on load transfer efficiency in concrete pavements, using Abaqus. In Proceedings of the 5th National Congress on Civil Engineering, Mashhad, Iran, 4–6 May 2010.
14. Shoukry, S.N.; William, G.W.; Riad, M.Y. *Evaluation of Load Transfer Efficiency Measurement*; West Virginia University, Department of Civil and Environmental Engineering: Morgantown, WV, USA, 2005.
15. Yu, H.T.; Tayabji, S. *Best practices for Dowel Placement Tolerances*; Federal Highway Administration: Washington, DC, USA, 2007.
16. ACPA. *Dowel Bar Alignment and Location*; American Concrete Pavement Association Guide specification: Rosemont, IL, USA, 2018.
17. ACPA. *Evaluating and Optimizing Dowel Bar Alignment*; American Concrete Pavement Association Guide specification: Rosemont, IL, USA, 2006.
18. Kivi, A.K. Dowel bar alignment in concrete pavements—21st century standards and methods. In Proceedings of the Transportation Association of Canada 2020 Conference and Exhibition—The Journey to Safer Roads, Toronto, ON, Canada, 21 September–8 October 2020.
19. Rao, S.; Hoegh, K.; Yu, T.; Khazanovich, L. Evaluation of dowel alignment constructability in Portland cement concrete pavements. *J. Transp. Res. Board* **2009**, *2098*, 86–93. [CrossRef]

20. Tayabji, S. Dowel placement tolerances for concrete pavements. *J. Transp. Res. Board.* **1986**, *10625*, 47–54.
21. Rens, L. *Guide for the Design of Jointed Plain Concrete Pavements*; EUPAVE (European Concrete Paving Association): Brussels, Belgium, 2003.
22. Grosek, J.; Zuzulova, A.; Brezina, I. Effectiveness of Dowels in Concrete Pavement. *Materials* **2019**, *12*, 1669. [CrossRef] [PubMed]
23. Mackiewicz, P. Analysis of stresses in concrete pavement under a dowel according to its diameter and load transfer efficiency. *Can. J. Civ. Eng.* **2015**, *42*, 845–853. [CrossRef]
24. Mackiewicz, P. Finite-element analysis of stress concentration around dowel bars in jointed plain concrete pavement. *J. Transp. Eng.* **2015**, *141*, 1–8. [CrossRef]
25. Khazanovich, L.; Hoegh, K.; Snyder, M. *Guidelines for Dowel Alignment in Concrete Pavements*; National Cooperative Highway Research Program, Transportation Research Board: Washington, DC, USA, 2009.
26. Odden, T.R.; Snyder, M.; Schultz, A.E. *Performance Testing of Experimental Dowel Bar Retrofit Designs, Part 1—Initial Testing*; University of Minnesota, Department of Civil Engineering. Transportation Research Board: Minneapolis, MN, USA, 2003.
27. Shoukry, S.N.; William, G.W.; Riad, M. Characteristics of concrete stresses in doweled transverse joints. *Int. J. Pavement Eng.* **2002**, *3*, 117–129. [CrossRef]
28. Davids, W.G.; Wang, Z.; Turkiyyah, G.; Mahoney, J.P.; Bush, D. Three-dimensional finite element analysis of jointed plain concrete pavement with EverFE2.2. *Transp. Res. Rec.* **2003**, *1853*, 92–99. [CrossRef]
29. Yancong, Z.; Lingling, G. Effect of dowel bar position deviation on joint load-transfer ability of cement concrete pavement. *Int. J. Pavement Res. Technol.* **2016**, *9*, 30–36.
30. Shoukry, S.N.; William, G.W.; Riad, M.Y. Motamarri, S.S. *Effect of Bonding Force on Stresses in Concrete Slab*; West Virginia Department of Transportation: Charleston, WV, USA, 2003.
31. Breeman, W.V. Experimental Dowel Installation in New Jersey. *Bull. Highw. Res. Board* **1955**, *34*, 8–33.
32. Mackiewicz, P. Thermal stress analysis of jointed plane in concrete pavements. *Appl. Therm. Eng.* **2014**, *73*, 1169–1176. [CrossRef]
33. Maitra, S.R.; Reddy, K.S.; Ramachandra, L.S. Load transfer characteristics of dowel bar system in JPCP. *J. Transp. Eng.* **2009**, *135*, 813–821. [CrossRef]
34. Davids, W.G. Effect of dowel looseness on response of jointed concrete pavements. *J. Transp. Eng.* **2000**, *126*, 50–57. [CrossRef]
35. Sii, H.W.; Chai, G.W.; Staden, R.V.; Guan, H. Effect of dowel looseness on response of jointed concrete pavements using three-dimensional finite element analysis. *Adv. Mater. Res.* **2014**, *900*, 435–444. [CrossRef]
36. Silva, E.R.; Balbo, J.T.; Cargnin, A.P. Effects of dowel bars misalignment in jointed plain concrete pavements—A numerical analysis considering thermal differentials and bonded slab-base interface. *IBRACON Struct. Mater. J.* **2021**, *14*, 1–14. [CrossRef]
37. Kim, K.; Chun, S.; Han, S.; Tia, M. Effect of dowel bar arrangements on performance of jointed plain concrete pavement (JPCP). *Int. J. Concr. Struct. Mater.* **2018**, *12*, 39. [CrossRef]
38. Nishizawa, T.; Koyanagawa, M.; Takeuchi, Y.; Kimura, M. Study on mechanical behaviour of dowel bar in transverse joint of concrete pavement. In Proceedings of the 7th International Conference on Concrete Pavements, Orlando, FL, USA, 9–13 September 2001.
39. Sadeghi, V.; Hesami, S. Investigation of load transfer efficiency in jointed plain concrete pavements (JPCP) using FEM. *Int. J. Pavement Res. Technol.* **2018**, *11*, 245–252. [CrossRef]
40. Sargand, S.M.; Hazen, G.A.; Bazeley, C.C.; Copley, J.R.; George, M.E. *Instrumentation of a Rigid Pavement System*; Ohio Department of Transportation, Federal Highway Administration: Columbus, OH, USA, 1997.
41. Leong, P.; Tighe, S.; Rothenburg, L.; Hein, D. Finite difference modelling of misaligned dowel bars and their effects on joint performance. *Transp. Res. Rec.* **2006**, *1946*, 101–110. [CrossRef]
42. Prabhu, M.; Varma, A.H.; Buch, N. Experimental and analytical investigation of mechanistic effects of dowel misalignment in jointed concrete pavements. *Transp. Res. Rec.* **2007**, *2037*, 12–29. [CrossRef]
43. Strömberg, L.; Silfwerbrand, J.; Ansell, A.; Hintze, S. Making concrete pavements competitive by using the standardised framework for comparisons of Infrastructure projects in terms of cost-efficiency and climate impact. *Nord. Concr. Res.* **2020**, *62*, 21–39. [CrossRef]
44. Yaqoob, S. Concrete Pavements' Repair Techniques and Numerical Assessment of Dowel Bar Load Transfer Efficiency. Licentiate Thesis, KTH Royal Institute of Technology, Stockholm, Sweden, 2024.
45. Silfwerbrand, J. Concrete pavements for modern Swedish highways. In Proceedings of the 14th International Symposium on Concrete Roads, Krakow, Poland, 25–28 June 2023.
46. Yaqoob, S.; Silfwerbrand, J. Rapid repair of concrete pavement using precast technology. In Proceedings of the 14th International Symposium on Concrete Roads, Krakow, Poland, 25–28 June 2023.
47. Tayabji, S. *Precast Concrete Pavement Technology Implementation*; Federal Highway Administration (FHWA): Washington, DC, USA, 2019.
48. Tayabji, S. *Load Transfer Systems for Jointed Precast Concrete Pavement: [techbrief]*; Federal Highway Administration (FHWA): Washington, DC, USA, 2015.
49. Smith, P.; Snyder, M. *Manual for Jointed Precast Concrete Pavement*, 3rd ed.; National Precast Concrete Association (NPCA): Washington, DC, USA, 2019.
50. Pierce, L.M.; Weston, J.; Uhlmeyer, J.S. *Dowel Bar Retrofit-Do's and Don'ts*; Washington (State). Department of Transportation—Office of Research and Library Services: Olympia, WA, USA, 2009.

51. Dhawale, A.W.; Vishwanath, M. Comparative study of wheel load stress and warping stress on concrete pavements. *Int. J. Civil Struct. Environ. Infrastruct. Eng. Res. Dev. (IJCSEIERD)* **2014**, *4*, 9–14.
52. Westergaard, H.M. Stresses in Concrete Pavements Computed by Theoretical Analysis. *Public Roads* **1926**, *7*, 25–35.
53. Teller, L.W.; Sutherland, E.C. The structural design of concrete pavements, Part 3—A case study of concrete pavement cross sections. *Public Roads* **1935**, *16*, 45–65.
54. Teller, L.W.; Sutherland, E.C. The structural design of concrete pavements, Part 4—A study of the structural action of several types of transverse and longitudinal joint designs. *Public Roads* **1936**, *17*, 67–114.
55. Teller, L.W.; Sutherland, E.C. The structural design of concrete pavements, Part 5—Experimental study of the Westergaard analysis of stress conditions in concrete pavement slabs of uniform thickness. *Public Roads* **1943**, *23*, 115–160.
56. Kelley, E.F. Application of the research results to the structural design of concrete pavements. *Public Roads* **1939**, *20*.
57. Eisenmann, J. *Betonfahrbahnen. Handbuch für Beton-, Stahlbeton-und Spannbetonbau*; Verlag von Wilhelm Ernst & Sohn: Berlin, Germany; München, Germany; Düsseldorf, Germany, 1979.
58. Petersson, Ö. *Svensk Metod för Dimensionering av Betongvägar*; Licentiatavhandlig, Kungl Tekniska Högsklan Institutioen för Byggkonstruktion: Stockholm, Sweden, 1996.
59. Silfwerbrand, J. *Dimensionering av Betongbeläggningar, (Design of Concrete Pavements)*; KTH Royal Institute of Technology, Department of Structural Mechanics and Engineering: Stockholm, Sweden, 1995.
60. Dassault Systems. Simulia User Assistance 2021. Available online: https://www.3ds.com/support/documentation/user-guides (accessed on 14 January 2024).
61. AB Svensk Byggtjänst. *Betong på Mark (Handbok) (Concrete slabs on grade [Handbook])*; AB Svensk Byggtjänst: Stockholm, Sweden, 2002; p. 253. (In Swedish)
62. STA. *ATBVÄG: Allmän Teknisk Beskrivning för vä Gkonstruktioner*; Swedish Transport Administration: Stockholm, Sweden, 2003. (In Swedish)
63. Chen, F.; Balieu, R.; Kringos, N. Thermodynamics-based finite strain viscoelastic-viscoplastic model coupled with damage for asphalt material. *Int. J. Solids Struct.* **2016**, *129*, 61–73. [CrossRef]
64. El-Matty, A.E.A.; Hekal, G.M.; El-Din, E.M.S. Modeling of dowel jointed rigid airfield pavement under thermal gradients and dynamic loads. *Civ. Eng. J.* **2016**, *2*, 38–51. [CrossRef]
65. Caliendo, C.; Parisi, A. Application of the results of research to the structural design of concrete pavements. *J. Transp. Eng.* **2010**, *136*, 664–667. [CrossRef]

buildings

Article

Natural Gradient Boosting for Probabilistic Prediction of Soaked CBR Values Using an Explainable Artificial Intelligence Approach

Esteban Díaz [1,*] and Giovanni Spagnoli [2]

[1] Department of Civil Engineering, University of Alicante, 03690 Alicante, Spain
[2] DMT GmbH & Co., KG, Am TÜV 1, 45307 Essen, Germany
* Correspondence: esteban.diaz@ua.es

Abstract: The California bearing ratio (CBR) value of subgrade is the most used parameter for dimensioning flexible and rigid pavements. The test for determining the CBR value is typically conducted under soaked conditions and is costly, labour-intensive, and time-consuming. Machine learning (ML) techniques have been recently implemented in engineering practice to predict the CBR value from the soil index properties with satisfactory results. However, they provide only deterministic predictions, which do not account for the aleatoric uncertainty linked to input variables and the epistemic uncertainty inherent in the model itself. This work addresses this limitation by introducing an ML model based on the natural gradient boosting (NGBoost) algorithm, becoming the first study to estimate the soaked CBR value from this probabilistic perspective. A database of 2130 soaked CBR tests was compiled for this study. The NGBoost model showcased robust predictive performance, establishing itself as a reliable and effective algorithm for predicting the soaked CBR value. Furthermore, it produced probabilistic CBR predictions as probability density functions, facilitating the establishment of reliable confidence intervals, representing a notable improvement compared to conventional deterministic models. Finally, the Shapley additive explanations method was implemented to investigate the interpretability of the proposed model.

Keywords: machine learning; CBR; soil index properties; subgrade; compaction characteristics; probabilistic model; explainable artificial intelligence

Citation: Díaz, E.; Spagnoli, G. Natural Gradient Boosting for Probabilistic Prediction of Soaked CBR Values Using an Explainable Artificial Intelligence Approach. *Buildings* **2024**, *14*, 352. https://doi.org/10.3390/buildings14020352

Academic Editors: Andrea Baliello and Di Wang

Received: 24 December 2023
Revised: 23 January 2024
Accepted: 24 January 2024
Published: 26 January 2024

1. Introduction

Most civil engineering projects requiring earthwork, such as highways, earth dams, airport runways, and pavements, typically need the establishment of a suitable subbase layer, i.e., subgrade. Subgrade soil plays a vital role in the long-term performance of pavements because it gives them essential foundational support. The subgrade must meet specific engineering criteria, which include factors like bearing capacity, settlement, and swell properties. The California bearing ratio (CBR) is one of the most frequently used tests to assess the stiffness and bearing capacity of subgrade [1]. The CBR is essentially an indirect measure that compares the strength of a specific soil to the strength of a standard crushed rock, typically expressed as a percentage value. The CBR test is considered a penetration test, initially introduced by the California State Highway Department, USA, to assess the subgrade in the process of flexible pavement design. The CBR test involves using soil previously compacted in a cylindrical mould to perform loading tests on the soil's surface with the aid of a plunger [2,3]. The CBR value of any soil sample can be established in soaked and unsoaked conditions. Nevertheless, the CBR value obtained for soaked soil samples is typically lower than the value obtained for unsoaked samples, and CBR values for soaked soil samples are often considered a conservative estimate for design purposes. On the other hand, the soaked CBR test is time-consuming and labour-intensive because it involves soaking the soil sample underwater for 96 h to simulate the worst-case scenario when the pavement is submerged. Additionally, the soaked CBR test demands a substantial quantity of materials, nearly 6 kg, and involves greater effort in preparing the test specimen. Furthermore, it is important to note that the subgrade soil's properties can exhibit substantial variations even over short distances, primarily due to the heterogeneous nature of soil. This variability can make it challenging to establish consistent design parameters. When building a new road or an earth dam, it becomes necessary to collect numerous soil samples and then estimate the CBR value for each of these samples using the CBR test. This process typically takes at least 4 days per sample (5 to 6 days, including the required modified Proctor test). This can lead to delays in obtaining test results, which may not be practical for making quick decisions related to pavement design, potentially resulting in a significant increase in construction costs. Moreover, the testing method involves expenses related to material transportation (from the construction site to the laboratory) and testing fees. Additionally, test results can occasionally be unreliable due to sample disturbance and issues with the quality of the laboratory conditions. Therefore, it can be challenging to ensure that the soil sample used for the CBR test accurately reflects the in situ conditions. Considering the points previously discussed, many works have tried to establish correlations between the CBR value and readily determinable index properties of soils. These tests are easy to conduct and are typically performed on soil samples as soon as they are brought to the laboratory. The study conducted by Jumikis [4] is one of the early attempts to establish a correlation between CBR values and Atterberg limits. Black [5] formulated a correlation to estimate the CBR value of cohesive soils, utilizing the plasticity index (PI) and the liquidity index (LI) as key parameters. Ring [6] focused on estimating the CBR value using Atterberg limits along with key compaction parameters, i.e., optimum moisture content (OMC) and maximum dry density (MDD). De Graft-Johnson, Bhatia [7] attempted to estimate the CBR value based on Atterberg limits and the soil's grain size distribution. Katte, Mfoyet [8] used the linear regression technique (LR) to establish a relationship between the CBR value and MDD. Additionally, they developed a group of multiple linear regression (MLR) models for estimating the CBR value, taking into account factors such as grain size distribution, Atterberg limits, OMC, and MDD. Patel and Desai [9] used MLR to establish correlations between CBR values and the liquid limit (LL), plastic limit (PL), PI, OMC, and MDD of cohesive soils. Hassan, Alshameri [10] proposed a set of MLR models for estimating the CBR value based on the grain size distribution, Atterberg limits, OMC, and MDD. Thus, numerous researchers have made several attempts to establish a relationship between the soaked CBR value and parameters associated with grain size distribution (such as gravel, sand, and fine content as well as the particle sizes corresponding to 10%, 30%, 50%, and 60% finer materials on the cumulative particle size distribution curve), Atterberg's limits (LL, PL, and PI), and compaction parameters (OMC and MDD). However, a comprehensive review of some conventional studies conducted by Taskiran [11] revealed that satisfactory correlations based on LR or MLR could not be attained in many cases because most of the proposed empirical equations lacked a high degree of accuracy and did not provide a generalized solution. Wang and Yin [12] analysed several traditional expressions and concluded that these expressions could not be considered highly reliable. This was primarily because most of these predictive equations were developed through regression analysis using a limited dataset, typically consisting of a small number of samples, ranging from 20 to 158, and were specific to certain soil types. As a result, it was noted that the degree of accuracy of these relationships was generally unsatisfactory in many cases, leading to poor predictions of CBR values. Therefore, more sophisticated and satisfactory approaches should be employed to estimate CBR values using soil properties. It should be noted that machine learning (ML), with its ability to perform nonlinear modelling, offers a viable tool for simulating complex processes [13,14]. Since the development of ML techniques, numerous studies have adopted them, considering the same input parameters as traditional correlations, and have achieved superior performance metrics compared to traditional methods (e.g., [15,16]). The study of Taskiran [11] marked the first application of ML techniques to predict the

CBR value in soils. A summary of these can be consulted in Bardhan, Gokceoglu [17], Khasawneh, Al-Akhrass [18], Othman and Abdelwahab [19], or Verma, Kumar [20]. It is worth emphasizing the study conducted by Bardhan, Gokceoglu [17], in which they employed four soft computing techniques and analysed 312 experimental results for soaked CBR values. They concluded that multivariate adaptive regression splines with piecewise linear models achieved the highest level of accuracy with a coefficient of determination (R^2) value of 0.969. On the other hand, this study analysed most of works where ML techniques were applied to predict the CBR value [11,21–35]. The study concluded that the works conducted with small datasets (consisting of fewer than 160 data points) exhibited higher predictive accuracies, with R^2 values between 0.81 and 1.00, in contrast to the studies conducted with larger datasets ranging from 358 to 389 soil test results [21,22]. When small datasets are utilized for ML models, they often yield exceptionally high performance; however, these models tend to suffer from overfitting and lack generalization ability. Conversely, the results also indicated that the prediction accuracy of ML models decreased as the number of samples increased, with accuracies ranging from 0.78 to 0.80 in terms of R^2, which is less than desirable. It is worth noting that studies based on larger datasets are generally considered to be more reliable than those conducted with a limited number of samples. Posteriorly, Hao and Pabst [36] compiled 39 samples for predicting CBR values and the resilient modulus of crushed waste rock. They used MLR, k-nearest neighbours (KNN), decision tree (DT), random forest (RF), multilayer perceptron (MLP), and neuroevolution of augmenting topologies (NEAT). The RF model provided better results with an R^2 value of 0.926. Khasawneh, Al-Akhrass [18] utilized 110 data points and investigated three ML techniques, namely, artificial neural networks (ANNs), M5P model tree, and KNN, alongside conventional methods such as MLR and nonlinear regression (NLR). The study concluded that the most effective model for predicting the CBR value was the ANN model, which achieved an R^2 value of 0.905, followed by KNN, MLR, M5P, and NLR in descending order of performance. Khatti and Grover [37] compiled 182 training, 15 testing, 36 validation, and 12 cross-validation data points for fine-grained soil. They employed gene expression programming (GEP) and relevance vector machine (RVM) models for predicting the soaked CBR value. The genetic algorithm-optimized Laplacian kernel-based SRVM model outperformed the others, achieving a correlation coefficient (r) of 0.9874. Othman and Abdelwahab [19] developed 240 ANNs with various hyperparameters. They achieved an R^2 value of 0.945, surpassing the performance of traditional MLR models. The size of the dataset was 77 samples. Verma, Kumar [20] collected 1011 soil samples and used kernel ridge regression, KNN, and Gaussian process regression (GPR) to predict the soaked CBR value of soils. GPR exhibited the highest performance, achieving an r value of 0.83 on the test set. Kamrul Alam and Shiuly [38] considered three distinct models developed using three different methodologies: a fuzzy inference system (FIS), an ANN, and an adaptive neuro-fuzzy inference system. To conduct the study, a dataset of 2000 soil samples was collected. The hybrid model of ANN and FIS (ANFIS) outperformed both ANN and FIS in terms of predictive accuracy, achieving an r value of 0.92. It should be noted that in some cases, when dealing with highly complex datasets, a better/more optimized network or model may not necessarily improve the model accuracy. Reducing the complexity of a dataset can also potentially improve the accuracy of the ML model [39,40]. Thus, in order to obtain good performance metrics, it is necessary to employ ML models to predict the CBR value based on a large number of samples, which allows the algorithm to generalize when it is tested with new samples.

However, ML methods have limitations for field applications, presenting two main challenges. Firstly, the uncertainty in predicted results makes reliability analysis difficult. Secondly, the models often lack interpretability. To enable reliability analysis, quantifying the uncertainty for each data point is necessary. There are two primary sources of uncertainty in data sets: (1) aleatoric uncertainty stems from the inherent variability in the data generation process, and (2) epistemic uncertainty is caused by limitations in the predictive capabilities of the model. Probabilistic forecasts hold greater value than deterministic ones because they offer additional information about prediction uncertainty. These data can be utilized by engineers to reduce risk and make more advantageous decisions. Thus, the deterministic perspective offers insight into the model's accuracy in making precise predictions, while the probabilistic perspective sheds light on the model's ability to understand and quantify the uncertainty inherent in those predictions. Together, they provide a nuanced and in-depth evaluation of the model's overall effectiveness and reliability in a regression setting. On the other hand, due to the growing complexity of these models, interpreting their results presents a significant challenge. A recently developed technique in the field of explainable artificial intelligence (XAI), known as Shapley additive explanation (SHAP) within the ML community, can be employed to interpret these models. Using the SHAP explainable ML model enables the determination of the importance and directional impact of each input variable on the predicted results, thereby providing insights into the relationships between input variables and the output.

Thus, this study attempted to predict the soaked CBR values through a probabilistic approach, using variables previously correlated with it. These variables encompassed gravel content (G), sand content (S), fine content (F) expressed in terms of silt and clay content, LL, PI, OMC, and MDD.

The main contributions of this study include the following: (1) the proposal and application of a reliable model for predicting the soaked CBR value considering a large database to ensure proper generalisation, (2) the ability of the proposed model to provide not only accurate point prediction but also estimation of the predictive uncertainty for reliable decision-making, and (3) an explanation approach based on explainable artificial intelligence to understand the importance of the input factors and estimate the quantitative results of the impact of every feature on the soaked CBR value.

2. Methodology

In the present investigation, the variables G, S, F, LL, PI, OMC, and MDD were used as input parameters to predict the soaked CBR value. For this aim, the study followed the framework illustrated in Figure 1. The adopted framework consists of five main parts:

(1) Data preparation. In this case, an outlier analysis and a normalization of the values were performed.
(2) ML model selection. Four probabilistic ML models were compared to select the best one.
(3) ML model optimization. The artificial bee colony algorithm was selected to obtain the definitive fine-tuned model.
(4) ML model evaluation. First, it is evaluated from a deterministic perspective and then from a probabilistic perspective to establish the confidence intervals and uncertainty in the predictions.
(5) Implementing an explanation approach based on XAI.

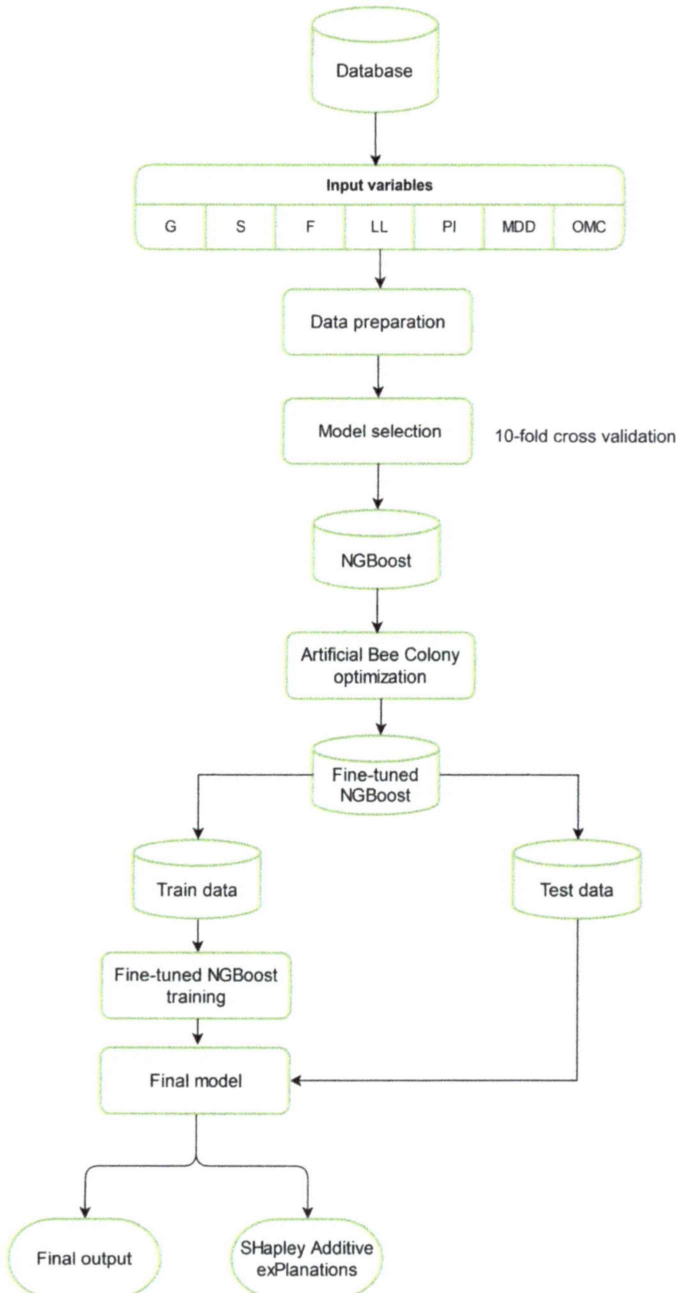

Figure 1. Framework of the methodology performed.

3. Database

A total of 2130 soil samples were collected from several construction projects in Spain. These soil samples were divided into training and testing sample sets in an 80/20 ratio. The laboratory tests were conducted in accordance with Spanish standard specifications, which are mostly equivalent to European standards. These tests were grain size distribution [41],

Atterberg limits [42], modified Proctor [43], and the soaked CBR [44]. Through these laboratory tests, numerous geotechnical parameters were collected such as the gravel content (G), sand content (S), silt and clay content termed fine content (F), liquid limit (LL), plasticity index (PI), maximum dry density (MDD), optimum moisture content (OMC), and soaked CBR value. The laboratory-obtained soil database comprised a wide spectrum of soils, ranging from soils classified as gravel according to the USCS, accounting for 33.5% of the total, sands at 29.4% of the total, and fine soils making up 37.1% of the total. Regarding plasticity characteristics, there were non-plastic soils or low-plasticity soils, i.e., LL < 50 (97.5%), and high-plasticity soils, i.e., LL ≥ 50 (2.5%). The main statistics of the considered properties are summarized in Table 1.

Table 1. Descriptive statistics of the dataset.

	G (%)	S (%)	F (%)	LL (%)	PI (%)	MDD (g/cm^3)	OMC (%)	CBR
Mean	21.08	38.18	40.75	28.18	9.89	1.97	10.28	29.01
Std. deviation	27.01	18.11	26.34	10.29	7.97	0.23	3.66	27.23
Minimum	0	1.80	0.70	0	0	1.46	4.10	0.90
25th percentile	0	26.20	15.40	20.60	4.00	1.79	6.50	7.10
Median	1.90	33.80	39.40	25.90	7.30	1.96	11.40	17.50
75th percentile	53.80	50.50	61.88	33.30	13.60	2.12	12.90	55.00
Maximum	73.90	90.20	98.20	82.30	53.80	2.52	20.00	144.00
Skewness	0.69	0.54	0.34	1.65	1.89	0.13	−0.17	0.91
Kurtosis	−1.38	−0.20	−1.04	5.04	5.82	−0.71	−0.93	−0.59

From the analysis of Table 1, it can be concluded that in the dataset, G had a mean of 21.08%, with a high standard deviation of 27.01%, suggesting substantial variability. The skewness and kurtosis indicated a moderate skew and a flatter than normal distribution. S had a mean of 38.18% and a standard deviation of 18.11% reflecting moderate variability. The data followed a slightly skewed, nearly normal distribution. F showed a mean of 40.75% and a standard deviation of 26.34%, indicative of a broad spread of data points. The skewness and kurtosis pointed to a slightly right-skewed and somewhat flatter distribution. LL had a mean of 28.18% and a standard deviation of 10.29%, showing lesser but significant variation. The data indicated a significantly skewed and leptokurtic distribution. The PI presented a mean of 9.89% and a standard deviation of 7.97%, suggesting moderate variability. Its high skewness and kurtosis indicated a highly skewed and leptokurtic distribution. MDD had a mean of 1.97 g/cm^3 and a low standard deviation of 0.23 g/cm^3, showing very little variation in values. MDD followed a nearly symmetric, slightly platykurtic distribution. OMC had a mean of 10.28% and a standard deviation of 3.66%, indicating moderate variability. Its skewness of −0.17 and kurtosis of −0.93 suggested a nearly symmetric, slightly flat distribution. Lastly, CBR values had a median of 17.5, a mode of 2.0, a mean of 29.01, and a high standard deviation of 27.23, showing a wide spread in these values. The distribution of CBR values exhibited a moderate right skew, indicating that there is a clustering of data points on the lower end of the scale with a few higher-value data points. Additionally, it displayed a flatter peak and lighter tails compared to a normal distribution, suggesting a lower occurrence of extreme values.

In Figure 2, scatter plots of the analysed parameters are depicted, providing a descriptive overview of the data distribution. Furthermore, each scatterplot included the best-fit line and the correlation coefficient (r) value for added context. These graphs are useful for detecting linear correlations between variables, as if it were a correlation matrix.

The analysis of these plots revealed several interesting relationships between variables. A strong negative correlation existed between G and F (−0.77), suggesting that as one increases, the other tends to decrease significantly, an obvious fact. Similarly, the F and CBR values showed a strong negative correlation (−0.76). This pattern was also seen with G and OMC (−0.83), indicating inverse relationships. Conversely, G positively correlated with MDD (0.73) and the CBR value (0.83), implying that these pairs of variables tend to increase or decrease together. LL and PI exhibited a very high positive correlation (0.90), suggesting they are likely to change in the same direction, which is quite normal as these variables are highly related. Other correlations, such as S with G (−0.37) and S with MDD (−0.31), were less pronounced but still notable.

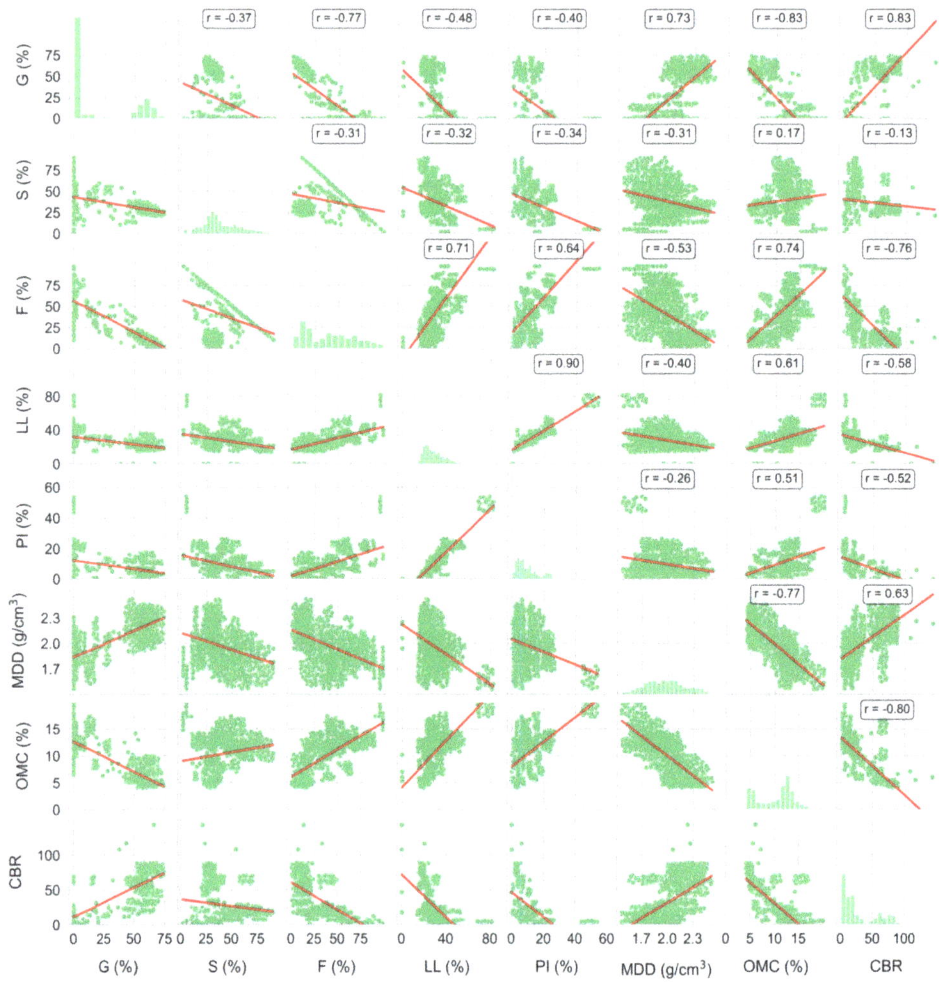

Figure 2. Correlation analysis including scatter plots and distribution histograms of the considered variables.

4. Machine Learning Process

4.1. Data Preparation

Before starting the ML process, it is necessary to conduct an outlier study on the dataset because outliers can significantly distort statistical analyses and predictive modelling, leading to inaccurate results. Identifying these anomalies ensures more reliable and generalizable insights from the data. The isolation forest (IF) algorithm [45] was employed for this purpose. This is an effective algorithm for anomaly detection and is particularly suited for high-dimensional datasets. It operates on the principle that anomalies are data points that are few and different, making them easier to 'isolate' than normal points. The algorithm selects a feature at random and then chooses a split value between the maximum and minimum values of the selected feature. This process creates 'isolation trees' or random decision trees. For each data point, the path length from the root to the node is calculated. Anomalies are expected to have shorter path lengths on average, as they are more susceptible to isolation. This path length forms the basis for the anomaly score, which is used to determine outliers: the shorter the path length, the more anomalous the data point. Thus, in an IF, anomalies are indeed 'isolated' more efficiently compared to normal points. After applying the IF algorithm to our dataset, 64 data points were removed. On the other hand, a normalization process was conducted to standardize the data values to a common scale while preserving the original differences in value ranges. This was achieved using min-max normalization, which scaled the data between 0 and 1.

4.2. Model Selection

Once the initial data were processed, four probabilistic ML algorithms were initially selected as candidates to choose the optimal prediction model through a cross-validation process with k = 10. The selection of k = 10 in k-fold cross-validation is a widely accepted practice that strikes a balance between bias and variance in the performance estimation of ML models [46]. Opting for a higher k value can decrease variance but may lead to increased bias. Conversely, a lower k tends to increase variance while reducing bias. Therefore, taking into account the dataset's size, k = 10 constitutes a practical compromise, ensuring that each fold includes a sufficiently large sample size and offering 10 distinct training-testing splits for comprehensive model evaluation. In this case, the following models were used: natural gradient boosting (NGBoost), quantile regression forest, Bayesian ridge regression, and Gaussian process regression.

NGBoost [47] represents a probabilistic model based on gradient boosted decision trees. However, in contrast to quantile regression or conditional mean estimation, NGBoost is designed to learn the complete probability distribution as an inherent part of its methodology. The fundamental concept behind NGBoost involves the utilization of natural gradients instead of traditional gradients, enabling the algorithm to effectively model a probability distribution across the outcome space, taking into account the predictor variables. Quantile regression forest [48,49] is a non-parametric method for estimating conditional quantiles, offering flexibility for modelling complex relationships and handling non-Gaussian distributions. It utilizes an ensemble of decision trees to predict different quantiles of the response variable, providing a comprehensive view of uncertainty in high-dimensional datasets. Bayesian ridge regression is a probabilistic technique employed to formulate and estimate statistical models [50]. Its probabilistic perspective offers significant advantages, particularly in scenarios where the data distribution is non-ideal or when the dataset is limited. Bayesian ridge regression leverages probability distributions to make predictions. Instead of yielding a single point estimate, this approach provides a full probability distribution over the predicted values. Gaussian process regression (GPR) [51] is a non-parametric approach that stands out in making probabilistic predictions by modelling data points as variables in a Gaussian process. By utilizing a covariance kernel, it effectively captures intricate non-linear relationships and dependencies between data points. Through Bayesian inference, GPR updates beliefs about the underlying function, providing predictions as Gaussian probability distributions, characterized by the mean and variance.

In Figure 3 are shown raincloud plots representing the distribution of R^2 and mean absolute error (MAE) values obtained in the cross-validation process. A raincloud plot combines elements of a half-violin plot, a boxplot, and a strip plot. This comprehensive visualization enables full statistical interpretation of the data. Specifically, it showcases the distribution's shape (half-violin plot), the five-number summary statistics including the minimum, first quartile (25th percentile), median, third quartile (75th percentile), and maximum (boxplot component), and the sample size and exact positions of individual data points (strip plot). Additionally, their means are highlighted in a darker colour. This format provides a nuanced view of the data, highlighting both its aggregate properties and individual variations. In this figure, to enhance visualization, Gaussian process regression is not included due to its low metrics (R^2 below 0.5).

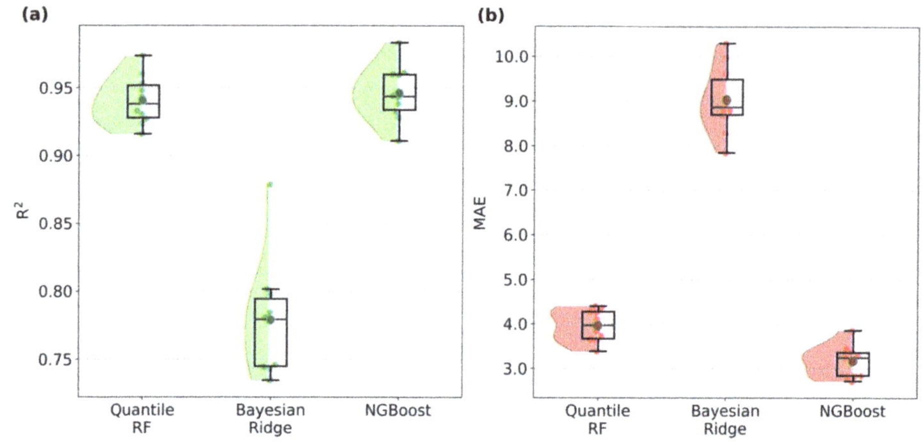

Figure 3. Raincloud plots of the initial probabilistic models considered. (**a**) R^2, (**b**) MAE.

Based on the analysis of Figure 3, it is evident that NGBoost exhibits the best overall performance among the algorithms, followed by quantile regression forest and Bayesian ridge. Conversely, Gaussian process demonstrates the poorest performance. Consequently, NGBoost was selected as the definitive model.

4.3. *Artificial Bee Colony Optimization*

After conducting the model selection study and determining that the algorithm to be used in this study for predicting the CBR value is NGBoost, an optimization of its hyperparameters was carried out. To achieve this, the artificial bee colony (ABC) algorithm was employed. The ABC algorithm [52] is a metaheuristic optimization algorithm inspired by the foraging behaviour of honey bee colonies. It works by iteratively updating a population of artificial bees, which represent potential solutions to the optimization problem. The ABC algorithm has three types of bees: employed bees, onlooker bees, and scouts. Employed bees explore the search space and evaluate the quality of the solutions they find. Onlooker bees then choose solutions to explore based on the quality of the solutions found by employed bees. Scouts explore new areas of the search space. The ABC algorithm is a powerful and versatile optimization technique that can be used to solve a wide range of complex optimization problems (e.g., [53]). This process was performed three times. The first time was used to choose the base model of NGBoost where the following were considered the next models: RF, extremely randomized trees (ET), and DT. The best base model was ET, and then this model was tuned by the ABC algorithm. The parameters to be adjusted in the ET base model were the number of trees in the forest (n_estimators), the maximum depth of them (max_depth), the minimum samples required for node splitting (min_samples_split), the minimum number of samples required in a leaf node (min_samples_leaf), and the maximum features to consider for node splitting (max_features). In Table 2, the search values and the selected values for each of them are included. In all optimization processes carried out, the search ranges for the hyperparameters were primarily determined through the collective expertise of the authors and common values in ML. Furthermore, an additional layer of validation in the selection process ensured that the chosen hyperparameters were not at the extremes of their respective ranges.

Once the base model was optimized, the process was repeated with NGBoost. For this purpose, a normal distribution was set as the probability distribution to model the predictions. The tuned ET base model was also used, and the optimization process was carried out using the ABC algorithm. In this case, the hyperparameters to be optimized were the number of trees in the forest (n_estimators), the fraction of data per mini-batch during training (minibatch_frac), and the step rate at which the model learns from the training data (learning_rate). The results of this optimization process are summarized in Table 3.

Table 2. Inputs for tuning hyperparameters of the ET base model using ABC.

Hyperparameter	Search Range	Optimal Value
n_estimators	1–200	51
max_depth	1–20	10
min_samples_split	2–15	10
min_samples_leaf	1–15	4
max_features	sqrt, log2, None	None

Table 3. Inputs for tuning hyperparameters for the NGBoost model using the ABC algorithm.

Hyperparameter	Search Range	Optimal Value
n_estimators	1–200	50
minibatch_frac	0–1	0.5
learning_rate	0.005–0.5	0.05

4.4. Fine-Tuned NGBoost

After tuning the hyperparameters, the mean prediction accuracy of the fine-tuned NGBoost was assessed using the test set. In this way, the training and testing sets were divided in an 80/20 proportion. The metrics obtained are included in Table 4. This analysis included R^2, MAE, root mean squared error (RMSE), and the a20-index [54]. The a20-index provides a clear and practical interpretation for engineering applications. It specifically denotes the percentage of samples where the predicted values are within a $\pm 20\%$ range of deviation from the experimentally measured values.

Table 4. Summary of performance metrics for the fine-tuned NGBoost.

Set	R^2	MAE	RMSE	a-20 Index
Train	0.975	1.97	3.68	0.9025
Test	0.967	2.53	4.26	0.8697

In Figure 4a are presented scatterplots comparing actual versus measured CBR values, showcasing the performance of the NGBoost model post hyperparameter fine-tuning for predicting the CBR mean values. From the analysis of Table 4 and Figure 4a, it can be observed that the NGBoost algorithm has been adequately trained, exhibiting similar performance on both the training and testing sets. On the testing set, the R^2 was 0.967, and the RMSE and MAE values were 4.26 and 2.53, respectively. Nearly 87% of the mean prediction results fell within the ± 20 relative error boundary for the testing set. This suggested that NGBoost's regression capacity for scalar prediction was satisfactory, and it offers the added benefit of probabilistic predictions. Figure 4b includes the residuals obtained by comparing the model's mean predictions with the actual measured values for both the training and testing datasets. The results from the analysis of residuals indicated that in the training set, the mean of residuals was 0.16, the median was -0.07, and the standard deviation of residuals was 3.25. In the testing set, the mean of residuals was 0.09, the median was -0.15, and the standard deviation of residuals was 3.74. Generally, the analysis of the residuals and error metrics for both the training and testing sets revealed that the metrics were quite similar across both sets. This similarity in metrics suggested that the model performed consistently on both the training and testing data. Regarding the Durbin–Watson statistic, it was 1.999 for the residuals in the training set and 2.121 for the testing set. The Durbin–Watson statistic ranges from 0 to 4, where a value close to 2 suggests there is no significant autocorrelation in the residuals. The obtained values, being approximately 2, indicated that there was likely no positive or negative autocorrelation in the residuals of both sets. This suggested that the model was effective in capturing correlation in the data and that the residuals were approximately independent of each other.

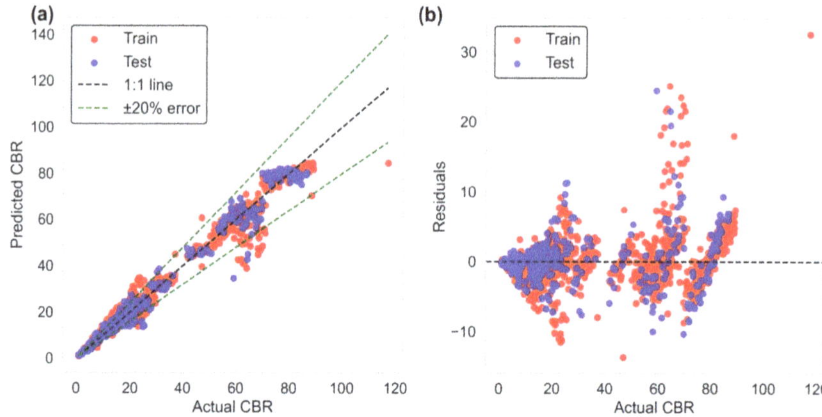

Figure 4. Fit between the mean predictions and the actual values for the training and testing sets in the NGBoost model. (**a**) Actual vs. predicted CBR values for the training and testing sets, (**b**) residuals of the predictions.

In Figure 5, the ratios between the test results and predictions are shown as functions of database variables. The NGBoost model demonstrated a consistently reasonable level of prediction accuracy across the entire range of database variables. The slopes of the linear regression lines (in pink) were nearly zero for all database features, confirming the suitability of the chosen input parameters. As mentioned earlier, almost all predictions fell within a ±20% error range, with no observed systematic deviations in the value ranges for any variables.

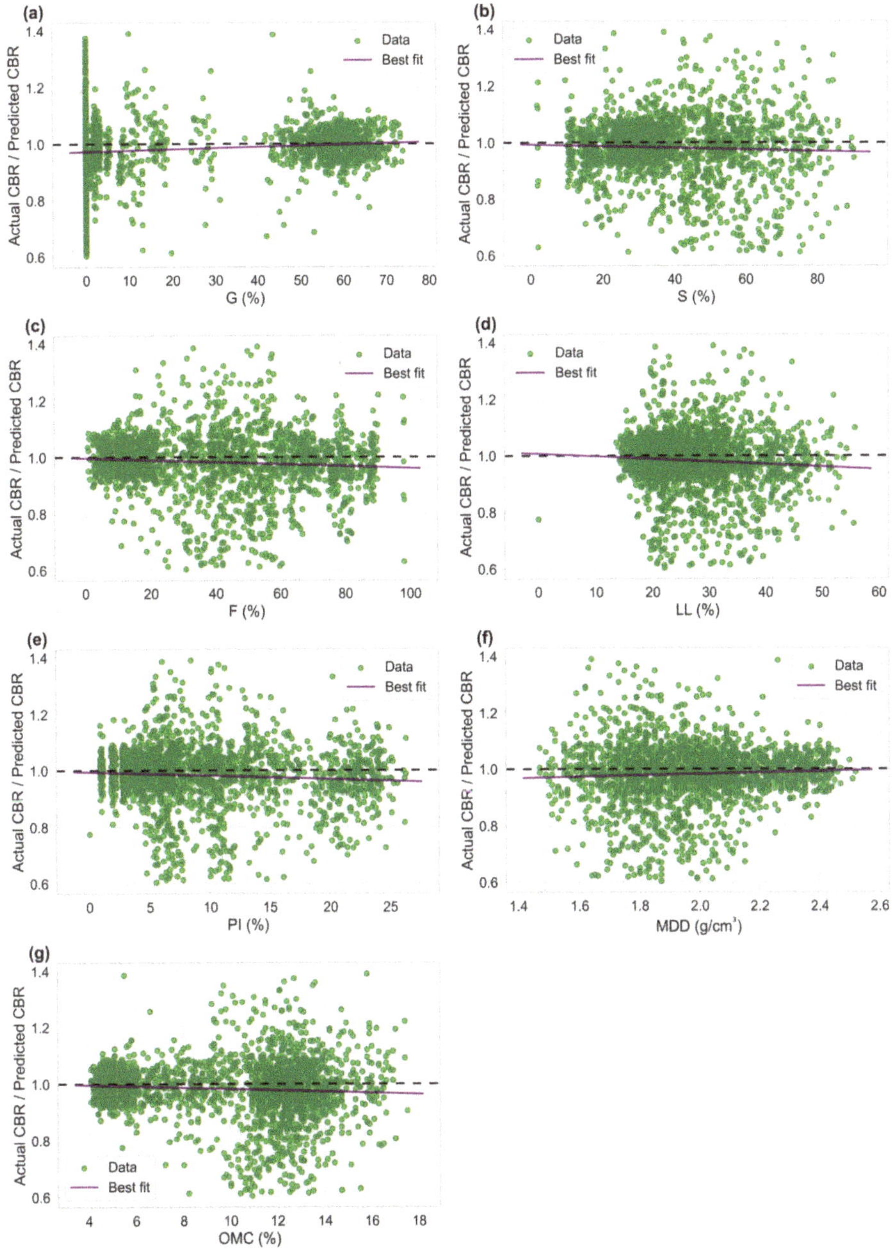

Figure 5. Prediction ratios versus database variables considering the entire dataset. (**a**) Gravel content, (**b**) sand content, (**c**) fine content, (**d**) liquid limit, (**e**) plasticity index, (**f**) maximum dry density, and (**g**) optimum moisture content.

5. Analysis of Uncertainty in the Predictive Model

In this section, the applicability of the probabilistic approach of NGBoost for predicting soaked CBR values in soils is analysed. It should also be highlighted that, in addition to providing reasonably accurate point estimates, the NGBoost-based probabilistic prediction model can also offer the variance of the point to account for the inherent uncertainty in the response. The NGBoost's probabilistic predictions, based on a normal distribution, are displayed in Figure 6. In addition to the mean prediction (μ), two confidence intervals, dependent on the standard deviation (std), were also provided using the NGBoost model. It can be observed that the majority of the test data (77.44%) fell within the interval $\mu \pm$ std, corresponding to a confidence level of 68.2%. Furthermore, 96.15% of the test data fell within the interval $\mu \pm 2$ std, corresponding to a confidence level of 95.4%. The evaluation of the NGBoost model for CBR values below 20 demonstrated a high level of reliability in the estimations, as evidenced by the average width of the confidence intervals, which was 4.88 for 68.2% and 9.75 for 95.4%. This suggested that the model provided estimations with a reasonably well-defined degree of uncertainty, which is crucial for users to understand the potential margins of error in their analyses. Furthermore, the model's reliability was reinforced by the fact that 92.41% of the actual data fell within the $\mu \pm$ std interval and that 98.66% fell within the $\mu \pm 2$ std interval. This indicated a strong ability of the model to predict accurately within this range of CBR values. This is significant because this range of values is where soil values are predominantly found, especially in soils where the greatest support capacity problems are encountered. On the other hand, the evaluation of the NGBoost model for CBR values above 40 exhibited distinct characteristics when compared to lower CBR values. The average width of the confidence intervals was 8.88 for 68.2% and 17.76 for 95.4%. These wider values suggested that the model provided estimations with a higher degree of uncertainty in this upper range of the CBR value, which is an important consideration for users when interpreting the results and understanding potential margins of error. In terms of reliability, it could be observed that 57.26% of the actual data fell within the $\mu \pm$ std interval and that 94.02% fell within the $\mu \pm 2$ std interval. Although most of the actual values still fell within these confidence intervals, the proportion was significantly lower compared to the CBR range below 20. This indicated a greater level of uncertainty in predictions for soils with higher bearing capacity. This fact can be explained by the scarcity of samples in the training dataset for CBR values > 40 (approximately half compared to CBR < 20), which means that the model had less information to learn the characteristics of that range. This could lead to more uncertain estimations because the model was not sufficiently exposed to data in that specific range. For the range of CBR values between 20 and 40, the behaviour was intermediate between the ones analysed. Obviously, these findings could not have been obtained by deterministic ML methods.

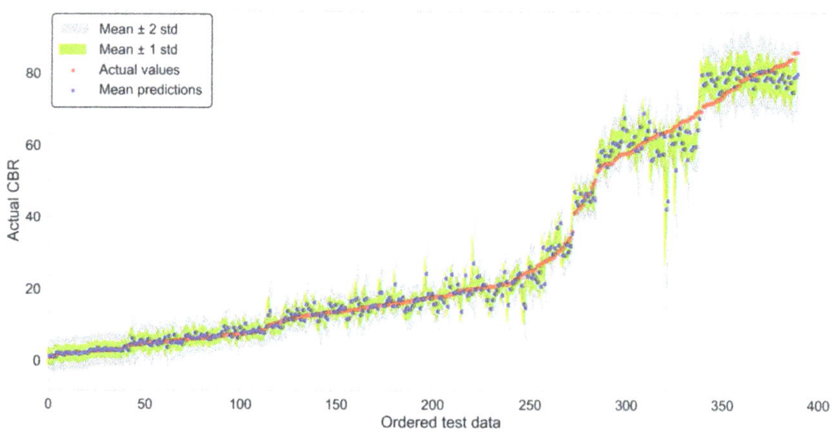

Figure 6. Performance of the NGBoost model for predicting the probabilistic distribution of CBR values. Note that the predicted CBR values are in ascending order. The grey and green areas indicate the $\mu \pm 2$ std (95.4%) and $\mu \pm$ std (68.2%) prediction intervals, respectively.

Additionally, to showcase the probabilistic prediction capabilities of the NGBoost model, Figure 7 includes three examples of density plots depicting the predicted probability distributions for three specific data points. These data points were the closest to the CBR values of (a) 5.0, (b) 25, and (c) 50. The true value (red) and mean (green) are marked along with shaded regions for ±1 std and ±2 std from the mean. The distributions characterize the uncertainty in the predictions at those data points.

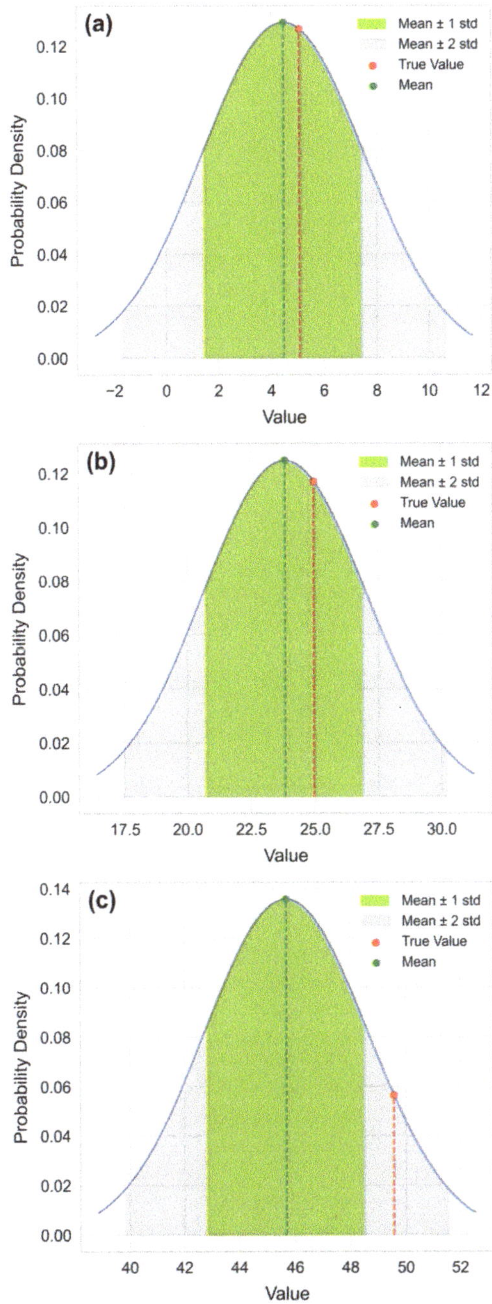

Figure 7. Probability density distributions for three selected data points in the test set. (**a**) CBR value closest to 5, (**b**) 25, and (**c**) 50.

These graphs provide a visual representation of how the NGBoost model generated probabilistic predictions for specific data points, which helps to understand the uncertainty associated with these predictions and how they relate to the real values. Consequently, engineers can have greater confidence in their decisions, especially when considering predetermined target reliability levels. These visualizations are valuable for the interpretation and evaluation of probabilistic models, something that is not achieved with average performance metrics as is usually used in deterministic ML models.

6. Interpretability Analysis of the Predictive Model

SHAP [55] was utilized to identify and interpret the impact of various factors on predicting the CBR value using the NGBoost probability model. SHAP is a method in ML that provides insights into how models make their predictions by utilizing concepts from cooperative game theory. Essentially, it breaks down a prediction to show the contribution of each feature. SHAP calculates this by considering the impact of including or excluding a particular feature on the model's output. It does this for every feature, across all possible combinations of other features, to ensure a fair attribution. This is based on the idea of Shapley values, which were originally developed to distribute payouts in cooperative games fairly. In the context of machine learning, these 'payouts' are the predictions, and SHAP assigns each feature a value that represents its contribution towards making the prediction. This method stands out for its ability to handle complex interactions between features and provide both global (across all predictions) and local (specific to one prediction) interpretations. By offering a detailed breakdown of feature contributions, SHAP helps in understanding, trusting, and improving ML models. SHAP enables a comprehensive analysis of how various soil properties and conditions impact the predicted CBR value, establishing SHAP as an indispensable tool for interpreting complex ML models. The SHAP values, representing the impact of a factor and a sample on the model output, are illustrated in the summary plot depicted in Figure 8. Additionally, in this plot, variables were ordered by their importance, with those having the greatest impact on the model's performance appearing at the top and those with lesser impact positioned towards the bottom. The summary plot serves to elucidate the ranking of factors by importance and their specific effects on the model's predictions. It visually depicts each factor's contribution, where the importance of a factor reflects its overall impact on the model's predictive performance, and its effect indicates the specific role in influencing individual predictions. The colour red signifies a positive influence where the factor increases the prediction value, while blue denotes a negative influence, reducing the prediction value. Furthermore, the positions of points further away from zero indicate stronger influences on the prediction outcome. Each point represents a SHAP value for an individual observation, offering a detailed view of how each factor alters the regression model's predictions. This dual interpretation of colour and distance from zero provides a comprehensive understanding of the factors' directional impact (positive or negative) and their magnitude of effect on the model's predictions.

From the analysis of the summary plot, it can be concluded that G was the most important variable for predicting the CBR value, and furthermore, an increase in G had a positive impact on the CBR value. Next in importance was the variable OMC, although in this case, high values of OMC tended to decrease the model's output. In the third position of importance was the variable F, indicating that high values of F decreased the model's output. The variable of greatest significance following F was PI, and its higher values decreased the model's output. Subsequently, was LL, which behaved similarly to OMC, F, and PI, as high values of LL tended to decrease the model's output. In the second-to-last position of importance was MDD, with a slight tendency for high values of MDD to positively impact the CBR value. Finally, the least important variable in the model was S, where the trend is not clear, with a fairly balanced distribution on both sides of zero. All of these findings were consistent with the established principles in soil mechanics and the known relationships among soil properties (e.g., [56,57]).

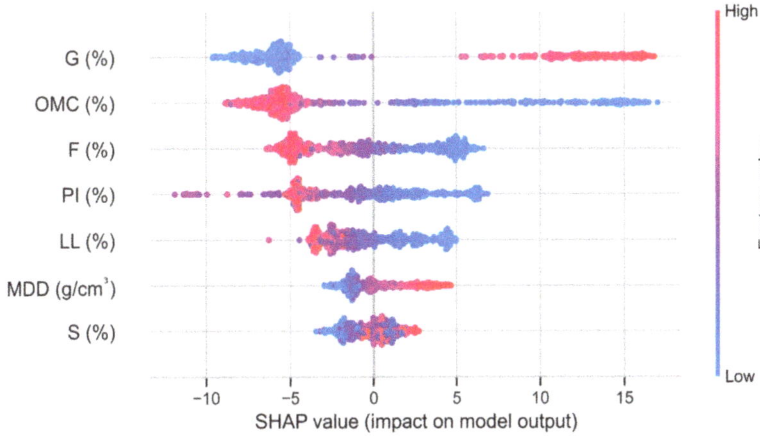

Figure 8. SHAP summary plot for the proposed model.

In Figure 9 are depicted the scatter plots of SHAP values, which aid in understanding the contribution of each feature to the model's predictions. Each point on the plots corresponds to an individual prediction from the model, with the position on the Y-axis reflecting the magnitude and direction of a feature's impact on that prediction. A higher SHAP value indicates a higher contribution to the model's prediction, while a lower value indicates a lower or negative contribution.

From the examination of the individual plots provided, it could be observed that G displayed a positive relationship that appeared slightly non-linear. In contrast, S showed widely dispersed points without a clear linear trend, indicating that the S feature may have influenced predictions in a complex, non-proportional way. F revealed a clearly negative and rather linear trend, where an increase in F correlated with a proportional decrease in the impact on the output. Turning to the LL, there was a negative trend that seemed non-linear. For PI, the trend was similar to that for LL. MDD showed a positive and linear relationship, with the impact on the output increasing in a proportional manner as MDD rose. Lastly, OMC indicated that the impact on the output significantly declined as OMC increased, suggesting a negative and slightly non-linear relationship, particularly stable at the higher range of OMC values. Considering the obtained results, it can be concluded that the algorithm has predominantly comprehended the relationship between each independent variable and the CBR value, aligning with the well-known relationships between properties of soils (e.g., [56,57]). In conclusion, a better understanding of the key factors affecting CBR values facilitates the optimization of construction practices for cost-effectiveness. This insight allows engineers to make informed choices regarding material selection, taking into account the variables' significance and impact direction. Such knowledge guides engineers in concentrating their efforts strategically to maximize benefits.

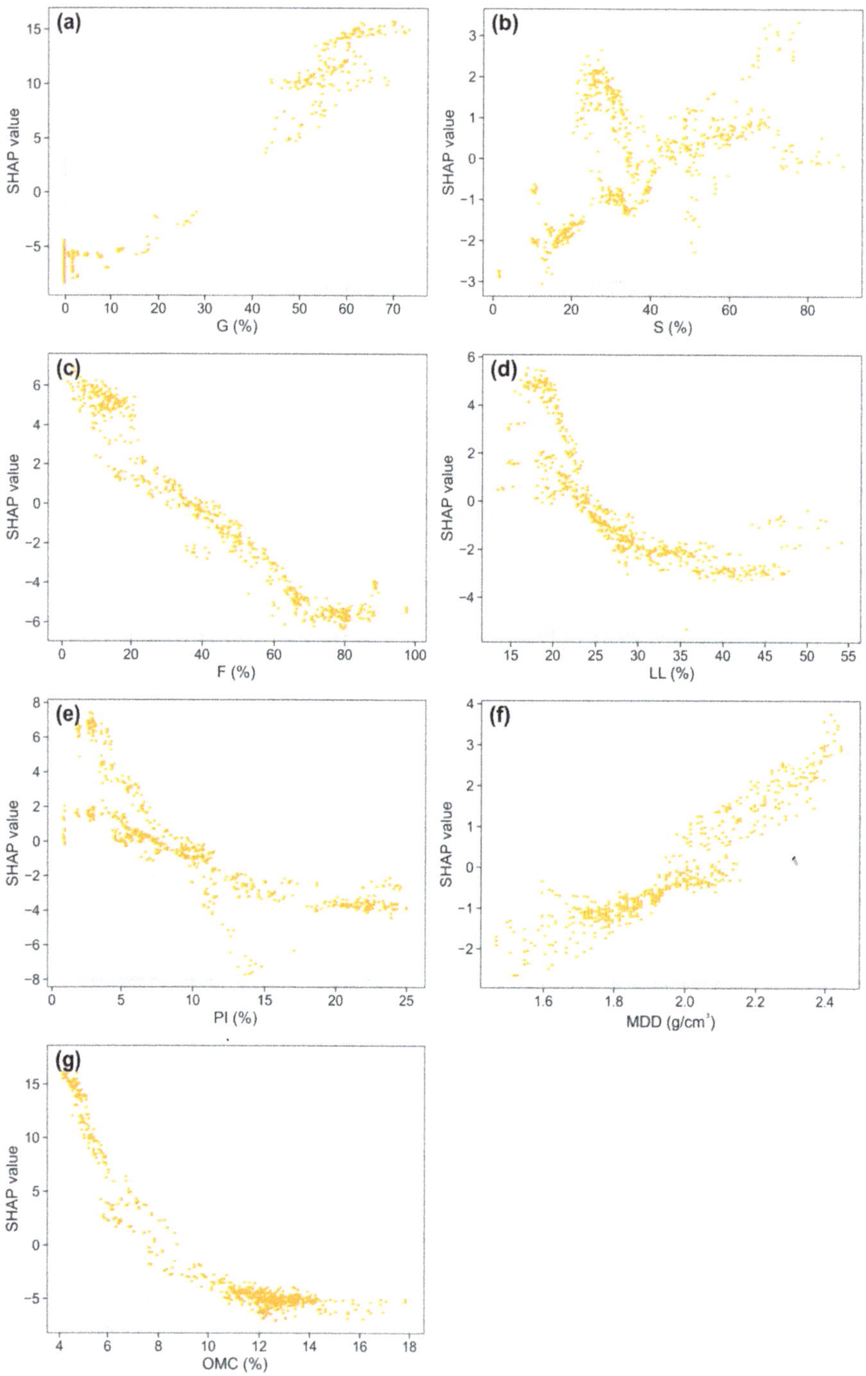

Figure 9. SHAP value scatter plots for the input variables. (**a**) Gravel content, (**b**) sand content, (**c**) fine content, (**d**) liquid limit, (**e**) plasticity index, (**f**) maximum dry density, and (**g**) optimum moisture content. Note that at the bottom of each plot, the distribution histogram of the data is displayed in light grey.

7. Conclusions

This paper presents an ML model based on the natural gradient boosting algorithm for probabilistic predictions of the soaked CBR value in soils. The model was developed employing a database of 2130 soaked CBR tests, which is considerably larger than the databases used to develop existing design models. A model selection study of probabilistic ML models revealed that NGBoost consistently achieved the best performance. Subsequently, the hyperparameters of the NGBoost model were extensively tuned and optimized using the artificial bee colony algorithm to enhance model predictions. The optimized NGBoost model demonstrated outstanding performance metrics on the testing set, with R^2, MAE, and RMSE values of 0.967, 2.53, and 4.26, respectively. With regard to the obtained performance metrics, when compared to studies that do not consider small datasets, they are very similar to those obtained by Bardhan, Gokceoglu [17] and above the rest of the works. Furthermore, NGBoost was utilized to estimate predictive uncertainty and demonstrated that 96.15% of the actual soaked CBR values in the test dataset fell within the 95.4% confidence level. Additionally, the predictions of CBR values lower than 20 showed a high degree of reliability and low uncertainty. This demonstrated the model's robust ability to provide precise predictions within the range of CBR values lower than 20, which encompasses the majority of soil types. However, for CBR values exceeding 40, the model exhibits increased uncertainty and lower reliability compared to CBR values below 20. This implementation demonstrated a robust and effective approach, not only providing point predictions but also quantifying the associated uncertainty caused by the quality of the data and the number of samples. This is crucial in data-driven decision-making within contexts of uncertainty. As a result, a more reliable evaluation of the CBR value can be carried out. The optimized NGBoost model was interpreted using the SHAP method, which indicated that the gravel content, optimum moisture content, and fine content are the most important parameters affecting the soaked CBR value. Finally, it is important to emphasize that the algorithm was trained on a large database, but the samples are localized to a specific site, giving it a strong local character. Extrapolation to other areas should be conducted with caution.

Author Contributions: Conceptualization, E.D. and G.S.; methodology, E.D. and G.S.; software, E.D.; validation, E.D. and G.S.; formal analysis, E.D. and G.S.; investigation, E.D. and G.S.; data curation, E.D.; writing—original draft preparation, E.D.; writing—review and editing, E.D. and G.S.; visualization, E.D.; supervision, E.D. and G.S.; project administration, E.D. All authors have read and agreed to the published version of the manuscript.

Funding: This research received no external funding.

Data Availability Statement: The data presented in this study are available on request from the corresponding author. The data are not publicly available due to their current utilization for future works involving the authors of this paper.

Acknowledgments: The authors would like to express their gratitude to ITC, S.L. for their valuable assistance with the sample data.

Conflicts of Interest: Author Giovanni Spagnoli was employed by the company DMT GmbH & Co., KG. The remaining authors declare that the research was conducted in the absence of any commercial or financial relationships that could be construed as a potential conflict of interest.

References

1. Davis, E. The California bearing ratio method for the design of flexible roads and runways. *Géotechnique* **1949**, *1*, 249–263. [CrossRef]
2. BS, B.S. *Methods of Test for Soils for Civil Engineering Purposes*; British Standards Institution: London, UK, 1990.
3. *ASTM D1883-21*; Standard Test Method for California Bearing Ratio (CBR) of Laboratory-Compacted Soils. ASTM: West Conshohocken, PA, USA, 2016.
4. Jumikis, A.R. Geology of soils of the newark (NJ) metropolitan area. *J. Soil Mech. Found. Div.* **1958**, *84*, 1646-1–1646-41. [CrossRef]
5. Black, W. A method of estimating the California bearing ratio of cohesive soils from plasticity data. *Geotechnique* **1962**, *12*, 271–282. [CrossRef]
6. Ring, G. Correlation of compaction and classification test data. *Hwy. Res. Bull.* **1962**, *325*, 55–75.

7. De Graft-Johnson, J.; Bhatia, H.; Gidigasu, D. The engineering characteristics of the laterite gravels of Ghana. In Proceedings of the 7th International Conference on Soil Mechanics and Foundation Engineering, San Fandila, Mexico, 21 July 1969; pp. 117–128.
8. Katte, V.Y.; Mfoyet, S.M.; Manefouet, B.; Wouatong, A.S.L.; Bezeng, L.A. Correlation of California Bearing Ratio (CBR) Value with Soil Properties of Road Subgrade Soil. *Geotech. Geol. Eng.* **2019**, *37*, 217–234. [CrossRef]
9. Patel, R.S.; Desai, M. CBR predicted by index properties for alluvial soils of South Gujarat. In Proceedings of the Indian Geotechnical Conference, Mumbai, India, 16–18 December 2010.
10. Hassan, J.; Alshameri, B.; Iqbal, F. Prediction of California Bearing Ratio (CBR) Using Index Soil Properties and Compaction Parameters of Low Plastic Fine-Grained Soil. *Transp. Infrastruct. Geotechnol.* **2022**, *9*, 764–776. [CrossRef]
11. Taskiran, T. Prediction of California bearing ratio (CBR) of fine grained soils by AI methods. *Adv. Eng. Softw.* **2010**, *41*, 886–892. [CrossRef]
12. Wang, H.-L.; Yin, Z.-Y. High performance prediction of soil compaction parameters using multi expression programming. *Eng. Geol.* **2020**, *276*, 105758. [CrossRef]
13. Díaz, E.; Salamanca-Medina, E.L.; Tomás, R. Assessment of compressive strength of jet grouting by machine learning. *J. Rock Mech. Geotech. Eng.* **2023**, *16*, 102–111. [CrossRef]
14. Díaz, E.; Spagnoli, G. Gradient boosting trees with Bayesian optimization to predict activity from other geotechnical parameters. *Mar. Georesources Geotechnol.* **2023**, 1–11. [CrossRef]
15. Venkatasubramanian, C.; Dhinakaran, G. ANN model for predicting CBR from index properties of soils. *Int. J. Civil Struct. Eng.* **2011**, *2*, 614–620.
16. Sabat, A.K. Prediction of California bearing ratio of a soil stabilized with lime and quarry dust using artificial neural network. *Electron. J. Geotech. Eng.* **2013**, *18*, 3261–3272.
17. Bardhan, A.; Gokceoglu, C.; Burman, A.; Samui, P.; Asteris, P.G. Efficient computational techniques for predicting the California bearing ratio of soil in soaked conditions. *Eng. Geol.* **2021**, *291*, 106239. [CrossRef]
18. Khasawneh, M.A.; Al-Akhrass, H.I.; Rabab'ah, S.R.; Al-Sugaier, A.O. Prediction of California Bearing Ratio Using Soil Index Properties by Regression and Machine-Learning Techniques. *Int. J. Pavement Res. Technol.* **2022**. [CrossRef]
19. Othman, K.; Abdelwahab, H. The application of deep neural networks for the prediction of California Bearing Ratio of road subgrade soil. *Ain Shams Eng. J.* **2023**, *14*, 101988. [CrossRef]
20. Verma, G.; Kumar, B.; Kumar, C.; Ray, A.; Khandelwal, M. Application of KRR, K-NN and GPR Algorithms for Predicting the Soaked CBR of Fine-Grained Plastic Soils. *Arab. J. Sci. Eng.* **2023**, *48*, 13901–13927. [CrossRef]
21. Tenpe, A.R.; Patel, A. Utilization of Support Vector Models and Gene Expression Programming for Soil Strength Modeling. *Arab. J. Sci. Eng.* **2020**, *45*, 4301–4319. [CrossRef]
22. Al-Busultan, S.; Aswed, G.K.; Almuhanna, R.R.A.; Rasheed, S.E. Application of artificial neural networks in predicting subbase CBR values using soil indices data. *IOP Conf. Ser. Mater. Sci. Eng.* **2020**, *671*, 12106. [CrossRef]
23. Yildirim, B.; Gunaydin, O. Estimation of California bearing ratio by using soft computing systems. *Expert Syst. Appl.* **2011**, *38*, 6381–6391. [CrossRef]
24. Kumar, S.A.; Kumar, J.P.; Rajeev, J. Application of machine learning techniques to predict soaked CBR of remolded soils. *IJERT* **2013**, *2*, 3019–3024.
25. Bhatt, S.; Jain, P.K.; Pradesh, M. Prediction of California bearing ratio of soils using artificial neural network. *Am. Int. J. Res. Sci. Technol. Eng. Math* **2014**, *8*, 156–161.
26. Varghese, V.K.; Babu, S.S.; Bijukumar, R.; Cyrus, S.; Abraham, B.M. Artificial Neural Networks: A Solution to the Ambiguity in Prediction of Engineering Properties of Fine-Grained Soils. *Geotech. Geol. Eng.* **2013**, *31*, 1187–1205. [CrossRef]
27. Sabat, A.K. Prediction of California bearing ratio of a stabilized expansive soil using artificial neural network and support vector machine. *Electron. J. Geotech. Eng.* **2015**, *20*, 981–991.
28. Erzin, Y.; Turkoz, D. Use of neural networks for the prediction of the CBR value of some Aegean sands. *Neural Comput. Appl.* **2016**, *27*, 1415–1426. [CrossRef]
29. Ghorbani, A.; Hasanzadehshooiili, H. Prediction of UCS and CBR of microsilica-lime stabilized sulfate silty sand using ANN and EPR models; application to the deep soil mixing. *Soils Found.* **2018**, *58*, 34–49. [CrossRef]
30. Suthar, M.; Aggarwal, P. Predicting CBR Value of Stabilized Pond Ash with Lime and Lime Sludge Using ANN and MR Models. *Int. J. Geosynth. Ground Eng.* **2018**, *4*, 6. [CrossRef]
31. Farias, I.G.; Araujo, W.; Ruiz, G. Prediction of California bearing ratio from index properties of soils using parametric and non-parametric models. *Geotech. Geol. Eng.* **2018**, *36*, 3485–3498. [CrossRef]
32. Kurnaz, T.F.; Kaya, Y. Prediction of the California bearing ratio (CBR) of compacted soils by using GMDH-type neural network. *Eur. Phys. J. Plus* **2019**, *134*, 326. [CrossRef]
33. Alam, S.K.; Mondal, A.; Shiuly, A. Prediction of CBR Value of Fine Grained Soils of Bengal Basin by Genetic Expression Programming, Artificial Neural Network and Krigging Method. *J. Geol. Soc. India* **2020**, *95*, 190–196. [CrossRef]
34. Islam, M.R.; Roy, A.C. Prediction of California bearing ratio of fine-grained soil stabilized with admixtures using soft computing systems. *J. Civil Eng. Sci. Technol.* **2020**, *11*, 28–44. [CrossRef]
35. Taha, S.; Gabr, A.; El-Badawy, S. Regression and Neural Network Models for California Bearing Ratio Prediction of Typical Granular Materials in Egypt. *Arab. J. Sci. Eng.* **2019**, *44*, 8691–8705. [CrossRef]

36. Hao, S.; Pabst, T. Prediction of CBR and resilient modulus of crushed waste rocks using machine learning models. *Acta Geotech.* **2022**, *17*, 1383–1402. [CrossRef]
37. Khatti, J.; Grover, K.S. Prediction of soaked CBR of fine-grained soils using soft computing techniques. *Multiscale Multidiscip. Model. Exp. Des.* **2023**, *6*, 97–121. [CrossRef]
38. Kamrul Alam, S.; Shiuly, A. Soft Computing-Based Prediction of CBR Values. *Indian Geotech. J.* **2023**, 1–15. [CrossRef]
39. Bolón-Canedo, V.; Remeseiro, B. Feature selection in image analysis: A survey. *Artif. Intell. Rev.* **2020**, *53*, 2905–2931. [CrossRef]
40. Kabir, H.; Garg, N. Machine learning enabled orthogonal camera goniometry for accurate and robust contact angle measurements. *Sci. Rep.* **2023**, *13*, 1497. [CrossRef] [PubMed]
41. *UNE UNE-EN ISO 17892-4:2019*; Investigación y Ensayos Geotécnicos. Ensayos de Laboratorio de Suelos. Parte 4: Determinación de la Distribución Granulométrica. Asociación Española de Normalización y Certificación: Madrid, Spain, 2019.
42. *UNE, UNE-EN ISO 17892-12:2019*; Investigación y Ensayos Geotécnicos. Ensayos de Laboratorio de Suelos. Parte 12: Determinación del Límite Líquido y del Límite Plástico. (ISO 17892-12:2018). Asociación Española de Normalización y Certificación: Madrid, Spain, 2019.
43. *UNE, UNE 103501:1994*; Ensayo de Compactación. Proctor Modificado. Asociación Española de Normalización y Certificación: Madrid, Spain, 1994.
44. *UNE, UNE 103502:1995*; Método de Ensayo para Determinar en Laboratorio el Índice C.B.R. de un Suelo. Asociación Española de Normalización y Certificación: Madrid, Spain, 1995.
45. Liu, F.T.; Ting, K.M.; Zhou, Z.-H. Isolation forest. In Proceedings of the 2008 Eighth IEEE International Conference on Data Mining, Pisa, Italy, 15–19 December 2008; IEEE: New York, NY, USA, 2008.
46. Kohavi, R. A study of cross-validation and bootstrap for accuracy estimation and model selection. In Proceedings of the 14th International Joint Conference on Artificial Intelligence (IJCAI), Montreal, QC, Canada, 20–25 August 1995; Volume 2, pp. 1137–1143.
47. Duan, T.; Anand, A.; Ding, D.Y.; Thai, K.K.; Basu, S.; Ng, A.; Schuler, A. Ngboost: Natural gradient boosting for probabilistic prediction. In Proceedings of the International Conference on Machine Learning, Virtual Event, 13–18 July 2020.
48. Meinshausen, N.; Ridgeway, G. Quantile regression forests. *J. Mach. Learn. Res.* **2006**, *7*, 983–999.
49. Koenker, R.; Hallock, K.F. Quantile regression. *J. Econ. Perspect.* **2001**, *15*, 143–156. [CrossRef]
50. McElreath, R. *Statistical Rethinking: A Bayesian Course with Examples in R and Stan*; Chapman and Hall/CRC: Boca Raton, FL, USA, 2018.
51. Rasmussen, C.E. *Gaussian Processes in Machine Learning, in Summer School on Machine Learning*; Springer: Berlin/Heidelberg, Germany, 2003; pp. 63–71.
52. Karaboga, D. *An Idea Based on Honey Bee Swarm for Numerical Optimization*; Technical Report-tr06; Erciyes University, Engineering Faculty, Computer Engineering Department: Kayseri, Turkey, 2005.
53. Asteris, P.G.; Nikoo, M. Artificial bee colony-based neural network for the prediction of the fundamental period of infilled frame structures. *Neural Comput. Appl.* **2019**, *31*, 4837–4847. [CrossRef]
54. Asteris, P.G.; Koopialipoor, M.; Armaghani, D.J.; Kotsonis, E.A.; Lourenço, P.B. Prediction of cement-based mortars compressive strength using machine learning techniques. *Neural Comput. Appl.* **2021**, *33*, 13089–13121. [CrossRef]
55. Lundberg, S.M.; Lee, S.-I. A unified approach to interpreting model predictions. *Adv. Neural Inf. Process. Syst.* **2017**, *30*, 4765–4774.
56. Hunt, R.E. *Geotechnical Engineering Investigation Handbook*; CRC Press: Boca Raton, FL, USA, 2005.
57. Bowles, J.E. *Physical and Geotechnical Properties of Soils*; McGraw-Hill: New York, NY, USA, 1979.

buildings

MDPI

Article

Performance Evaluation on the Application of MAST and RCC on a Cambodian Rural Road: A Case Study

Carlo Elipse, Je Won Kim * and Byung-Sik Ohm

Department of Highway and Transportation Research, Korea Institute of Civil Engineering and Building Technology, 283 Goyangdae-ro, Ilsanseo-gu, Goyang-si 10223, Republic of Korea; carloelipse@kict.re.kr (C.E.); bseom@kict.re.kr (B.-S.O.)
* Correspondence: jewonkim@kict.re.kr

Abstract: Cambodia's transportation sector has exhibited tremendous growth in the past decades together with its economic advancement. However, these improvements are only focused on the national and provincial roads, leaving the rural roads underdeveloped. A large percentage of the rural roads in Cambodia are unpaved while those paved roads are in deteriorated condition, making these rural roads prone to structural failure when subjected to heavy truck traffic and massive rainfall. Therefore, an innovative pavement material shall be used in the construction and rehabilitation of Cambodian rural roads. In this study, a composite pavement testbed consisting of a roller compacted cement (RCC) base layer and a modified double bituminous surface treatment (DBST), namely multi-layered bituminous surface treatment (MAST), was constructed in Cambodia and its performance was evaluated. To have a point of comparison, the testbed was divided into three sections with varying combinations of surface and base layer: (1) MAST surface with an RCC base, (2) MAST surface with an aggregate base, and (3) DBST surface with an aggregate base. Initially, a visual inspection was conducted to investigate the surface condition of the testbed. To assess the structural capacity, a light-falling weight deflectometer device was used. Meanwhile, the surface roughness was evaluated through the sand patch test. Lastly, the pass-by test was performed to quantify the noise level of the pavement. Based on the results, the proposed composite pavement performed significantly superior to the typical pavement in Cambodia, which is the third section. In summary, the composite pavement with MAST as the surface layer and RCC as the base layer was observed to be suitable for heavy truck traffic loading and the environmental conditions of Cambodia.

Keywords: Cambodia; double bituminous surface treatment; multi-layered asphalt surface treatment; light-falling weight deflectometer; sand patch test; pass-by test

Citation: Elipse, C.; Kim, J.W.; Ohm, B.-S. Performance Evaluation on the Application of MAST and RCC on a Cambodian Rural Road: A Case Study. *Buildings* **2023**, *13*, 2993. https://doi.org/10.3390/buildings13122993

Academic Editors: Andrea Baliello and Di Wang

Received: 1 November 2023
Revised: 24 November 2023
Accepted: 28 November 2023
Published: 30 November 2023

1. Introduction

The economic growth of a developing country, such as Cambodia, greatly relies on its national and rural roads that connect the whole country for the exchange of goods. Over the past few decades, Cambodia has made a significant improvement in its transportation system [1,2]. As of 2022, the percentages of pavements covered with Portland cement concrete (PCC), asphalt concrete (AC), or double bituminous surface treatment (DBST) for the national and provincial roads are 84% and 39%, respectively [3]. However, these developmental projects were merely focused on the connectivity of economically strategic areas, such as major cities, in the country. Moreover, national and provincial roads are only a quarter of the whole road network of Cambodia, leaving rural roads, which include the majority of Cambodia's road network, underdeveloped.

Out of the 48,000 km of rural roads in Cambodia, around 45,600 km, which amounts to 95% of the rural road network of Cambodia, is unpaved [3]. These roads have soil or laterite as a surface layer. Being situated along the Mekong River and Tonle Sap basin, Cambodia's geographical location makes it prone to periodic river floods and massive rainfall [4]. These unpredictable and inevitable natural hazards that subject the unpaved rural roads to a high volume of water cause a decrease in the road's structural capacity, making it prone to deformation [5]. In the presence of deformation, the road's serviceability and ride quality or comfort of the road users are reduced.

Furthermore, the rise of the economy also implies an increase in the traffic on its road network. In the case of rural roads, the number of heavy truck traffic that delivers products from the rural areas to the cities is also increasing [6]. Due to the overloading of these heavy trucks, longitudinal unevenness and permanent deformations may occur on unpaved rural roads [5]. In the long run, rural roads will not be able to support the heavy truck traffic load it experiences. Therefore, improving the unpaved rural roads of Cambodia is necessary.

The improvement that needs to be applied on the unpaved roads shall be able to support the heavy traffic loading and overcome the weakening of the soil due to the high volume of rainfall and floods. Furthermore, being a developing country, the budget is limited to the Cambodian government's capacity for the construction; thus, choosing a cost-effective material for paving the rural roads is necessary [7,8]. Typical pavement materials, such as reinforced concrete or asphalt, are designed to support heavy traffic loading and extreme environments. However, in the case of Cambodia, paving a lengthy span of rural roads with the typical pavement materials is costly [9–11]. Thus, an alternative pavement material with a lower cost than the typical pavement materials but also able to withstand heavy traffic under a tropical environment is needed.

Among the pavement materials, rigid pavement, a specifically reinforced concrete pavement, is most effective in withstanding heavy truck traffic loading [12]. However, due to the restrictions in the construction budget, the construction of reinforced concrete pavement would not be feasible. Therefore, a cheaper alternative to reinforced concrete pavement shall be used, which is the roller compacted concrete (RCC).

Unlike reinforced concrete pavement, RCC is a non-reinforced type of concrete pavement and, as stated in its name, is compacted using a vibratory roller [13]. Having no reinforcements needed, RCC is definitely cheaper, simpler, and less time-consuming to construct than the typical reinforced concrete pavement without compromising its structural strength. Moreover, the absence of reinforcements makes the RCC more appropriate for high moisture environmental conditions [14]. Furthermore, RCC has a lower water-to-cement ratio, which gives it the advantage of lowering the amount of Portland cement compared to the typical concrete pavement [15]. The reduction in the Portland cement in the mix design makes RCC more cost-efficient than the typical concrete pavement. However, due to RCC's lower water-to-cement ratio, the surface of the pavement tends to be rougher than that of the typical concrete pavement, which makes it unsuitable for general roads [16]. Hence, a pavement material suitable as a surface course for RCC shall be applied.

The typical surface course for a composite pavement with cement concrete as the base course is AC [17]; although, using this material for Cambodian rural roads will amount to high construction expenses. A cheaper alternative for the surface course is DBST, which is commonly used in Cambodia [3]. DBST promotes the appropriate skid resistance for a typical surface course and serves as a layer for waterproofing the layers underneath [18]. However, general DBST alone is prone to early stages of wearing, and if there is an insufficient adherence in the binder–aggregate interface, aggregate stripping may occur. Improvements in the general DBST shall be applied to make it more suitable for Cambodian rural roads.

In 2005 [19], North Carolina DOT observed that the enhancement in the gradation of aggregate used in the bituminous surface treatment from finer-graded to coarser and more uniformly graded minimizes aggregate loss. That study suggests an aggregate gradation of larger than 2.36 mm with less than 1.5% of fine aggregates less than 0.075 mm. Moreover, there is a study regarding the modification of asphalt emulsion used in the bituminous surface treatment by adding polymer to increase the mixture's resistance to rutting [20]. It was found that polymer-modified emulsions enhance the rutting resistance of the mixture at high temperatures, reduce the loss of aggregate coverage, and prolong the life by at least two years more compared to the unmodified emulsion. For the improvement of DBST for Cambodia, these factors were considered and the final mixture was coined multi-layered asphalt surface treatment (MAST).

Before the implementation of the MAST-RCC composite pavement, the cost-effectiveness of the proposed pavement system was validated in 2022 [21]. A simple life-cycle cost analysis was conducted on the MAST-RCC composite pavement and the typical AC pavement. A developed analysis program by the Asphalt Paving Association was used, and the theories and calculations were based on FHWA procedures. The analysis was set to a 30-year service life for both pavements, and the prices for construction, maintenance, rehabilitation, and user costs were considered. The periods at which the rehabilitation and maintenance will be implemented on the MAST-RCC pavement were set to 10 years and 4 years, respectively, while for the AC pavement, the periods were set to 15 years and 5 years, respectively. Furthermore, the costs of the rehabilitation and maintenance of the MAST-RCC pavement were determined to be 42% and 5.1% of the initial construction cost, respectively. Meanwhile, the costs of the rehabilitation and maintenance of the typical AC pavement were set to 42% and 5.1% of the initial construction cost, respectively. After running the analysis, it was found that the 30-year life-cycle cost of MAST-RCC pavement is lower by more than 30% compared to that of the typical AC pavement. This finding showed that the MAST-RCC composite pavement is a viable option for a more cost-effective pavement material. However, the performance of the proposed composite pavement on the actual field condition of the rural roads of Cambodia must be investigated.

In this study, the feasibility of the MAST-RCC composite pavement system for rural roads in Cambodia was evaluated. Initially, laboratory tests were conducted to achieve the best mix design for RCC and MAST. To observe the RCC and MAST's in situ performance, a 200-m testbed was constructed on a typical rural road in Phnom Penh, Cambodia. The testbed was divided into three different sections: MAST surface layer with an RCC base course, MAST surface layer with a general aggregate base course, and normal DBST with a general aggregate base course. Three months after opening the testbed to traffic, tests were conducted to assess each section's performance. Firstly, a visual inspection was performed to observe the surfaced distresses present in each section. Secondly, a light-falling weight deflectometer (LFWD) was used to evaluate each section's structural condition. Thirdly, a sand patch test was conducted to assess the surface roughness of the pavement. Lastly, since rural roads are also located near residential areas in Cambodia, the noise level produced by the tire–pavement interaction on MAST was compared to the typical concrete pavement present in most of the residential areas in Cambodia using the pass-by test. The general flow of this research is presented in Figure 1.

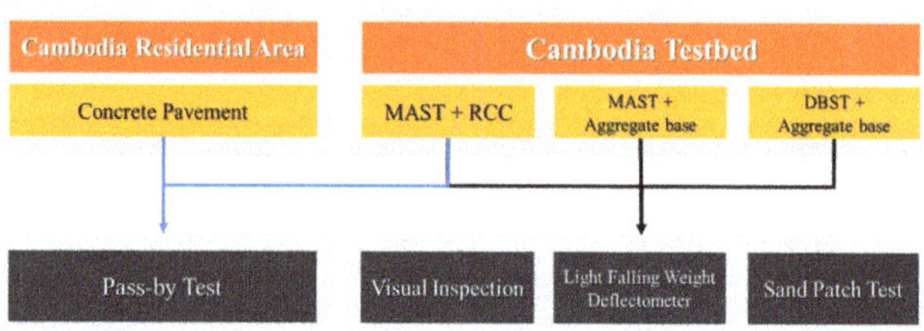

Figure 1. Research flowchart.

2. Materials and Methods

2.1. Materials

2.1.1. Multi-Layered Asphalt Surface Treatment (MAST)

Bituminous surface treatment, also known as chip seal, is a low-cost surfacing material commonly used in road maintenance and is a low-budget surface layer with a structurally reliable base layer. Basically, it is a layer of aggregates coated with emulsified asphalt as a binding agent [22]. In Cambodia, two layers of bituminous surface treatment, also known as DBST, is commonly used. However, DBST in Cambodia is prone to early surface wearing due to insufficient adherence between the binder–aggregate interface. Therefore, a modified asphalt emulsion is used as a binding agent for the DBST, called the modified mixture MAST.

In this study, to improve the performance of MAST, latex and glass fibers were used as an additive to the mixture. Latex is a type of polymer additive that improves the elasticity of the asphalt emulsion, making it more resistant to aggregate stripping [23]. Meanwhile, glass fibers increase MAST's resistance to moisture damage [24]. MAST is developed to improve DBST's resistance to surface wearing and aggregate stripping.

The gradation for the aggregate used in MAST is shown in Table 1. The amount of glass fibers is between 60 g/m^2 and 12 g/m^2 while the amount of latex is between 3% and 5% of the weight of the emulsified bitumen used. The amount of asphalt emulsion and aggregates on the first layer of application was set to 1.15 kg/m^2 and 15.2 kg/m^2, respectively. Meanwhile, for the second application, the amount of asphalt emulsion and aggregates was set to 1.55 kg/m^2 and 6.15 kg/m^2, respectively. These values on the application of MAST were in accordance with the construction standards of the Kingdom of Cambodia [25].

Table 1. Percent by weight passing of MAST aggregates.

Nominal Size of Material	Percent by Weight Passing AASHTO Sieve Sizes							
	37.5 mm	25.0 mm	19.0 mm	12.5 mm	9.5 mm	4.75 mm	2.36 mm	1.18 mm
25.0 mm	100	90–100	0–45	0–10	0–5	-	0–2	0–0.05
19.0 mm		100	90–100	0–30	0–8	-	0–2	0–0.05
12.5 mm			100	90–100	0–40	0–8	0–2	0–0.05
9.5 mm				100	90–100	0–30	0–2	0–2.00

To assess the MAST's resistance to aggregate stripping in the laboratory, a simple sweep test was conducted and additional DBST specimens were included for comparison [26]. At 3-h curing time, the percent mass loss of MAST was 1.4% while the percent mass loss of DBST was 2.9%, which means that the aggregate loss on the surface of MAST specimen was 50% less than the aggregate loss of DBST. Figure 2 shows the comparison of the appearance of the surface of MAST and DBST after the sweep test.

(a) (b)

Figure 2. Laboratory tests performed on RCC. (**a**) MAST; (**b**) DBST.

2.1.2. Roller Compacted Concrete (RCC)

Roller compacted concrete (RCC) is basically a traditional concrete mixture with a slightly adjusted mixture design to achieve a zero-slump. The components of RCC are the same as the traditional concrete mixtures but with reduced water content, making it stiff enough to support itself while compacting it using a roller. Unlike typical concrete structures, RCC is constructed in the absence of forms, dowels, joints, or any reinforcement. In recent years, RCC has been used in highly loaded infrastructures and the most notable is its use in dams. Its usage in dams significantly reduced the time of construction, making it more efficient and economical. Moreover, RCC has been present in the construction of horizontal infrastructures that include heavy loading, such as haul roads, loading docks, intermodal port facilities, parking lots, airports, and, recently, highways [27].

In this study, the aggregates used are a combination of coarse and fine aggregates to achieve a uniform particle size distribution with close proximity to the Fuller curve. The aggregate gradation is shown in Figure 3, which is in accordance with the RCC gradation limit of ACI PRC-327-14 [28]. The target water-to-cement ratio of the mix design is 48%, as calculated from the mix design by unit weight summarized in Table 2.

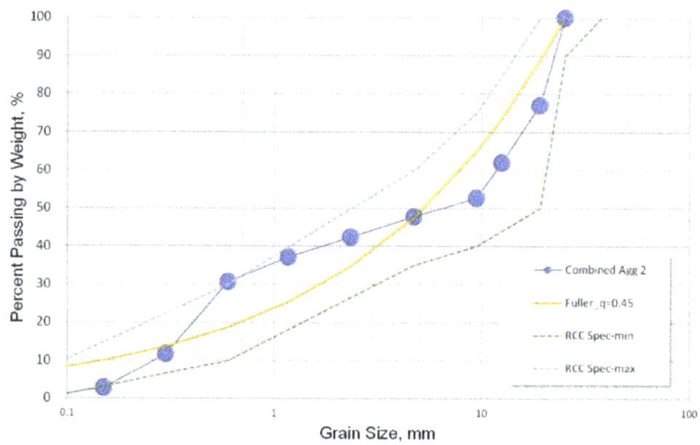

Figure 3. RCC aggregate gradation.

Table 2. RCC mix design by unit weight.

	Water	Cement	Fine Aggregate	Coarse Aggregate	Admixture
Unit Weight (kg/m^3)	132.0	274.3	913.4	1120.4	1.9

Using the gradation and mix design stated, laboratory specimens with a curing age of 28 days were fabricated to determine the RCC's compressive strength and chloride ion permeation resistance using the laboratory tests shown in Figure 4. Based on the results, the values of the compressive strength and chloride ion permeation resistance were 39.4 MPa and 2813 Coulomb, respectively, which classified the RCC as a normal grade.

2.2. Testbed Details

The testbed was constructed on the rural road just before the boundary of Phnom Penh. Figure 5 shows an aerial view of the testbed location, as well as the span of the road where the testbed was constructed. The testbed's total length was 200 m and was divided into three sections as shown in Figure 6. The first section with a length of 100 m was the proposed composite pavement, which consisted of a MAST surface layer and an RCC base layer. The second section was a MAST surface layer with an aggregate base layer and had a length of 50 m. Lastly, the third section was a 50-m long typical DBST surface layer and aggregate base layer. The third section served as the point of reference in this study. The thicknesses of the surface and base layers of all the sections were 10 mm and 150 mm, respectively, which were in accordance with the typical road structure in Cambodia [25].

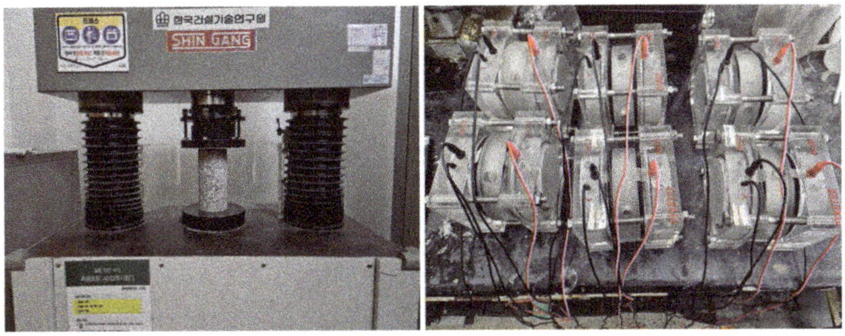

(a) (b)

Figure 4. Laboratory tests performed on RCC. (**a**) Compressive strength test; (**b**) chloride ion permeation resistance test.

Figure 5. Aerial view of the testbed in Cambodia.

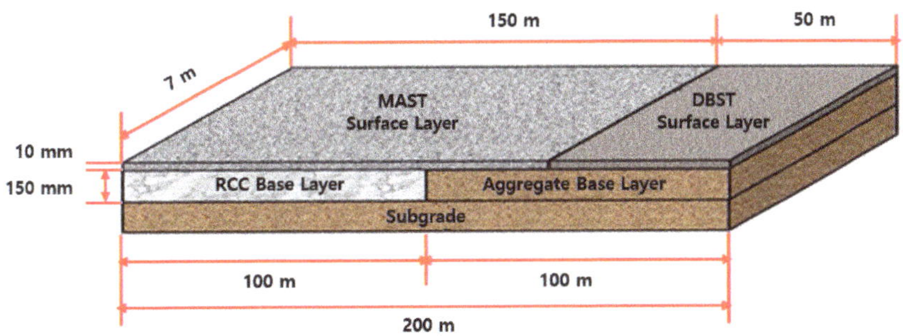

Figure 6. Structure and different sections of the testbed.

2.3. Methods

2.3.1. Light-Falling Weight Deflectometer

Light-falling weight deflectometer (LFWD) is a non-destructive pavement testing device that is mainly used to determine the stiffness modulus of the pavement or soil from the deflection caused by the application of an impulse load. It is also used to determine the compaction degree and evaluate the bearing capacity of the pavement or soil for road construction and earthworks. The device is derived from the static plate loading test and improved the testing procedure by replacing the static load with dynamic load, which closely represents the traffic loading generated by passing vehicles [29–31].

In this study, the test procedure using the LFWD was performed in accordance with ASTM E2583-07 [32], and Figure 7 shows the specific device used in the test, Dynatest LWD 3031. The LFWD device is composed of several parts:

- Guide rod;

- Release handle;
- Drop weight;
- Rubber buffer;
- Loading plate;
- Load cell;
- Velocity transducer.

The guide rod connects all of the parts into one device. In the initial phase of the testing, the release handle holds the drop weight on top of the guide rod with a magnetic plate underneath and a security lock. In the loading phase of the test where the drop weight is released, the drop weight makes an impulse load that is received by the rubber buffer. The presence of a rubber buffer makes the lower part of the device a spring-damper system [30]. The impulse load received by the rubber buffer is transferred to the pavement or soil through the loading plate. The load cell records the impulse load created by the drop weight while the deflection due to the impulse load is obtained through the velocity transducer underneath the loading plate.

Figure 7. LFWD testing device.

To calculate the stiffness modulus of the testing point, the Boussinesq half-space equation is used [33]:

$$E_{LFWD} = \frac{2(1-\mu^2)\rho R}{s},$$

where E_{LFWD} is the stiffness modulus of the testing location in MPa, μ is the Poisson ratio, ρ is the applied stress on the testing surface in kPa, R is the plate radius in mm, and s is the deflection in micron.

The *LFWD* test can be conducted by only one person, but for ease of testing, three people are highly suggested. While testing, the guide rod should be as perpendicular to the surface as possible to avoid variations in the impulse load due to inclination. Moreover, the device should be as firm as possible to the ground and the surface should be flat to achieve a uniform loading to the whole area of the loading plate. Lastly, the load cell and deflection sensor shall be periodically calibrated to make sure that the results are accurate [33].

2.3.2. Sand Patch Test

Sand patch test was conducted to quantify the roughness or macrotexture of the pavement surface. The pavement's surface texture correlates to the pavement's skid resistance, tire-pavement noise, and tire wearing. Moreover, it also affects the road user's ride quality and safety. A highly deteriorated pavement with a very smooth surface is prone to slippage, splash, and spray, especially after heavy rain or floods [34].

In this study, the sand patch test was conducted according to ASTM E965-15 standards [35], and the test procedure is presented in Figure 8. Basically, the sand patch test determines the pavement surface's roughness by volumetric approach. A known volume of sand was placed on the clean and dry pavement surface. The sand was gradually spread on the pavement surface in a circular manner using a smooth and flat disk until the entire amount of sand was spread evenly on the pavement surface. Assuming that the sand uniformly fills the full depth of the crevices on the pavement surface, at least four diameters of the formed circle were measured on different angles, and the average diameter was used to calculate the mean text depth (*MTD*) as follows:

$$MTD = \frac{4V}{\pi D^2},$$

where V is the volume of sand in mm^3 and D is the average diameter of the circular area covered by the sand in mm.

In this procedure, it is important that the pavement surface is thoroughly cleaned and dried before testing. Moreover, the sand patch test is not applicable on pavement surfaces with visible cracks. The presence of debris, water, and cracks on the pavement surface affects the volume of space the sand needs to fill on the pavement surface, resulting in values not representative of the pavement's roughness [36].

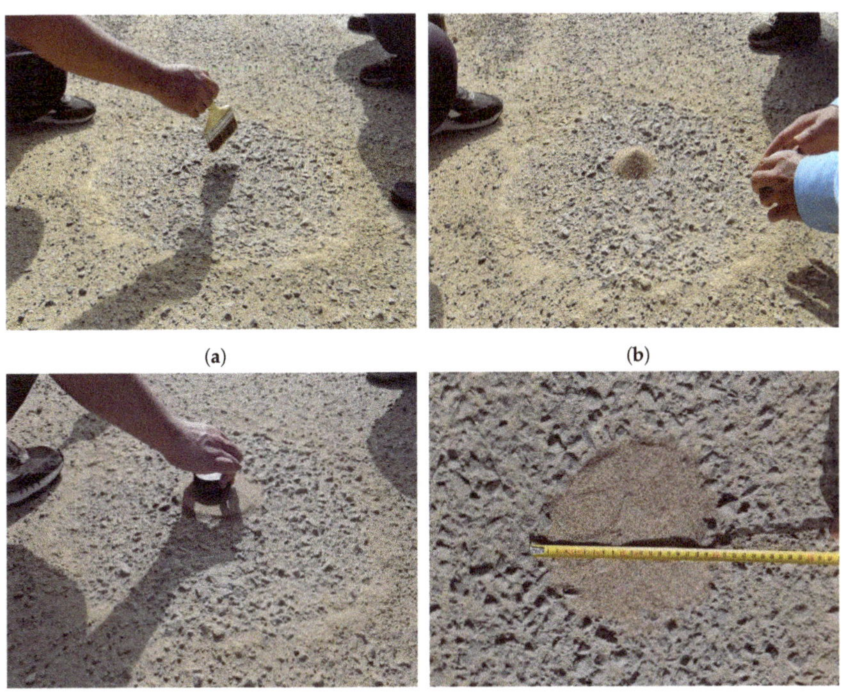

(a) (b) (c) (d)

Figure 8. Sand patch test procedure. (**a**) Cleaning of the surface to be tested; (**b**) pouring of the sand on the cleaned surface; (**c**) spreading of the sand; (**d**) measuring of the diameter.

2.3.3. Pass-By Test

Pass-by test aims to measure the noise level of a passing vehicle at a certain point at the roadside. This testing can be utilized to analyze the sound propagation generated by a certain vehicle at a specific speed. Through this, the comparison of sound levels from different types of vehicles, tire properties, and speed can be conducted [37]. Furthermore, it can also be used to compare the noise mitigation properties of different types of pavement materials.

For this study, the pass-by test was conducted according to EN ISO 11819-1 standard [38]. As shown in Figure 9, the pass-by test consisted of a microphone that was placed 7.5 m from the middle of the vehicle path with a height of 1.2 m and a vehicle driven at a uniform speed [39]. The test should be repeated at least three times for a single location and vehicle speed to observe precision among the results. Due to the pass-by test's simplicity, this test is the most commonly used among the noise test methods.

In conducting the pass-by test, a controlled environment is needed to minimize the unnecessary noise from different sources other than the interaction of the vehicle to the pavement and the vehicle engine. Moreover, there shall be no obstructions around the microphone that may hinder the travel of sound from the road to the microphone, such as trees, bushes, and traffic signs. The maximum sound level in one vehicle pass is the representative result of each test. Under the assumption of a constant speed with the passing vehicle, a log-linear trend shall be observed in the maximum sound level with respect to the vehicle speed [40].

Figure 9. Pass-by test microphone set up.

3. Results and Discussion

After three months of opening the testbed to traffic, the performance of MAST and RCC was evaluated. The surface condition of the pavement was observed through visual inspection and the sand-patch test. Moreover, the structural condition of the pavement was assessed using the LFWD test. Lastly, the noise level of the pavement was determined by conducting the pass-by test.

3.1. Visual Inspection

To assess the current surface condition of the testbed, a visual inspection was conducted. The three sections of the testbed were thoroughly inspected and Figure 10 shows the surface condition of each section.

It can be observed that a significant amount of depression is present in Section 3 while Sections 1 and 2 still maintain an evenly leveled surface. These depressions were mainly caused by the heavy truck traffic, shown in Figure 11 that passes on the testbed. The location of the testbed is on the outskirts of Phnom Penh, which is commonly the route of the heavy truck that carries goods from rural areas to the city and vice versa. The structural capacity of Section 3 was not sufficient to support the high frequency of the heavy loading of the truck. Moreover, the tropical climate of Cambodia weakened the aggregate base layer of the pavement. Frequent rainfall and poor drainage promote a long duration of water retention in the aggregate base, thus decreasing its structural capacity.

(**a**)

(**b**)

(**c**)

Figure 10. Surface condition of the testbed in Cambodia. (**a**) Section 1: MAST surface layer with RCC base layer; (**b**) Section 2: MAST surface layer with aggregate base; (**c**) Section 3: general DBST surface layer with aggregate base.

Figure 11. Heavy truck passing on the testbed.

The RCC base layer in Section 1 increased the stability of the underlying layer, making it less prone to deformation. Meanwhile, even though Section 2 has the same aggregate base layer as Section 3, the MAST layer on top of the aggregate base layer assisted the whole pavement system in making it intact due to a stiffer binder than that of the general DBST. This proves that replacing the typical DBST with a MAST surface layer improves the performance of the pavement. It served as an effective waterproofing layer to the other pavement layers underneath. Moreover, the RCC base layer is more suitable for heavy truck traffic than the typical aggregate base layer, especially in an environment with frequent heavy rainfall, such as tropical countries.

Furthermore, edge deterioration is present in all of the sections of the testbed, as shown in Figure 12. As observed in Figure 10, the whole span of the testbed has an unbounded edge. The lack of a concrete curb or any stable edge that prevents the deformation of the pavement on its side makes the pavement prone to edge deterioration. Moreover, the settlement at the edge of the pavement was also intensified by the absence of a proper drainage system at the testbed site. As mentioned earlier, Cambodia is a tropical country with a tremendous amount of rainfall. The retention of water on and under the pavement weakens the edge of the pavement, thus forming a settlement. As the deteriorated edge of the pavement is subjected to cycles of rainfall and water retention, its structural condition worsens making it easier to be damaged.

Figure 12. Deterioration on the edge of the testbed.

3.2. LFWD Test

Three testing points were chosen for each section, and these points are evenly spaced to represent the whole span of the section. In Section 3, one point was intentionally located on the depression to investigate the structural condition of the deteriorated location. At each point, three repetitions of LFWD testing were conducted. The stiffness modulus per repetition was calculated using the Boussinesq half-space equation, and the average value of the three stiffness moduli was considered as the representative stiffness modulus of the testing point. Figure 13 shows the values of the representative stiffness modulus of the testing points in all the sections.

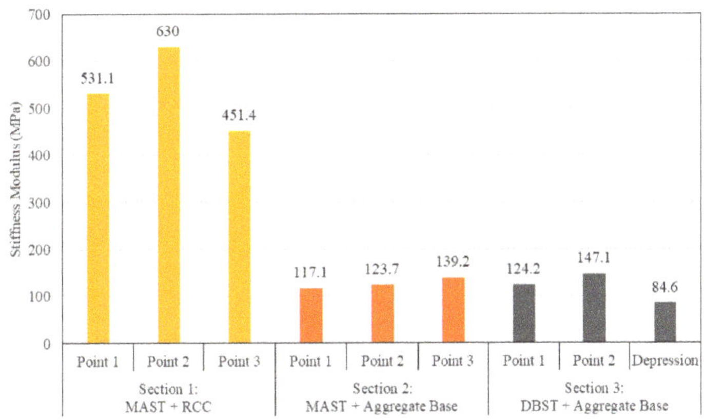

Figure 13. LFWD test results.

As observed in the LFWD test results, all the testing points in Section 1 exhibit a significantly high stiffness modulus compared to Sections 2 and 3, ranging from 450 MPa to 650 MPa. This is expected due to the structural strength contributed by the RCC on the pavement system. Moreover, the dominant stiffness modulus of Section 1 correlates to the absence of noticeable surface distress or deformation observed in the visual inspection.

Meanwhile, the stiffness moduli for Sections 2 and 3 are almost similar, ranging from 100 MPa to 150 MPa, with the exception of the point in Section 3 with the depression having a stiffness modulus of less than 100 MPa. Having the aggregate base, it is expected that Sections 2 and 3's stiffness moduli are in the same range.

Moreover, the lowest value of the stiffness modulus on the point of depression shows the reason behind the deterioration in that area. With a low stiffness modulus and a high traffic volume to support, the pavement is most likely to fail given those conditions.

In summary, it is evident that the main factor for the stiffness modulus in this testbed is the base layer. A base layer with a higher stiffness modulus, like the RCC, is more suitable for heavy traffic than the base layer with a lower stiffness modulus, like the aggregate base.

3.3. Sand Patch Test

To assess the surface roughness of each section of the testbed, the sand patch test was performed. One testing point per section was chosen, and a uniform volume of sand amounting to 5000 mm^3 was used for the test. Four diameter readings were obtained, and the average diameter was calculated to compute the MTD. The results of the sand patch test are summarized in Figure 14.

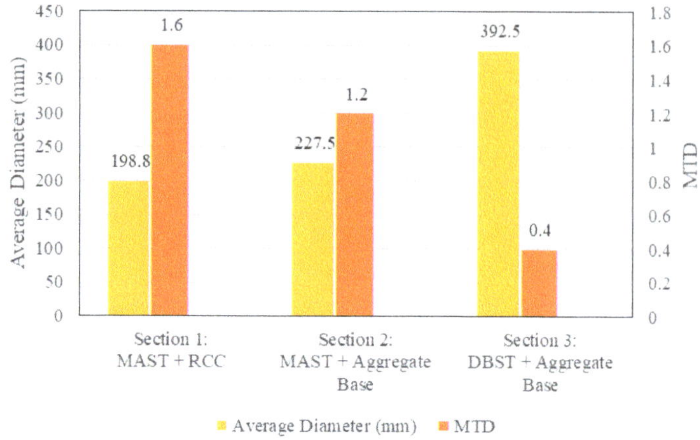

Figure 14. Sand patch test results.

According to the principle of the sand patch test, the circular area covered by the sand on a surface with low roughness is bigger since the sand has fewer crevices to fill in. Meanwhile, surfaces with high roughness tend to form circular areas with smaller diameters since more spaces should be filled in by the sand. In the results of the sand patch test, it is evident that Section 3, having the highest average diameter, has the lowest surface roughness among the three sections since it is the only section with a general DBST surface layer. Meanwhile, Sections 1 and 2 have average diameters closer to each other with measurements of 198.8 mm and 227.5 mm, respectively, since both sections have the same MAST surface.

Based on [41], the acceptable MTD for bituminous surface treatment is at least 0.7. Anything lower than that would not be sufficient to provide appropriate traction to the tire of the vehicle. As shown in the results, Sections 1 and 2 performed well in the sand patch test with values higher than 0.7. Meanwhile, the surface roughness in Section 3 failed the sand patch test, with a MTD value of 0.4. It is implied that the surface roughness of the general DBST declines abruptly in just a few months of exposure to traffic.

Both the results on average diameter and MTD suggest that MAST is significantly superior to the general DBST in terms of surface roughness after being subjected to several months of traffic loading. The poor surface roughness of DBST was caused by the degradation due to the insufficient strength of the pavement system, as well as the low resistance to aggregate stripping based on the sweep test conducted in the laboratory. Therefore, the latex and glass fibers additive of the MAST surface layer effectively increased the material's resistance to surface wear. Given that information, it is more appropriate to use the MAST surface layer on Cambodian rural roads since it can withstand heavy truck traffic loading without compromising its surface roughness.

3.4. Pass-By Test

The future utilization of MAST in residential areas of Cambodia is also being taken into consideration. Thus, the noise level that it produces should be taken into account. In this study, the noise level of a typical concrete pavement in the residential area of Cambodia was compared to that of the MAST pavement. Figure 15 shows the MAST pavement and concrete pavement where pass-by tests were conducted.

(a) (b)

Figure 15. (**a**) MAST pavement and (**b**) concrete pavement where the pass-by tests were conducted.

For each location, three vehicle speeds were tested: 40 kph, 50 kph, and 60 kph. The ambient noise was also measured for each location and deducted from the testing results to isolate the noise produced by the tire–pavement interaction and the vehicle engine. The final values of the noise level at different vehicle speeds are summarized in Figure 16.

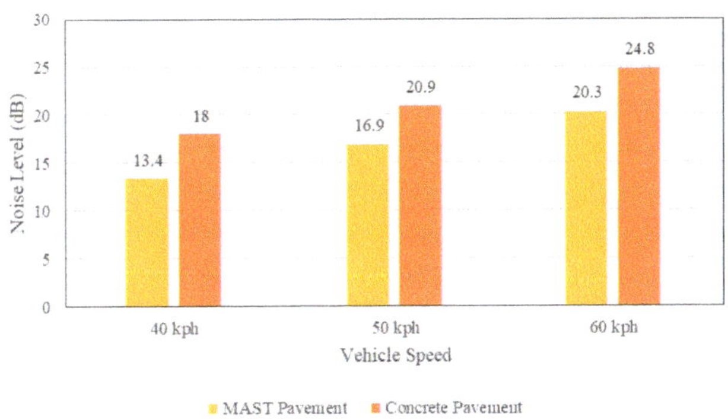

Figure 16. Pass-by test results.

Based on the results, the noise level of the MAST pavement is lower than that of the concrete pavement by at least 4 dB in all of the testing vehicle speeds. Therefore, the MAST pavement is an appropriate replacement for concrete pavement in residential areas of Cambodia in terms of noise level.

4. Conclusions

This study focused on determining the performance of MAST as a surface layer and RCC as a base layer for Cambodian rural roads. To evaluate the proposed pavement material, a testbed on a rural road in Phnom Penh was constructed. The testbed consisted of three sections with varying surface layers and base layers: (1) MAST surface layer with an RCC base layer, (2) MAST surface layer with an aggregate base layer, and (3) DBST surface layer with an aggregate base layer. The third section served as the control in the experiment since it is the typical pavement material used in Cambodia. The testbed was subjected to traffic load for several months and the condition of the pavement after opening it to traffic was assessed. A visual inspection was performed to investigate the surface condition of the pavement and take into account the notable surface distresses present on the pavement. For the structural capacity of the pavement, the stiffness modulus of each section was determined using the LFWD device. Moreover, the surface roughness of the pavement was evaluated using the sand patch test. Lastly, the noise level of MAST as a surface layer was compared to the typical concrete pavement in residential areas of Cambodia using the pass-by test. The following are the conclusions drawn from this study:

- Through the visual inspection of the sections, it was found that the section with the typical rural road design in Cambodia, which is the DBST surface layer with an aggregate base layer, is susceptible to depression. Meanwhile, the other two sections with MAST as the surface layer preserved its evenly leveled surface. For Section 1, the RCC base layer sustained the pavement system's structural stability under heavy truck traffic loading. Moreover, its resistance to moisture damage is exhibited in this study with the help of MAST as a waterproofing layer on the surface.
- Edge deterioration was also observed in all the sections of the testbed. This is due to the unbounded nature of the testbed all throughout its span. The lack of concrete curbs or any edge stabilizing structures causes the sideways deformation of the pavement, resulting in settlement and water retention at the edge. This problem must be addressed in future constructions of testbed or actual roads in Cambodia.
- The LFWD test results confirmed the structural strength of the RCC base layer compared to the aggregate base layer. The results showed that all the testing results in Section 1 are significantly higher than that of Sections 2 and 3, with values ranging from 450 MPa to 650 MPa. Meanwhile, having the same aggregate base layer, the stiffness modulus of the points obtained in Sections 2 and 3 are in close proximity to each other, with values ranging from 100 MPa to 150 MPa.
- Moreover, the depression in Section 3 was also included in the LFWD testing points to investigate the reason behind the failure. The results correlated with the visual inspection with the depression point having the lowest stiffness modulus among all the testing points. The stiffness modulus at the depression was under 100 MPa. Given these findings, the combination of low structural capacity, heavy traffic loading, and massive rainfall was found to be the most probable reason for the occurrence of the depressions in Section 3.
- The sand patch test was conducted on the three sections to compare each section's surface roughness after experiencing traffic for several months. Based on the results, Sections 1 and 2, both with a MAST surface layer, sustained their surface roughness and are still suitable for traffic. However, the DBST surface layer of Section 3 was deemed to be susceptible to surface wearing with an MTD lower than 0.7. This implies that the addition of latex and glass fiber to the MAST effectively elevated the material's resistance to surface wearing.
- The feasibility of MAST being an alternative pavement material for Cambodian residential areas in terms of noise levels was investigated. The pass-by test was performed on both the MAST pavement and concrete pavement to compare the noise levels. It was found that in all three vehicle speeds tested, the noise level of the MAST pavement is lower by at least 4 dB compared to the noise level produced by the concrete pavement. Therefore, it can be concluded that MAST is an appropriate alternative in terms of noise levels for pavements in Cambodian residential areas.
- In summary, the MAST surface layer and RCC base layer showed promising results in the evaluation of the testbed in the rural roads of Cambodia. Further tests are suggested to fully investigate the feasibility of the proposed composite pavement as an alternative for Cambodian rural roads.

Author Contributions: Conceptualization, C.E., J.W.K. and B.-S.O.; methodology, C.E., J.W.K. and B.-S.O.; validation, C.E., J.W.K. and B.-S.O.; formal analysis, C.E.; investigation, C.E., J.W.K. and B.-S.O.; resources, C.E.; data curation, C.E.; writing—original draft preparation, C.E.; writing—review and editing, J.W.K. and B.-S.O.; visualization, C.E.; supervision, J.W.K. and B.-S.O.; project administration, J.W.K. and B.-S.O.; funding acquisition, J.W.K. and B.-S.O. All authors have read and agreed to the published version of this manuscript.

Funding: This research was conducted under the KICT Research Program (project no. 20230202-001, Development of road network sustainability strategy and localization-specific technology for improvement of unpaved roads in Mekong River Neighboring Countries) funded by the Ministry of Science and ICT.

Data Availability Statement: Data will be provided upon request. The data are not publicly available due to privacy.

Conflicts of Interest: The authors declare no conflict of interest.

References

1. Dainichi, S. *Cambodia: Provincial Roads Improvement Project*; Technical Report; ADB: Mandaluyong, Philippines, 2016.
2. Bunna, Y.; Sophort, P.; Seila, C.V. *How to Design a Pavement with an Overload Environment*; Technical Report; MPWT: Phnom Penh, Cambodia, 2004.
3. IRITWG. *Overview of the Transport Infrastructure Sector in the Kingdom of Cambodia*, 7th ed.; Technical Report; JICA: Tokyo, Japan, 2023.
4. Wang, B.; Miyagi, M. A Pathway to Resilient Rural Roads in Cambodia: Sharing "The Tested and Tried" in Japan. Available online: https://www.gfdrr.org/en/feature-story/pathway-resilient-rural-roads-cambodia/ (accessed on 15 October 2023).
5. Salpisoth, H. Simple Evaluation Methods for Road Pavement Management in Developing Country. Ph.D. Thesis, Kyoto University, Kyoto, Japan, 2014.
6. Sisovanna, S. Improvements and Challenges Associated with the Facilitation of Road Transport in Cambodia. In *Cross-Border Transport Facilitation in Inland ASEAN and the ASEAN Economic Community*; Wiley: Hoboken, NJ, USA, 2019; pp. 45–89.
7. JICA. *The Study on the Road Network Development in the Kingdom of Cambodia: Final Report*; Technical Report; JICA: Tokyo, Japan, 2006.
8. Bank, T.W. *Cambodia Road Connectivity Improvement: Project Proposal Document*; Technical Report; The World Bank: Carroll, NH, USA, 2020.
9. ADB. *Kingdom of Cambodia: Road Network Improvement Project, Project Administration Manual*; Technical Report; ADB: Mandaluyong, Philippines, 2017.
10. ADB. *Cambodia: Primary Roads Restoration Project, Performance Evaluation Report*; Technical Report; ADB: Mandaluyong, Philippines, 2009.
11. Thach, R.; Jang, H.; Thu, N. The Problem of Economic Development Strategy in Cambodia: The Case of the Road Network. *Vnu J. Econ. Bus.* **2018**, *34* , 29–38.
12. Shaban, A.M.; Alsabbagh, A.; Wtaife, S.; Suksawang, N. Effect of Pavement Foundation Materials on Rigid Pavement Response. *IOP Conf. Ser. Mater. Sci. Eng.* **2020**, *671*, 012085. [CrossRef]
13. Piggott, R.W. *Roller-Compacted Concrete Pavements-A Study of Long Term Performance*; Technical Report Report RP366; PCA: Washington, DC, USA, 2021.
14. Algin, Z.; Gerginci, S. Freeze-thaw resistance and water permeability properties of roller compacted concrete produced with macro synthetic fibre. *Constr. Build. Mater.* **2020**, *234*, 117382. [CrossRef]
15. Harrington, D.; Abdo, F.; Adaska, W.; Hazaree, C. *Guide for Roller-Compacted Concrete Pavements*; Technical Report; Iowa State University: Ames, IA, USA, 2010.
16. Zollinger, D. *Roller-Compacted Concrete Pavement*; Technical Report FHWA-HIF-16-003; FHWA: Sevierville, TN, USA, 2016.
17. Rao, S.P. *Composite Pavement Systems: HMA/PCC Composite Pavements*; Transportation Research Board: Washington, DC, USA, 2013; Voulme 1.
18. Bituminous Surface Treatments. Available online: https://pavementinteractive.org/reference-desk/maintenance-and-rehabilitation/maintenance/bituminous-surface-treatments/ (accessed on 28 October 2023).
19. Kim, Y.R.; Lee, J. *Optimizing Gradation for Surface Treatments*; Technical Report FHWA/NC/2005-15; Development of Transportation, Research and Analysis Group: Chapel Hill, NC, USA, 2005.
20. Kim, Y.R.; Lee, J. *Performance-Based Analysis of Polymer-Modified Emulsions in Asphalt Surface Treatments*; Technical Report FHWA/NC/2007-06; Development of Transportation, Research and Analysis Group: Chapel Hill, NC, USA, 2009.
21. Kim, Y. *Development of High-Performance, Low-Cost MAST (Multi-Layered Asphalt Surface Treatments) Composite Pavement System for Improving Unpaved Roads in Vietnam (2/3)*; Technical Report KICT 2022-147; Korea Institute of Civil Engineering and Building Technology: Goyang, Republic of Korea, 2022.
22. Board, T.R. *Chip Seal Best Practices*; The National Academies Press: Washington, DC, USA, 2005. [CrossRef]
23. King, G.; King, H.; Johnston, J.; Galehouse, L.; Voth, M. *Performance-Based Analysis of Polymer-Modified Emulsions in Asphalt Surface Treatments*; Technical Report FHWA-CFL/TD-12-004; Federal Highway Administration: Washington, DC, USA, 2011. [CrossRef]
24. Wang, X.; Wu, R.; Zhang, L. Development and performance evaluation of epoxy asphalt concrete modified with glass fibre. *Road Mater. Pavement Des.* **2019**, *20*, 715–726. [CrossRef]
25. *CAM PW 03-102-99*; Road Design Standard—Part 2-Pavement. Ministry of Public Works and Transport: Phnom Penh, Cambodia, 1999.

26. *ASTM D1000-19a*; Standard Test Method for Sweep Test of Emulsified Asphalt Surface Treatment. American Society for Testing and Materials: West Conshohocken, PA, USA, 2020.
27. Williams, S.G.; McFarland, A.M. *Roller Compacted Concrete for Roadway Pavement*; Technical Report (Report No. TRC1005); TRC: Little Rock, AR, USA, 2013.
28. *ACI PRC-327-14*; Guide to Roller-Caompacted Concrete Pavements. American Concrete Institute: Indianapolis, IN, USA, 2015.
29. Ayyanchira, M.M. Introduction of Light Weight Deflectometer. *Int. J. Eng. Res. Technol.* **2014**, *3*, 303–305. [CrossRef]
30. Adam, C.; Adam, D. Modelling of the dynamic load plate test with the light falling weight device. *Asian J. Civ. Eng.* **2003**, *4*, 73–89.
31. Roksana, K.; Nowrin, T.; Hossain, S. A Detailed Overview of Light Weight Deflectometer (LWD). In Proceedings of the International Conference on Planning, Architecture and Civil Engineering, Rajshashi, Banglasdesh, 7–9 February 2019.
32. *ASTM E2583-07*; Standard Test Method for Measuring Deflections with a Light Weight Deflectometer. American Society for Testing and Materials: West Conshohocken, PA, USA, 2020.
33. Heathcote, J.H. Towards Using the Light Falling Weight Deflectometer as a Construction Control Device. Master's Thesis, University of Pretoria, Pretoria, South Africa, 2016.
34. Hibbs, B.O.; Larson, R.M. *Tire Pavement Noise and Safety Performance, PCC Surface Texture Technical Working Group*; Technical Report; Federal Highway Administration, Office of Engineering: Washington, DC, USA, 1996.
35. *ASTM E965-15*; Standard Test Method for Measuring Pavement Macrotexture Depth Using a Volumetric Technique. American Society for Testing and Materials: West Conshohocken, PA, USA, 2015.
36. Pavements—Sand Patch Test (SPT). Available online: https://infotechnology.fhwa.dot.gov/sand-patch-test/ (accessed on 28 October 2023).
37. Freitas, E.F.; Paulo, J.; Coelho, J.; Pereira, P.A. Towards Noise Classification of Road Pavements. 2008. Available online: https://www.academia.edu/33725316/Towards_noise_classification_of_road_pavements (accessed on 15 October 2023).
38. *EN-ISO 11819-1*; Acoustics-Measurement of the Influence of Road Surfaces on Traffic Noise—Part 1: Statistical Pass-by Method. International Organization for Standardizations: Geneva, Switzerland, 2015.
39. Hanson, D. Tire/Pavement Noise Measurement and Results. *Natl. Superpave News* **2005**, *4*, 3–10.
40. Moreno, R.; Bianco, F.; Carpita, S.; Monticelli, A.; Fredianelli, L.; Licitra, G. Adjusted Controlled Pass-By (CPB) Method for Urban Road Traffic Noise Assessment. *Sustainability* **2023**, *15*, 5340. [CrossRef]
41. Pierce, L.M.; Kebede, N. *Chip Seal Performance Measures: Best Practices*; Technical Report Report WA-RD 841.16; Washington State Department of Transportationp: Washington, DC, USA, 2015.

buildings

MDPI

Article

Application of Steel-Fiber-Reinforced Self-Stressing Concrete in Prefabricated Pavement Joints

Yun-Feng Xi [1], Sheng-Jun Ren [2,*], Bao-Ling Chen [3], Bing Yang [4], Jin Lee [1,*], Guang-Hao Zhu [1], Tian-Cheng Zhou [1] and Hao Xu [1]

1 School of Transportation Civil Engineering, Shandong Jiaotong University, Jinan 250357, China
2 Dezhou Highway Engineering Corporation, Dezhou 253011, China
3 Dezhou Highway Development Center, Dezhou 253011, China
4 Shandong Zhixing Consulting Survey and Design Institute, Dezhou 253011, China
* Correspondence: z2501120405@163.com (S.-J.R.); 21107042@stu.sdjtu.edu.cn (J.L.)

Abstract: Prefabricated pavement, with its advantages of a high paving speed, low material consumption, low carbon emissions, high strength, and easy construction, has gradually been used to address issues associated with traditional cement pavement construction. However, even under the long-term combined effects of vehicle loads and environmental loads, the joints between pavement slabs remain prone to various problems. This paper proposes the use of steel-fiber-reinforced self-stressing concrete (SFRSSC) with a certain level of self-stress for joint pouring and connection to control the development of cracks in the joints and achieve seamless integration between the slabs. Additionally, the self-stress generated by SFRSSC is utilized to enhance the continuity of the prestressed design in precast slabs, thereby extending their service life. Through laboratory experiments and field tests, the self-stress magnitude, mechanical strength, and long-term applicability of SFRSSC were studied. The results indicate that SFRSSC can achieve self-stress levels of over 7 MPa under standard curing conditions, but the values decrease significantly when removed from the standard curing environment. SFRSSC exhibits a compressive strength of over 60 MPa and a flexural strength of over 9 MPa, both of which exceed the requirements of the relevant standards, making it suitable for use as a pavement joint material. During long-term monitoring in the field, SFRSSC demonstrates favorable expansion effects and high stability. The longitudinal expansion remains stable at 100 $\mu\varepsilon$, while the transverse expansion exhibits minor shrinkage, maintained at around 25.2 $\mu\varepsilon$. Therefore, the application of SFRSSC in assembly-type prestressed pavement joints shows high applicability.

Keywords: steel fiber concrete; prefabricated pavement; self-stressing concrete; long-term monitoring

Citation: Xi, Y.-F.; Ren, S.-J.; Chen, B.-L.; Yang, B.; Lee, J.; Zhu, G.-H.; Zhou, T.-C.; Xu, H. Application of Steel-Fiber-Reinforced Self-Stressing Concrete in Prefabricated Pavement Joints. *Buildings* **2023**, *13*, 2129. https://doi.org/10.3390/buildings13092129

Academic Editors: Andrea Baliello and Di Wang

Received: 29 July 2023
Revised: 18 August 2023
Accepted: 21 August 2023
Published: 22 August 2023

1. Introduction

The technology of prefabricated prestressed concrete pavement refers to the prefabrication of the pavement structure layer in a factory, followed by transportation to the construction site for assembly, joint treatment, and other subsequent processes, thereby achieving rapid construction and repair of pavement structures [1]. This technology offers several advantages, including a fast paving speed, low material consumption, low carbon emissions, high reuse rate, and long structural lifespan. As a result, it has been widely applied in low-speed traffic roads, sidewalks, and urban roads [2–5]. The main difference between prefabricated prestressed concrete pavement and continuously cast concrete pavement lies in the composition of the pavement panels. In prefabricated pavement, the panels are primarily assembled. After assembly, joints unavoidably form, typically appearing as open cracks through the surface layer. In usual cases, some joints are preserved as expansion joints for the pavement, while others are sealed and waterproofed to form an integral connection. Currently, there are two main connection methods: dry and wet connections, with wet connections being the primary approach for prefabricated structures.

However, since each pavement panel is independently installed and influenced by various construction factors, there can be height variations between adjacent panels. Additionally, with the passage of time and inadequate maintenance, factors such as cracking, water seepage, and aging often occur at these joint locations, significantly impacting the service level and driving comfort of the pavement. The vulnerability of the joint area as a weak point in the structure has been widely recognized [6–8]. Therefore, how to treat the joint interface, reduce the number of joints, or even achieve seamless connections has been a hot topic in research on prefabricated pavement engineering.

Currently, both domestic and international approaches to pavement panel connections mainly involve the use of transfer bars and suitable joint materials to enhance the load-carrying capacity of the joints, aiming to make the pavement structure as close to a continuous panel without joints [9–11]. Transfer bars transmit loads through their own shear and bending stiffness. Relevant studies have shown that the transmission efficiency between pavement panels with transfer bars is between 85% and 95% [12,13]. Although transfer bars possess relatively high transmission efficiency, there are still concerns regarding their rust prevention treatment, especially during the concrete compaction stage where loss is most evident. Moreover, if there is an excessive amount of anti-rust coating on the surface of the transfer bars, the contact stress generated by the wheel load transmission may lead to excessive damage to the concrete surrounding the transfer bars, resulting in loosening and a reduced load-carrying capacity [14–18]. Therefore, the selection of joint materials becomes crucial, aiming to find a material that not only improves the bonding effect between transfer bars and pavement panels but also prevents water from infiltrating and damaging the various structural layers. Based on this concept, this study uses a steel-fiber-reinforced self-stressing concrete (SFRSSC) [19,20] material to address the aforementioned issues.

2. Experiment

2.1. Experimental Materials and Ratio

The basic mix proportion used in this experiment was (cement:fly ash:water:coarse aggregate:fine aggregate:expansion agent) 1:0.28:0.42:1.87:2.81:0.11. The gel material selected was ordinary Portland cement (P.O.42.5), and the chemical composition of the Class II fly ash is shown in Table 1.

Table 1. Main chemical composition of cement and fly ash (mass fraction,%).

Composite	CaO	Al_2O_3	MgO	SiO_2	Fe_2O_3	SO_3	LOI
Cement	65.1	3.1	2.8	23.2	2.6	1.6	1.6
Fly ash	6.2	32.6	0.7	48.5	8.7	1.2	2.1

The coarse aggregate consisted of natural aggregates with particle sizes of (5~10) mm and (10~20) mm, mixed in a 4:6 ratio. The fine aggregate was natural sand with a fineness modulus of 2.6. The superplasticizer used had a dosage of 1% and was a polycarboxylate-based high-efficiency water reducer with a water reduction efficiency of 25%. The steel fiber used was the milling type produced by Shanghai Harix Fiber as shown in Figure 1.

The main physical properties of the materials used are listed in Table 2. The expansion agent used in the experiment was a dual-expansion-source expansion agent produced by Wuhan Sanyuan Special Building Materials Co., Ltd. (Wuhan, China), and its chemical composition is shown in Table 3.

Figure 1. Steel ingot milling profile steel fiber.

Table 2. Physical properties of materials used.

Material	Physical Properties
Cement	Ordinary Portland cement (OPC, density: 3.15 g/cm^3, specific surface area: 3440 cm^2/g)
Fly ash	(Density: 2.35 g/cm^3, specific surface area: 4110 cm^2/g)
Fine aggregate	(River sand, size: 0.35~0.45 mm, density: 2.58 g/cm^3, absorption: 1%)
Coarse aggregate	(Basalt gravel, 5~20 mm, apparent density 2.933 g/cm^3)
Superplasticizer	Polycarboxylic-based superplasticizer (specific gravity: 1.05 ± 0.05, pH: 5.0 ± 1.5)
Steel fiber	(Milling type, size: 32 mm ± 2, length–diameter ratio: 35–45, tensile strength ≥ 700 MPa)

Table 3. Main chemical composition of expansion agent (mass fraction,%).

Components	CaO	Al_2O_3	MgO	SiO_2	Fe_2O_3	SO_3	LOI
HCSA	65.21	4.88	1.45	1.81	1.22	22.33	3.10

2.2. Experiment Method

2.2.1. Stress and Mechanical Response Measurement of Steel-Fiber-Reinforced Self-Stressing Concrete

The measurement of concrete self-stress utilized a custom-designed apparatus, as shown in Figure 2. The dimensions of the specimens were primarily determined based on the steel reinforcement ratio within the concrete, with a reverse calculation performed for the cross-sectional dimensions using a 1% reinforcement ratio. Due to the smallest commercially available steel bar having a diameter of 16 mm, this study employed prismatic specimens of reinforced fiber concrete with a length of 400 mm and a cross-sectional dimension of 141 mm × 141 mm. A 16 mm diameter steel bar was centrally positioned within the specimen, equipped with a strain gauge. Both ends of the steel bar were connected to 141 mm × 141 mm × 10 mm steel plates by welding, providing constraints to the overall deformation of the specimen and transmitting feedback on changes in tensile and compressive stresses during the testing process. The strain gauge used in this study was an intelligent chord-type digital steel reinforcement stress gauge (range: ±200 MPa, sensitivity: 0.1 MPa). When the strain gauge experienced axial forces, it caused tension or compression changes in the elastic steel chords, altering the vibration frequency of the steel chords. By measuring the frequency change using a frequency meter, the magnitude of the forces acting on the steel reinforcement could be determined. This, in turn, allowed for the estimation of the forces experienced by the concrete structure.

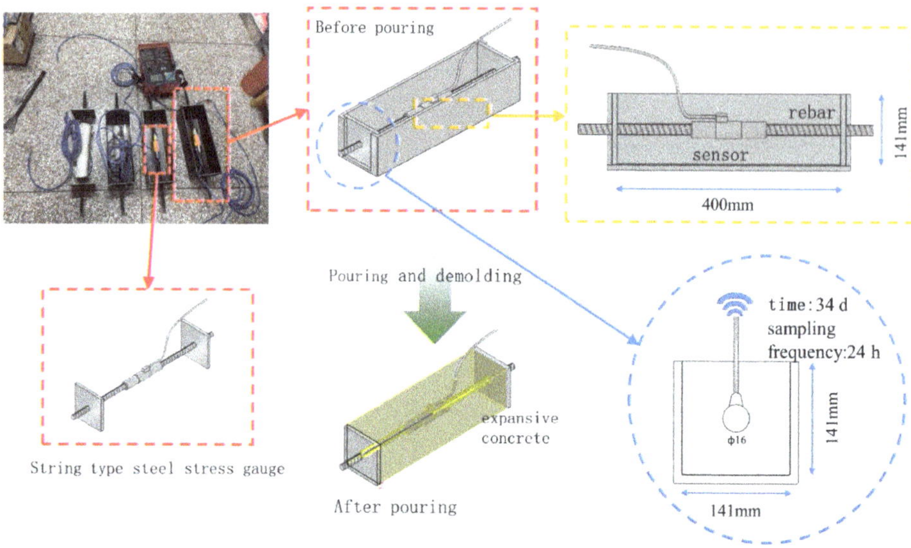

Figure 2. Self-developed concrete self-stress measurement mold and equipment.

For the experiment, a common 1% reinforcement ratio was used as the baseline. Three types of specimens were prepared by adding different volume percentages of steel fibers: 0%, 1%, and 1.5%. These specimens were labeled as G0, G1, and G1.5, respectively. After molding, the specimens were demolded after 24 h and then moved to a standard curing room for a curing period of 28 days (with a temperature of 20 ± 2 °C and relative humidity of 60 ± 5%). Due to the relatively low water-to-cement ratio used in the design, while this design increased the strength of the concrete, it was not conducive to the performance of the expansion agent in the concrete. This significantly affected the self-stress values of the micro-expansion fiber-reinforced concrete. To address this issue, during the curing period, the specimens were observed, and water was appropriately replenished on the surface to keep the specimens moist. After completion, the JMZX-3001 comprehensive testing machine was used to track and measure the stress changes in the specimens.

The compressive and flexural strength were determined according to the standard GB/T 50082-2009 "Standard Test Methods for Mechanical Properties of Ordinary Concrete" [21]. The testing equipment is shown in Figure 3. The dimensions of the flexural and compressive test specimens were 150 mm × 150 mm × 515 mm and 150 mm × 150 mm × 150 mm, respectively. After molding for 24 h, the specimens were demolded and placed in a standard curing room for curing until the ages of 3 days, 7 days, and 28 days.

Figure 3. Compression and bending resistance measurement equipment.

2.2.2. Field Test and Long-Term Observation of Steel-Fiber-Reinforced Self-Stressing Concrete

To analyze the long-term performance of steel-fiber-reinforced self-stressing concrete in prefabricated prestressed pavement joints, a prefabricated prestressed concrete pavement was laid at a certain intersection in Dezhou City, Shandong Province. The installation process involved the use of remote real-time monitoring equipment, as shown in Figure 4.

Figure 4. Installation diagram of long-term monitoring equipment.

Figure 5a shows the design drawing of the prefabricated prestressed pavement panels. The connection between the panels was achieved through steel bar welding and concrete bonding, as illustrated in Figure 5b. The length of the joint was 7500 mm, with a width of 200 mm and a thickness of 240 mm. The designed reinforcement ratio was 1%. Concrete strain gauges were embedded in the joints for continuous observation, with data recorded at 10-min intervals. The positions of the strain gauge measurement points are indicated in Figure 5c, with four sensors embedded in two joints, two in the transverse direction (MR-01 and MR-04), and two in the longitudinal direction (MR-02 and MR-03), arranged at the 1/4 and 1/2 positions of each joint. The production of steel-fiber-reinforced self-stressing concrete involves on-site mixing, and after the assembly of the panels, the concrete is poured. The proportion of steel fibers in the mix was selected based on the 28-day self-stress values of the concrete measured in the experiment mentioned in Section 2.2.1 of this paper, using the optimal numerical proportion for pouring. After 24 h of casting, a geotextile was placed on the surface, and construction personnel arranged for a 10-day water curing. Subsequently, the pavement was reopened to traffic, and long-term monitoring was conducted.

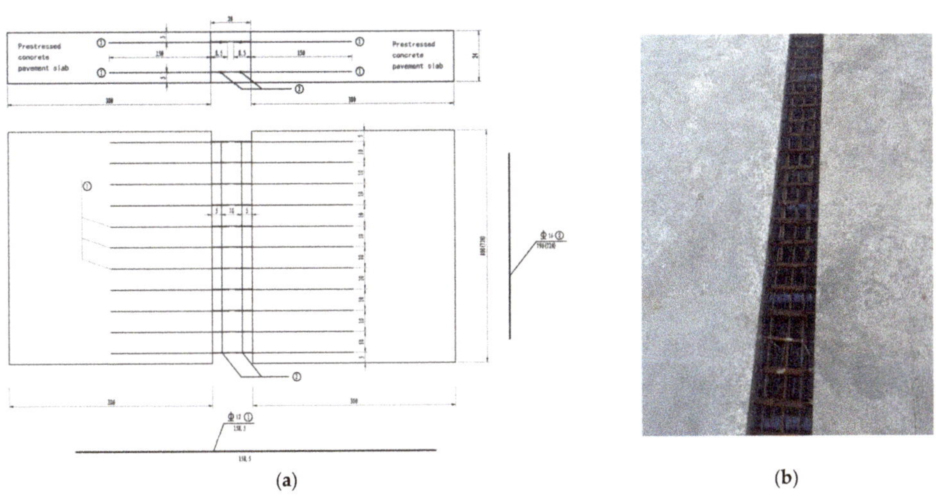

(**a**) (**b**)

Figure 5. *Cont.*

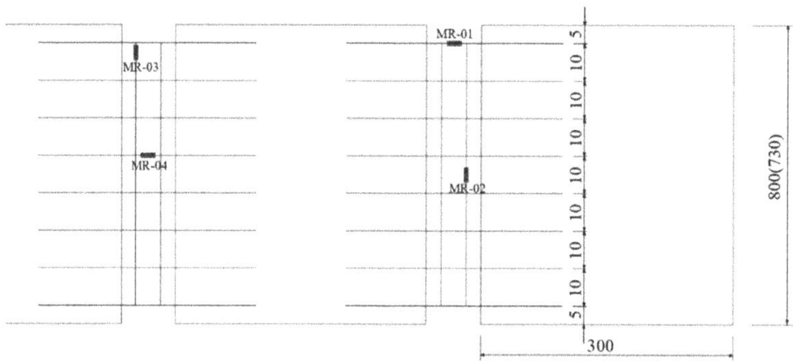

(**c**)

Figure 5. Schematic diagram of the prefabricated prestressed pavement slabs (unit: cm). (**a**) Design drawing of a prefabricated prestressed road panel joint; (**b**) Joint connection of prefabricated prestressed pavement panels; (**c**) Embedding of prefabricated prestressed pavement panel joint equipment.

3. Results and Discussion

3.1. Analysis of Self-Stress Influence of Micro-Expansion Fiber Concrete

Figure 6 shows the self-stress results of concrete with different steel fiber contents before and after 3 days under the same expansion agent content. The SFRSSC control sample (G0) showed a higher self-stress before and after 3 days (mainly due to the action of concrete on the steel plates at both ends under the constraint of steel bars). The self-stress increases rapidly before 3 d, the growth rate slows down after 3 d, and the concrete self-stress value begins to decrease slowly after reaching the first peak at about 7 d. When the curing specimen reached about 14 d, the concrete self-stress began to rise again and reached the maximum value of 7.4 MPa in the control sample (G0) after 28 d. This delayed self-stress growth phenomenon may be attributed to the current low water–binder ratio design, which makes the concrete lack water available for the expansion agent reaction. This makes part of the expansive agent unreacted in the early stage of the concrete hydration process. Since this experiment was conducted in a standard curing room, and the surface was properly sprinkled with water, the water entered the concrete through the pores and continued to hydrate the expansive agent to form a stable expansion element, so the concrete self-stress began to rise again.

Compared with G0, the specimens G1 and G1.5 with added steel fibers exhibited different phenomena before 3 days (3 d), and it is particularly noteworthy that G1 showed a significant negative self-stress before 3 d. This could be attributed to the fact that the steel fibers restricted the expansion of the hydration products of the expansion agent in the concrete. Previous research has indicated that nearly 50% of concrete's self-induced shrinkage occurs before 3 d. The lower expansion in the early stages cannot compensate for the higher shrinkage of the concrete, resulting in an apparent negative self-stress. After 3 d, the trend of change in self-stress was the same as observed in the G0 group. Self-stress increased with time and around 14 days (14 d), the self-stress generated in the G1 group surpassed that of the G0 group. This can be attributed to the constraint effect of the steel fibers on the concrete. While the steel fibers restricted the concrete's expansion in the early stages, they also partially limited its shrinkage, thereby improving the concrete's pore structure and making it more compact in the early stages. When the expansion agent's hydration products were formed, the volume of these products did not change under the higher constraints, resulting in more apparent macroscopic self-stress. The stress variation trend in the G1.5 group before 3 d was consistent with that of the G0 group, but the values were slightly smaller than those in the G1 group, which could be due to the increased amount of steel fibers providing higher constraints on the concrete. The higher constraints caused a significant slowing of the stress growth in the G1.5 group, and at 28 d, the self-stress value exhibited by the G1.5 group was the same as that of the G0 group. After 28 d, when the specimens were removed from the standard curing room, the self-stress values began to show a more apparent decrease, which may be attributed to the loss of moisture in the SFRSSC due to the lack of water supplementation in the ambient temperature environment. From the temperature change results obtained from the instrument, it was found that the internal temperature of the concrete stabilized after 3 d during the hydration process, and no significant correlation was found between the self-stress values and the temperature changes within the concrete.

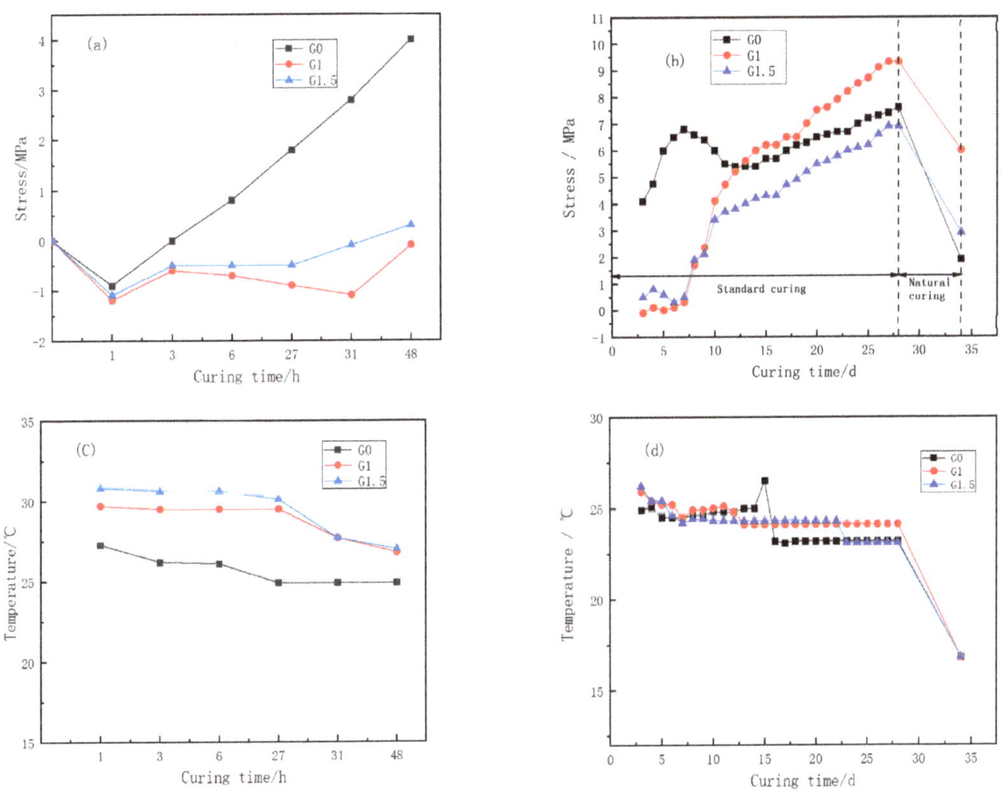

Figure 6. Effect of steel fiber content on self-stress and temperature change of SFRSSC. (**a**) Self-stress before and (**b**) after 3 days; (**c**) Temperature change before and (**d**) after 3 d.

In summary, the analysis of the concrete self-stress results shows that the addition of steel fibers in SFRSSC has no significant impact on the concrete's deformation before demolding. Within 3 days after demolding, due to the concrete's high shrinkage, increasing the amount of steel fibers would limit the concrete's expansion. After 3 days, during the standard curing period, the deformation also shows the restricting effect of steel fibers on the concrete's expansion. After leaving the curing room, the non-fiber-reinforced concrete undergoes rapid shrinkage, while the steel fibers, due to their limiting effect on the concrete, exhibit a significantly slowed rate of shrinkage. By comparing the rate of shrinkage between G1 and G1.5, it was found that the two rates were essentially the same. Therefore, compared to ordinary concrete, SFRSSC has a noticeable resistance to shrinkage, and using it at the joints between prefabricated prestressed concrete pavement panels can effectively avoid the risk of concrete cracking due to shrinkage. Furthermore, through detection, it was found that after the concrete leaves the curing room, there is still some residual stress, which can partially counteract the pavement arching caused by temperature gradients, thereby prolonging the durability of the joints.

3.2. Mechanical Response Analysis of Micro-Expansion Fiber Concrete

In the design of prestressed concrete pavement, according to the Highway Cement Concrete Pavement Design Code JTG D40-2011 [22], the compressive strength of concrete prefabricated blocks should not be less than 50 MPa. Therefore, the selected joint material should also have a compressive strength not lower than this value. Additionally, after the hydration of the joint material is completed, it typically takes on a rectangular prism shape, and when vehicles stop at the joint, they exert a certain force on both ends of the joint. This requires the selected joint material to have good flexural strength. Figure 7 shows the compressive and flexural strength values of SFRSSC with different steel fiber contents. After the addition of steel fibers, the compressive and flexural strengths of SFRSSC were improved at various ages. Compared to the blank control group, when the steel fiber content was 1%, the compressive and flexural strengths of SFRSSC after 28 days of curing increased by 11.5% and 32.6%, respectively. When the steel fiber content was 1.5%, the compressive and flexural strength of SFRSSC after 28 days of curing increased by 28.3% and 41.6%, respectively. The addition of steel fibers increased the flexural-to-compressive strength ratio of concrete, which had a beneficial effect on the development of concrete strength. It is worth noting that the increase in steel fiber content did not have a significant effect on the compressive strength of SFRSSC at 3 days. This is because the early concrete strength is not high, and under the action of the expansion agent, the concrete exhibits micro-expansion. After demolding, the surrounding constraints are reduced, resulting in some loosening of the early concrete structure. After 28 days of curing, the high content component (G1.5) showed an increase of approximately 9.44 MPa compared to the G1 group.

Based on the Highway Cement Concrete Pavement Design Code [21], it can be observed that SFRSSC (steel-fiber-reinforced self-stressing concrete) exhibits a significant increase in flexural performance due to the addition of steel fibers while maintaining excellent compressive strength. The mechanical behavior of SFRSSC enables it to serve as a stable joint material, thereby avoiding damage caused by insufficient strength at the joints.

3.3. Influence Analysis of Long-Term Changes in the Use of Micro-Expansion Fiber Concrete

Figure 8 shows the recorded graphs of joint strain variations and temperature changes detected by the on-site buried monitoring devices. Due to a delay in installing the real-time recording devices, there was no real-time recording with 10-min intervals for the first 3 days, and manual recording was performed instead. From the curves in the figure, it is evident that the two joints poured with SFRSSC (steel-fiber-reinforced self-stressing concrete) exhibit completely different behaviors in the transverse and longitudinal directions. The longitudinal measuring points, MR-01 and MR-04, experienced expansion, with their highest strain values, occurring at approximately 3 days, of 238.6 $\mu\varepsilon$ and 206.7 $\mu\varepsilon$, respectively. On the other hand, the transverse measuring points, MR-02 and MR-03, consistently experienced shrinkage, reaching a minimum value of -59.2 $\mu\varepsilon$ at around 6 days. This phenomenon is attributed to the outward expansion of the SFRSSC along the longitudinal direction, creating inward compressive stresses at the two ends of the concrete. These compressive stresses, combined with the shrinkage stresses inherent in the concrete itself, make it difficult for the expansion stresses induced by the expansive agent to compensate, resulting in a relatively obvious shrinkage phenomenon at the macroscopic level. Regarding the trend of changes, the longitudinal SFRSSC joints exhibited higher initial expansion at 3 days, which gradually decreased with curing time, stabilizing at ± 25 $\mu\varepsilon$ from 7 days onwards. Furthermore, in analyzing the temperature curve changes, it was observed that the ambient temperature in the surrounding environment decreased by 10 °C within a short period of 5 days during the 33-day inspection. However, the strain values of the SFRSSC did not show significant changes. This indicates that SFRSSC exhibits relatively high stability after completing its hydration process and is less influenced by temperature variations. As for the transverse trend, it was noted that the SFRSSC underwent continuous shrinkage during the first 6 days, but the shrinkage rate gradually reduced over time, with a shrinkage value of only −25.2 με at 120 days. It is worth noting that MR-03 showed a slight increase after 3 days, and MR-02's slope also decreased slightly. This could be attributed to non-uniform reactions of the internal expansive agent within the concrete, where the expansion induced by the agent exceeded the natural shrinkage of the concrete at 3 days. Comparing the numerical values of MR-01 and MR-04, MR-02 and MR-03, it is evident that the strain variations in the SFRSSC joints were not significantly affected by the measurement point locations but rather by the measuring direction. This indicates relatively high overall integrity of SFRSSC joints, with relatively uniform strain behavior in the same direction.

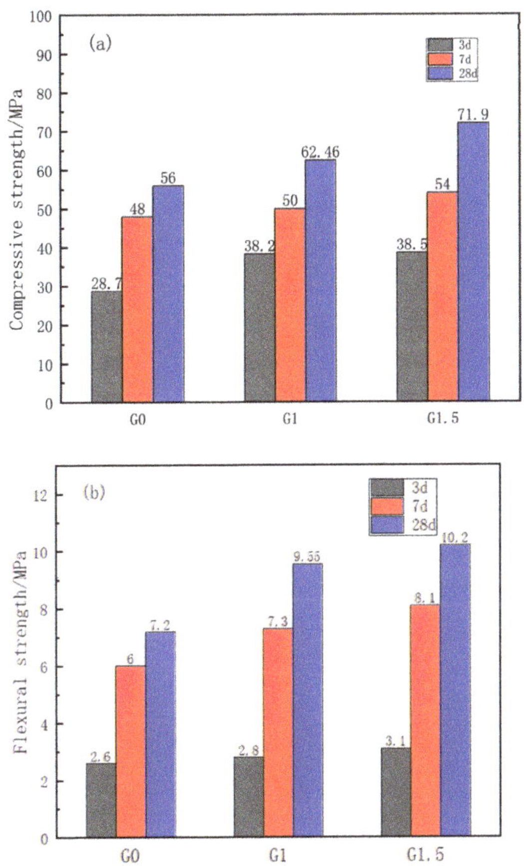

Figure 7. The effect of steel fiber content on the compressive and flexural strength of SFRSSC. (**a**) Compressive strength; (**b**) Flexural strength.

In conclusion, long-term monitoring of SFRSSC has revealed that its shrinkage deformation is more stable compared to other types of concrete, with shrinkage values ranging from −210 με to −600 με at 28 days [23–25]. Particularly, SFRSSC exhibits relatively small shrinkage deformation, which helps to address the issue of concrete cracking caused by its shrinkage. This characteristic offers the potential for achieving seamless connections between concrete slabs and joints, providing a viable solution for minimizing cracks and achieving seamless integration in concrete structures.

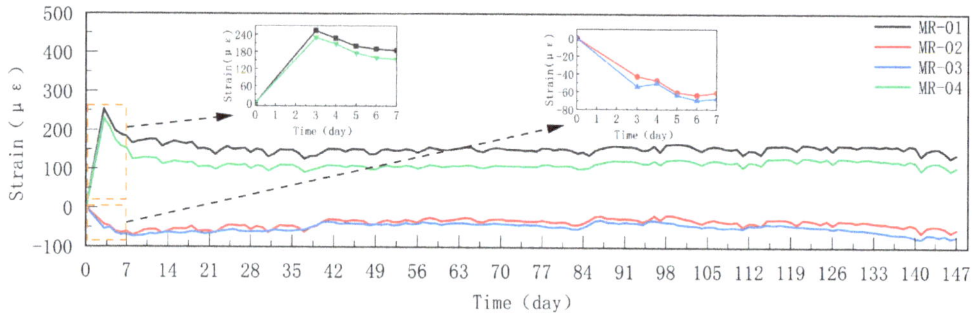

(**a**)

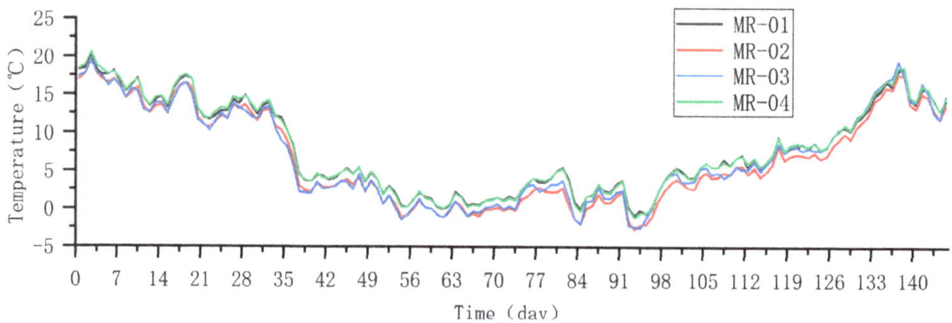

(**b**)

Figure 8. Variation of joint strain and temperature. (**a**) Joint strain variation diagram; (**b**) Joint temperature variation diagram.

4. Conclusions

This study proposes the use of steel-fiber-reinforced self-stressing concrete (SFRSSC) as an assembly-type prestressed pavement joint material to improve the connection between slab segments and precast panels, thereby avoiding joint-related problems and increasing the overall service life. The research investigates the trends of self-stress and mechanical strength variation of SFRSSC and evaluates its stability through long-term testing on experimental road sections. The specific research conclusions are as follows:

(1) SFRSSC exhibits higher self-stress compared to ordinary concrete, and this self-stress is mainly influenced by the moisture content and the amount of added steel fibers. A higher moisture content allows the expansive agent in SFRSSC to fully hydrate, leading to significant early expansion. However, a significant portion of this expansion is ineffective and decreases noticeably as the concrete leaves the moisture curing stage. The addition of steel fibers primarily limits the development of self-stress in the early stage, and this restriction reduces with increasing steel fiber content. However, in the later stages of concrete development, this limitation on self-stress also partially restricts the shrinkage of the concrete, enabling higher self-stress to be retained in the concrete over the long term, thereby improving the pavement's crack resistance. It should be noted that excessive steel fiber content is not ideal during usage, as a high content of steel fibers (1.5%) limits early self-stress development without retaining high self-stress in the later stages.

(2) SFRSSC overcomes the brittle fracture characteristics of ordinary concrete and exhibits high resistance to deformation under compressive and flexural loads. At the age of 28 days of curing, the compressive strength and flexural strength of SFRSSC can reach 62.46 MPa and 9.55 MPa, respectively, meeting the design strength requirements for assembly-type prestressed pavement materials.

(3) Through long-term monitoring of the experimental road section, SFRSSC demonstrates high stability in the joints. After 28 days, SFRSSC deformation remains in a relatively stable state, and its strain values do not vary significantly with changes in temperature. Even after 120 days, the longitudinal strain of the road remains in an expanded state. Although the transverse strain of the road is in a shrinking state, the shrinkage value slowly decreases, with only a −25.2 με variation observed at 120 days.

Author Contributions: Conceptualization, Y.-F.X., J.L., B.-L.C. and S.-J.R.; methodology, G.-H.Z.; software, B.Y.; validation, J.L. and B.-L.C.; formal analysis, T.-C.Z.; investigation, B.Y.; resources, H.X.; data curation, Y.-F.X.; writing—original draft preparation, S.-J.R.; writing—review and editing, Y.-F.X. and J.L.; visualization, J.L; supervision, B.-L.C.; project administration, Y.-F.X.; funding acquisition, J.L. and B.-L.C. All authors have read and agreed to the published version of the manuscript.

Funding: This work was supported by the research fund of the 2022 Postgraduate Science and Technology Innovation Project of Shandong Jiaotong University (No. 2022YK027); the Science and Technology Plan of Shandong Provincial Department of Transportation (No. 2022B29); and the Science and Technology Plan of Shandong Provincial Department of Transportation (2023B91).

Data Availability Statement: Not applicable.

Conflicts of Interest: The authors declare no conflict of interest.

References

1. Editorial Department of China Journal of Highway and Transport. Review on China's Pavement Engineering Research·2020. *China J. Highw. Transp.* **2020**, *33*, 1–66. [CrossRef]
2. Liu, M.; Zhan, H.; Zeng, P.; Gong, J.; Li, X.; Sun, Z.; Zhang, W. Design Optimization and Experimental Study of Assembled Prestressed Concrete Pavement. *Build. Sci.* **2020**, *36* (Suppl. 1), 245–252.
3. Rangkang, J.; Samang, L.; Adisasmita, S.; Hustim, M. The Infiltration Capacity of Eco-concrete Paving on Different Block Shapes. *IOP Conf. Ser. Earth Environ. Sci.* **2020**, *419*, 012086. [CrossRef]
4. Jamshidi, A.; Kurumisawa, K.; White, G.; Nishizawa, T.; Igarashi, T.; Nawa, T.; Mao, J. State-of-the-art of Interlocking Concrete Block Pavement Technology in Japan as a Post-modern Pavement. *Constr. Build. Mater.* **2019**, *200*, 713–755. [CrossRef]
5. Zhang, J.; Weng, X.; Yang, B.; Li, Y.; Liu, J.; Jiang, L. Bonding characteristics of grouting layer in Prefabricated Cement Concrete Pavement. *Constr. Build. Mater.* **2017**, *145*, 528–537. [CrossRef]
6. Qu, B.; Weng, X.-Z.; Zhang, J.; Mei, J.-J.; Guo, T.-X.; Li, R.-F.; An, S.-H. Analysis on the deflection and load transfer capacity of a prefabricated airport prestressed concrete pavement. *Constr. Build. Mater.* **2017**, *157*, 449–458. [CrossRef]
7. Tao, L.; Ma, X.; Peng, X.; Huang, S. Comparison of Pouring Schemes for PC Stiffening Beams Wet Joints of Yunnan Moon Bay Bridge. *Bridge Constr.* **2019**, *49* (Suppl. 1), 121–126.
8. Li, G.; Hu, H.; Ren, C.; Zhou, S.; Li, J. Durability Performance of Bridge Concrete Structural Joints. *J. Civ. Eng.* **2018**, *51*, 98–103. [CrossRef]
9. Verger-Leboeuf, S.; Charron, J.-P.; Massicotte, B. Design and Behavior of UHPFRC Field-Cast Transverse Connections between Precast Bridge Deck Elements. *J. Bridge Eng.* **2017**, *22*, 7. [CrossRef]
10. Fang, M.; Zhou, R.; Ke, W.; Tian, B.; Zhang, Y.; Liu, J. Precast System and Assembly Connection of Cement Concrete Slabs for Road Pavement: A Review. *J. Traffic Transp. Eng.* **2022**, *9*, 208–222. [CrossRef]
11. Murugan, R.B.; Natarajan, C.; Chen, S.-E. Material Development for a Sustainable Precast Concrete Block Pavement. *J. Traffic Transp. Eng.* **2016**, *3*, 483–491. [CrossRef]
12. Li, J.; Zhang, Q. Test Research on Load Transfer of Tie Bar of Concrete Pavement. *J. Zhengzhou Univ. Eng. Sci.* **2015**.
13. Peng, P.; Tian, B.; Niu, K. A Study on the Working Performance of Dowel Bars with Horizontal Installation Errors. *J. Highw. Transp. Res. Dev.* **2012**, *6*, 33–38. [CrossRef]
14. Tan, Y.Q.; Li, L.K.; Cao, P.; Mi, F.Y.; Gong, X.B. Analysis of Corrosion Mechanism and Deterioration Process for Dowel under Deicing Salt Environment. *Eng. Mech.* **2013**, *30*, 199–205.
15. Suta, M.; Lukasenoks, A.; Cepuritis, R. Determination of Material Design Values for Steel Fibre Reinforced Self-Stressing Concrete (SFRSSC) Structures. In Proceedings of the 19th International Scientific Conference Engineering for Rural Development, Jelgava, Latvia, 26–28 May 2020; pp. 631–637.
16. Suta, M.; Cepuritis, R.; Zegelis, A. Determination of Material Design Values for Steel Fibre Reinforced Self-Stressing Concrete (SFRSSC) and Regular Steel Fibre Reinforced Concrete (SFRC) in Statically Indeterminate Round Panel Tests. *Mater. Sci. Forum* **2022**, *1053*, 297–302. [CrossRef]
17. Xi, Y.F.; Lee, J.; Chen, B.L.; Yang, B.; Yu, M.Z.; Yan, X.Z.; Zhu, L. Impact of High-Performance Expansion and Shrinkage-Reducing Agents on the Mechanical Properties and Shrinkage Compensation of High-Strength Concrete. *Buildings* **2023**, *13*, 717. [CrossRef]

18. Liu, L.; Fang, Z.; Huang, Z.; Wu, Y. Solving Shrinkage Problem of Ultra-High-Performance Concrete by a Combined Use of Expansive Agent, Super Absorbent Polymer, and Shrinkage-Reducing Agent. *Compos. Part B Eng.* **2021**, *230*, 109503. [CrossRef]
19. Wang, T.; Gong, J.; Chen, B.; Gong, X.; Luo, H.; Zhang, Y.; Qu, Z. Effects of Shrinkage Reducing Agent and Expanded Cement on UHPC Fluidity, Mechanical Properties, and Shrinkage Performance. *Adv. Mater. Sci. Eng.* **2021**, *2021*, 9045754. [CrossRef]
20. Li, H.; Wang, Y.; Wang, Y.; Liu, J.; Tian, Q. Effect of CaO and MgO Based Expansive Agent on Deformation and Mechanical Properties of Concrete-Filled Steel Tubes—ScienceDirect. *Constr. Build. Mater.* **2020**, *250*, 118723. [CrossRef]
21. *GB/T 50081-2009*; China National Standard for Test Methods for Mechanical Properties of Ordinary Concrete. China Architecture & Building Press: Beijing, China, 2009. (In Chinese)
22. *JTG D40-2011*; China Communications Highway Planning and Design Institute Co., Ltd. Design Specification for Highway Cement Concrete Pavement: JTG D40-2011. People's Communications Press: Beijing, China, 2011.
23. Li, Y.; Ruan, X.; Huo, N. Experimental and Numerical Analysis on Early Shrinkage of Microporous Concrete. *J. Tongji Univ. Nat. Sci.* **2023**, *51*, 696–705.
24. Sun, H.; Ye, L.; Ding, J.; Guo, Y. Shrinkage and Creep Test of High-Strength Lightweight Aggregate Concrete. *Tsinghua Univ. J. Nat. Sci. Ed.* **2007**, 765–767+780. [CrossRef]
25. Zhang, J.; Yi, J.; Zhang, M.; Wang, K.; Wang, L. Experimental Study on Shrinkage Performance of All-Coral Concrete. *Concrete* **2023**, *399*, 165–168+177.

buildings

MDPI

Article

Feasibility and Sustainable Performance of RAP Mixtures with Low-Viscosity Binder and Castor Wax–Corn Oil Rejuvenators

Kyungnam Kim [1] and Tri Ho Minh Le [2,*]

1 Department of Highway & Transportation Research, Korea Institute of Civil Engineering and Building Technology, 283 Goyangdae-Ro, Ilsanseo-Gu, Goyang-si 10223, Gyeonggi-Do, Republic of Korea; kimkyungnam@kict.re.kr

2 Faculty of Civil Engineering, Nguyen Tat Thanh University, 300A Nguyen Tat Thanh Street, District 4, Ho Chi Minh City 70000, Vietnam

* Correspondence: lhmtri@ntt.edu.vn

Abstract: The utilization of Recycled Asphalt Pavement (RAP) mixtures in pavement construction is an environmentally friendly approach that promotes sustainable development by reducing energy consumption and material waste. However, the high cost of conventional rejuvenators limits the widespread use of RAP mixtures. In this study, a novel approach is proposed to enhance the performance of RAP mixtures by incorporating a combination of high-penetration asphalt binder and rejuvenators, namely Castor wax and Corn oil. The newly developed rejuvenator consists of 8.5% Castor wax oil, 3% Corn oil, 3% fatty acid amine surfactant, 0.2% additive, and 79.8% water. The test results demonstrate that the modified mixture exhibits superior properties compared with conventional RAP mixtures. The Multiple Stress Creep Recovery test results showed a 20% reduction in cumulative strain rate for the RAP mixture with the new rejuvenators compared with that for the conventional ones. Furthermore, the Tensile Strength Ratio test indicated a notable 9.47% improvement in the rejuvenated RAP mixture's resistance to moisture-induced damage compared with the conventional mixture. Evaluation of viscoelastic behaviors revealed a slight reduction in dynamic modulus for the rejuvenated binder, but a significant improvement in elastic behavior. In terms of rutting resistance, the Hamburg wheel tracking rut depths of the rejuvenated binder were significantly lower, representing reductions of 21.83% for specific binder compositions. Additionally, the absence of the stripping phenomenon further confirmed the superior moisture resistance of the modified mixture. The rejuvenated binder exhibited a remarkable 28.55% increase in fatigue load cycles to failure compared with the reference RAP binder, demonstrating substantial resistance to fatigue cracking. These quantitative comparisons not only confirm the superior performance of the modified mixture over conventional RAP mixtures, but also highlight the potential cost savings achieved through the utilization of Castor wax and Corn oil rejuvenators.

Keywords: recycled asphalt pavement; sustainable development; high penetration asphalt binder; castor wax; corn oil; rutting resistance; fatigue life

Citation: Kim, K.; Le, T.H.M. Feasibility and Sustainable Performance of RAP Mixtures with Low-Viscosity Binder and Castor Wax–Corn Oil Rejuvenators. *Buildings* **2023**, *13*, 1578. https://doi.org/10.3390/buildings13071578

Academic Editors: Andrea Baliello and Di Wang

Received: 7 June 2023
Revised: 18 June 2023
Accepted: 19 June 2023
Published: 21 June 2023

1. Introduction

The application of Reclaimed Asphalt Pavement (RAP) has gained considerable attention in the field of pavement engineering [1]. RAP is obtained by milling and crushing existing asphalt pavements, and it can be reprocessed and used as a valuable resource in new pavement construction [2]. The use of RAP offers numerous benefits, including reduced energy consumption, conservation of natural resources, and decreased waste generation [3,4]. Moreover, the incorporation of RAP in pavement construction aligns with the principles of sustainable development and supports the circular economy concept [5].

Over the years, several innovative technologies have emerged to enhance the utilization of RAP materials and improve the overall performance of recycled asphalt [6–8]. One notable technology is the use of warm-mix asphalt (WMA) additives in RAP mixtures [9]. WMA technologies enable the production and placement of asphalt at lower temperatures, reducing energy consumption and emissions during construction [10]. Studies have shown that incorporating WMA additives in RAP mixtures can enhance workability, reduce moisture susceptibility, and improve pavement durability [11]. This technology not only contributes to sustainable practices, but also offers economic advantages by lowering production costs [12,13]. Another promising technology is the introduction of recycling agents or modifiers specifically designed for RAP [14]. These additives, such as polymers, chemical agents, and bio-rejuvenators, aim to restore the aged binder in RAP materials and improve their rheological properties [15]. By rejuvenating the binder, these technologies can enhance the stiffness, rutting resistance, and fatigue performance of RAP mixtures [16]. Additionally, the use of bio-based rejuvenators offers an eco-friendly alternative to traditional petroleum-based products, aligning with sustainable development goals [17]. Advancements in milling and processing techniques have also contributed to the efficient use of RAP materials [18]. Innovations in milling machines, such as drum designs, automation features, and temperature control allow for better control and precision in RAP processing [19]. This results in a higher quality of recycled materials, improved particle size distribution, and enhanced compatibility with virgin binders. The observed increase in performance can be attributed to various factors, and recent studies by Xing et al. have highlighted the significant contribution of the degree of blending between virgin and RAP binders [20]. It has been noted that an effective blending of these binders noticeably contributes to the improvement in performance. This finding emphasizes the importance of achieving optimal blending between the two types of binders to enhance the overall performance of RAP mixtures. Furthermore, the integration of intelligent systems and artificial intelligence in RAP processing has the potential to optimize milling operations, reduce waste, and improve overall productivity [21].

The current status of expensive rejuvenators in RAP reflects the challenges and limitations associated with their use [4]. Traditional rejuvenators, which are commonly derived from petroleum-based products, can be costly and contribute to the overall expenses of RAP mixtures [22,23]. This high cost hinders the widespread adoption and application of RAP in pavement construction, particularly in regions with limited financial resources. The expense of these rejuvenators poses a significant barrier to the sustainable utilization of RAP materials, as it can undermine the economic viability and feasibility of using recycled asphalt in road projects [24]. As a result, there is a need to explore alternative and more affordable rejuvenator options that can maintain or enhance the performance of RAP mixtures while reducing overall project costs.

In developing countries, the application of RAP has shown great promise in addressing the challenges of limited resources and infrastructure development [25]. These countries often face constraints in terms of budget, construction materials, and environmental sustainability [26]. The utilization of RAP provides a cost-effective solution by recycling existing asphalt pavements and reducing the reliance on virgin materials [27]. Additionally, RAP offers the opportunity to enhance the performance and durability of road pavements without compromising on quality [28]. The use of RAP in developing countries not only helps in conserving natural resources, but also contributes to the reduction of carbon footprint and waste generation. Despite the potential benefits, the widespread adoption of RAP in these countries is still in its early stages and requires further investigation, optimization, and implementation strategies to overcome technical, economic, and policy challenges. By exploring and promoting the application of RAP, developing countries can achieve sustainable and resilient road infrastructure development while addressing their unique socio-economic and environmental needs.

While the use of RAP has demonstrated numerous advantages in sustainable road construction, its application in developing countries is often hindered by cost-related challenges [29]. One significant limitation lies in the high cost associated with conventional rejuvenators and additives used in RAP mixtures [30]. These countries, already grappling with limited financial resources and infrastructure budgets, find it challenging to afford expensive rejuvenators that are commonly used in developed nations [31]. As a result, the utilization of RAP materials becomes constrained, hindering the potential benefits of recycling and sustainability that RAP can offer [26]. Addressing this limitation requires innovative approaches and cost-effective solutions that cater to the specific economic constraints of developing countries, enabling them to fully leverage the potential of RAP in their road construction and maintenance endeavors.

Given the current limitations and cost implications surrounding the application of RAP in developing countries, there is a pressing need for research to address these challenges and develops cost-effective solutions. This study aims to fill this research gap by investigating the effectiveness of low-cost rejuvenators, specifically Castor wax and Corn oil, in replenishing the stiffness and hardened properties of RAP mixtures. The novelty of this research lies in the formulation of a newly developed rejuvenator that combines a high-penetration asphalt binder with Castor wax and Corn oil. By quantitatively comparing the performance of the modified RAP mixture with conventional RAP mixtures through various laboratory tests, this study provides valuable insights into the feasibility and practical application of low-cost rejuvenators in enhancing the performance and sustainability of RAP pavements. The incorporation of a low-viscosity binder and Castor wax with Corn oil rejuvenators holds significant importance in the field of asphalt pavement rejuvenation. Traditional methods of rejuvenation often rely on petroleum-based additives, which not only raise environmental concerns but also face limitations in terms of availability and cost. By exploring alternative rejuvenation strategies, such as the use of a low-viscosity binder and Castor wax with Corn oil rejuvenators, the research can address these challenges and fulfill the growing need for sustainable and cost-effective solutions in the asphalt industry. These innovative approaches offer the potential to enhance the performance of recycled asphalt mixtures by improving their quality and durability. Furthermore, by reducing the dependence on virgin asphalt binders, the research can contribute to the preservation of natural resources and mitigate the environmental impact associated with traditional rejuvenation methods.

In this research, a comprehensive mixed-design approach was employed to evaluate the effectiveness of low-cost rejuvenators in RAP mixtures. The mixed-design process involved combining a high-penetration asphalt binder with Castor wax, Corn oil, and other additives in specific proportions. The newly developed rejuvenator composition consisted of 8.5% Castor wax oil, 3% Corn oil, 3% fatty acid amine surfactant, 0.2% additive, and 79.8% water. The mixture's performance was assessed through a range of laboratory tests, including the Multiple Stress Creep Recovery (MSCR) test, the Indirect Tensile Strength (ITS) test, Hamburg wheel tracking (HWT) test, and dynamic modulus test. These testing methods allowed for a comprehensive evaluation of the modified RAP mixture's stiffness, resistance to deformation, moisture sensitivity, and overall performance. The results obtained from these tests were compared with those of conventional RAP mixtures, providing valuable insights into the potential of the developed rejuvenators in enhancing the performance of RAP pavements. A general summary of the research is briefly presented in Figure 1.

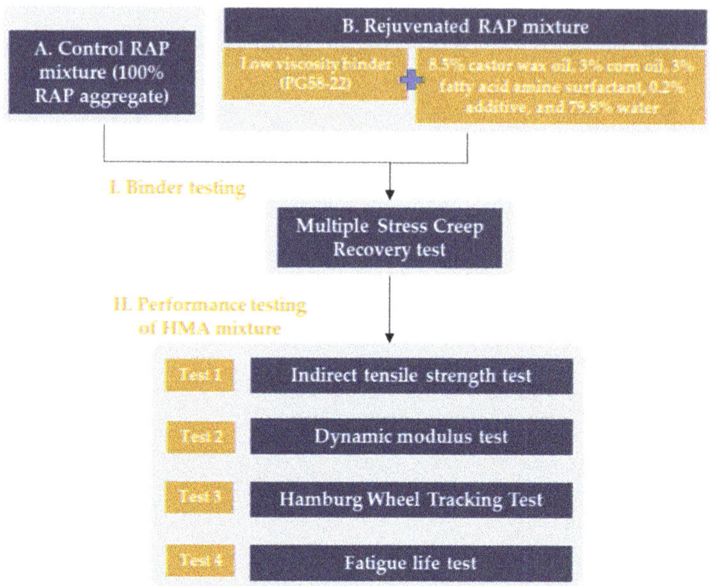

Figure 1. Research flowcharts.

2. Materials and Methods

2.1. Materials

The materials used in this study included RAP aggregates, asphalt binders, and rejuvenator additives. The RAP aggregates were obtained from local road construction projects and were processed to meet the required specifications [27]. The asphalt binder selected for the mix design was a high penetration grade binder, chosen for its compatibility with the RAP materials [4]. As for the rejuvenator additives, Castor wax and Corn oil were utilized as the primary rejuvenators due to their potential to replenish the aged binder and enhance the performance of the RAP mixture. These materials were sourced and prepared according to established guidelines to ensure consistency and accuracy throughout the testing and analysis phases.

2.1.1. RAP Aggregates

The RAP aggregate used in this study was obtained from local road construction projects (see Figure 2). It consisted of asphalt pavement materials that were milled and removed from existing road surfaces. The RAP material was processed to meet the required specifications for the experimental mixtures [27]. Initially, the RAP material underwent a thorough screening process to remove oversized particles and debris. Subsequently, it was crushed to achieve the desired nominal maximum aggregate size of 13 mm. To ensure the quality of the RAP aggregate, it was subjected to rigorous cleaning procedures. Any contaminants, such as soil, vegetation, or other foreign materials, were removed to minimize their adverse effects on the performance of the asphalt mixture. The resulting RAP aggregate exhibited satisfactory properties in terms of particle shape, size distribution, angularity, and texture, making it a suitable component for the experimental mixtures in this research. The gradation and general properties of the employed RAP aggregate in this research are presented in Figure 3 and Table 1, respectively.

Figure 2. Collecting of RAP aggregate in the asphalt plant.

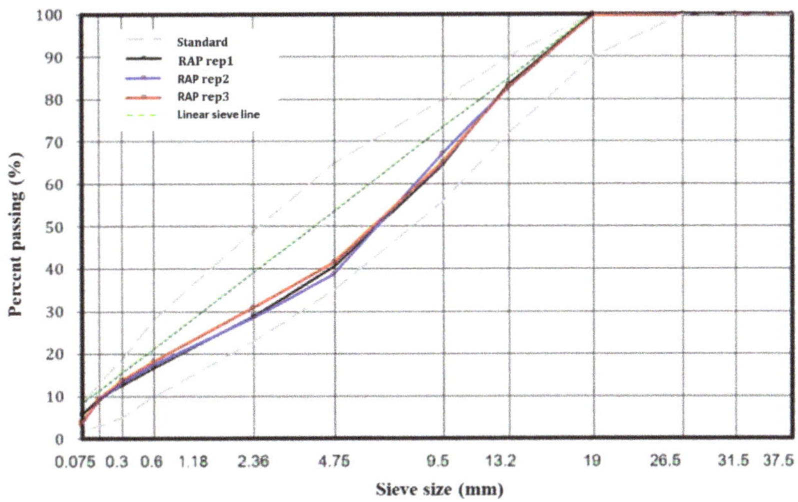

Figure 3. RAP aggregate gradation.

Table 1. RAP aggregate properties.

Properties	Value
Concrete asphalt content [32]	≥3.7%
Asphalt penetration [33]	≥21 (25 °C, 1/10 mm)
The amount lost in the wash test [34]	≤4.8%
Moisture content [35]	≤5%

2.1.2. Extracted RAP Binders

The extracted RAP binder refers to the asphalt binder obtained from the reclaimed asphalt pavement (RAP) material through a solvent extraction process [3]. In this study, the RAP aggregate was crushed to expose the asphalt binder within the RAP particles. Then, a suitable solvent, such as trichloroethylene or chloroform, was used to dissolve and extract the binder from the RAP aggregate. The extraction process involved mixing the crushed RAP material with the solvent and agitating the mixture to facilitate the dissolution of the binder. After a specified extraction period, the solvent and dissolved binder were separated from the solid residue through settling and subsequent separation techniques. The solvent was further removed from the extracted binder using methods like vacuum distillation or filtration to obtain a purified binder. The extracted RAP binder served as an important component in this research, as it allowed for the evaluation of its properties and its potential for reuse in asphalt mixtures. By utilizing the extracted binder, this study aimed to assess its performance characteristics and determine its suitability as a sustainable alternative to virgin asphalt binder in pavement construction. RAP binder exhibited key properties, including a penetration of 84.1 (1/10 mm) at 25 °C, a softening point of 68.2 °C, and a ductility of 102 cm/min at 5 °C. It demonstrated minimal mass loss (0.02%) and penetration loss (67) after thin film oven aging. Rheologically, it showed $G^*/\sin\delta$ values of 1.72 kPa (original), 2.41 kPa (after RTFO), and 1486 kPa (after PAV) at 76 °C. At −22 °C, the binder exhibited a stiffness of 182 MPa and an m-value of 0.32. These properties contributed to its performance in various asphalt applications.

2.1.3. New Low-Viscosity Rap Binder

The new low-viscosity RAP binder used in this study was a recently developed asphalt binder specifically designed for the rejuvenation of reclaimed asphalt pavement (RAP) materials [4]. This binder exhibited a high penetration grade of PG 58−22, indicating its low stiffness and improved workability at higher temperatures. The low viscosity of the binder was achieved through the careful selection of asphalt feedstocks and modification techniques during the refining process. This resulted in a binder with enhanced flow properties, allowing for better coating and binding of RAP aggregates in asphalt mixtures. The use of a low-viscosity binder in RAP mixtures offered several advantages. It promoted easier mixing and compaction of the asphalt mixture, leading to improved workability during construction. Additionally, the low-viscosity binder enhanced the coating and adhesion properties, facilitating the incorporation of RAP aggregates and ensuring better overall performance of the asphalt pavement.

2.1.4. Castor Wax Oil and Corn Oil Rejuvenators

The rejuvenator used in this study was a specially formulated mixture comprising Castor wax oil, Corn oil, a fatty acid amine surfactant, an additive, and water. The rejuvenator was designed to replenish the aging and stiffened properties of the extracted RAP binder, promoting the restoration of its original performance characteristics. The composition of the rejuvenator consisted of 8.5% Castor wax oil, 3% Corn oil, 3% fatty acid amine surfactant, 0.2% additive, and 79.8% water. Castor wax oil, derived from Castor beans, is known for its rejuvenating properties, helping to soften and improve the workability of the aged binder. Corn oil, on the other hand, acted as a natural rejuvenator, enhancing the elasticity and rejuvenation potential of the binder. The fatty acid amine surfactant was included to enhance the compatibility and dispersion of the rejuvenator within the binder matrix. The additive was incorporated to optimize the rejuvenation process and improve the overall performance of the RAP mixture. The newly developed rejuvenator composition aimed to provide an effective and cost-efficient solution for the rejuvenation of RAP materials. The research will evaluate the performance of a RAP mixture treated with the rejuvenator through various laboratory tests, including the MSCR test, to assess its ability to restore the desired rheological properties and improve the long-term performance of the asphalt pavement.

2.1.5. Binder Content for RAP Mixture

The percentage of the binder for the RAP mixture involved considering both the aged binder present in the RAP and the low-viscosity binder content added to rejuvenate the mixture. After careful analysis and testing, it was determined that a suitable percentage for the aged binder in the RAP was 3.7% by weight. This percentage accounted for the residual binder present in the reclaimed asphalt pavement. In addition to the aged binder, a low-viscosity binder was incorporated as a rejuvenator to replenish the stiffness and enhance the performance of the RAP mixture. Through extensive research and experimentation, it was found that a percentage of 2.3% by weight for the low-viscosity binder was effective in achieving the desired properties. Therefore, the recommended total binder content for the RAP mixture was 6% by weight, comprising 3% aged binder and 2% low-viscosity binder. This proportion ensured the proper balance of binder materials to promote adequate coating, bonding, and improved performance of the RAP mixture. The selection of these percentages took into account the specific characteristics of the RAP, the desired performance goals, and the anticipated traffic and environmental conditions the pavement will experience.

2.1.6. Mixing Process

The mixing process for the RAP mixture was conducted using the Superpave compaction gyratory method, which includes careful attention to temperature control at each step to ensure an effective blend of the materials [36]. Firstly, the RAP aggregates were preheated to a temperature of 160 °C to remove any moisture content. Next, the extracted RAP binder and the new low-viscosity RAP binder, along with the rejuvenator consisting of 8.5% Castor wax oil, 3% Corn oil, 3% fatty acid amine surfactant, 0.2% additive, and 79.8% water, were heated to their respective mixing temperatures. The preheated RAP aggregates were then combined with the binders and rejuvenator in a mechanical mixer, following the Superpave compaction gyratory method, which involved multiple cycles of mixing and compaction to simulate the field conditions. Throughout the mixing process, the temperature was maintained at a specified level, as specified by the Superpave compaction gyratory method, to facilitate proper blending and optimal workability of the mixture. After completion of the mixing, the RAP mixture was allowed to gradually cool to a suitable temperature for subsequent testing and analysis. The adherence to the Superpave compaction gyratory method during the mixing process ensured the proper compaction and performance of the RAP mixture. To conduct various laboratory tests, specimens of appropriate sizes were prepared from the RAP mixture. The specific specimen sizes were utilized for each specific laboratory test.

2.2. Methods

2.2.1. Multiple Stress Creep Recovery (MSCR) Test

The MSCR test was conducted to evaluate the performance of the asphalt binder used in the RAP mixture (see Figure 4). The test was performed in accordance with the standard AASHTO T350 procedure [37]. The MSCR test provided valuable insights into the rutting and healing potential of the binder under varying stress and temperature conditions [38–40]. The test involved subjecting the asphalt binder to a series of controlled stress and temperature cycles. The binder was loaded into a dynamic shear rheometer (DSR) equipped with a parallel plate geometry. The test was conducted at temperatures ranging from 40 °C to 70 °C and stress levels of 100 Pa, 200 Pa, and 300 Pa. During the test, the binder was subjected to a stress load for a specified period, followed by a recovery period. The creep compliance and recovery compliance of the binder were measured to assess its ability to resist deformation and recover its original properties. The MSCR test provided crucial data on the rutting potential, rutting resistance, and permanent deformation characteristics of the asphalt binder. The results of this test played a vital role in understanding the performance of the binder and its suitability for use in the RAP mixture.

The indirect tensile strength test, following the ASTM D6931 standard [41], was conducted to evaluate the mechanical properties of the RAP mixture. Cylindrical specimens with a diameter of 150 mm and a height of 75 mm were prepared from the compacted RAP mixture. The test was performed at a controlled temperature of 25 °C, simulating typical ambient conditions. Prior to testing, the specimens were conditioned in a controlled environment for 24 h to ensure moisture equilibration and consistent test results. During the test, the specimens were loaded at a constant displacement rate until failure occurred, and the load and deformation values were recorded. The indirect tensile strength was calculated based on these measurements. To compare the performance of the RAP mixture with that of the control, which consisted of virgin asphalt and aggregate without RAP or rejuvenators, the Tensile Strength Ratio (TSR) was determined. The TSR was calculated as the ratio of the average wet Indirect Tensile (IDT) strength to the average dry IDT strength. Higher TSR values indicated improved resistance to cracking and enhanced tensile strength in the RAP mixture. By conducting the indirect tensile strength test and evaluating TSR ratios, the effectiveness of the rejuvenators and their impact on the mechanical properties of the RAP mixture was assessed.

Figure 4. MSCR testing apparatus.

2.2.2. Indirect Tensile Strength Test

2.2.3. Dynamic Modulus Test

To assess the viscoelastic properties of RAP mixtures, an extensive experiment was conducted using an advanced MTS testing machine, as shown in Figure 5. The dynamic modulus test, based on AASHTO TP 62 [42], was performed to evaluate the viscoelastic properties of the RAP mixture. The test was conducted using a Universal Testing Machine (UTM) equipped with a dynamic modulus apparatus. Specimens in a cylindrical shape, with a diameter of 100 mm and a height of 150 mm, were prepared from the compacted RAP mixture. The test was performed at various temperatures ranging from −10 °C to 54 °C to assess the temperature-dependent behavior of the mixture.

Figure 5. Dynamic modulus test.

The test involved subjecting the specimens to sinusoidal loading at different frequencies (f) ranging from 0.1 Hz to 20 Hz. The resulting complex modulus (E^*) was calculated using the equation:

$$log|E^*| = a + \frac{b}{1 + \frac{1}{e^{d+g*\log(f_R)}}} \quad (1)$$

where f_R is the reduced frequency, calculated as

$$f_R = f \times a \quad (2)$$

To determine the temperature dependency of the dynamic modulus, the logarithm of the time-temperature shift factor (a_T) was plotted against temperature. The relationship was modeled using a quadratic equation:

$$\log(a_T) = \alpha_1 T^2 + \alpha_2 T + \alpha_3 \quad (3)$$

where α_1, α_2, and α_3 are coefficients that characterize the temperature dependency.

By conducting the dynamic modulus test and analyzing the obtained data, the viscoelastic behavior of the RAP mixture under varying temperatures and frequencies was assessed. This information provided insights into the material's stiffness and ability to resist deformation under different loading conditions, contributing to the understanding of its performance in pavement applications.

2.2.4. Fatigue Crack Resistance in Mixtures

An extensive investigation into the fatigue crack resistance of RAP mixtures was conducted using a standardized testing method outlined in AASHTO T 321-07 [43]. The primary objective of this dedicated fatigue cracking test was to assess the RAP asphalt mixture's ability to withstand repeated loading cycles and resist crack formation under cyclic loading conditions. Test specimens, prepared according to standard procedures, were subjected to repeated loading cycles at a predetermined strain level, using a beam subjected to bending action with a continuous 10 Hz haversine or sinusoidal loading. The specimens were fabricated to adhere to precise measurements, boasting dimensions of 380 mm in length, 50 mm in height, and 63 mm in width. Fatigue failure was determined when the material's stiffness was reduced to 50% of its original value after enduring repeated and continuous loading. Three replicates were employed in this test to generate the average result.

To establish a crucial baseline for stiffness, the material underwent 50 repetitive load cycles under controlled conditions at a constant room temperature of 20 °C. Throughout the test, the specimens' response was continuously monitored, meticulously observing any visible crack initiation and propagation. The fatigue life of the RAP asphalt mixture was determined by analyzing several performance indicators, including stiffness degradation, the number of cycles required for crack initiation, and the extent of crack propagation. These parameters enabled the evaluation of the resistance of the RAP asphalt mixture to fatigue cracking, comparing it against established performance criteria. The ultimate aim was to replicate the cumulative effects of traffic loading and environmental factors on the fatigue performance of the RAP asphalt mixture. By comprehensively assessing stiffness degradation and crack development, the test results provided valuable insights into the material's resistance to fatigue cracking.

2.2.5. Hamburg Wheel Tracking Test

In order to assess the vulnerability of the RAP asphalt mixture to moisture-induced damage, a comprehensive Hamburg wheel tracking test was conducted following the guidelines specified in AASHTO T 324 [44], as shown in Figure 6. The main purpose of this test was to evaluate the resistance of the mixture against potential issues caused by moisture. During the testing process, the RAP asphalt mixture specimens were submerged in water at a precisely controlled temperature of 50 °C while being subjected to a constant wheel load of 705 ± 4.5 N. The test consisted of a predetermined number of loading cycles designed to simulate real-world traffic loads and environmental conditions. It should be noted that the test results from each mix design were averaged from three replicates. Throughout the test, close attention was paid to the appearance of stripping points, which are indicative of the material's susceptibility to moisture-related damage.

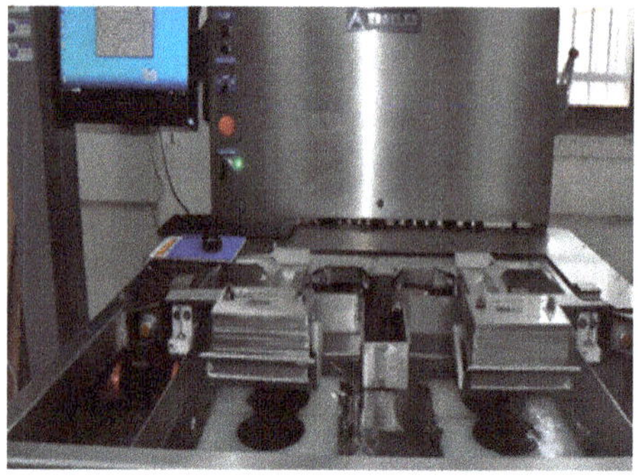

Figure 6. HWT results.

The determination of the stripping point in the HWT test involved analyzing the relationship between the stripping inflection point and stripping slope, which are indicators of the moisture resistance of HMA. The stripping inflection point represented the number of passes at the intersection of the creep slope and stripping slope curves. It served as a critical parameter in assessing the susceptibility of the HMA to moisture damage, providing valuable insights into its resistance to the detrimental effects of water infiltration and potential stripping phenomena. To ensure accurate and reliable results, certain criteria were set. Specifically, the detection of stripping points was carefully observed after a minimum of 10,000 cycles. Additionally, the settlement of the specimens was monitored, and it was ensured that the settlement did not exceed 20 mm after 20,000 repetitions of the test. These specific thresholds were established to provide valuable insights into the performance of the RAP asphalt mixture under moisture-induced conditions. The fabrication process of each sample was carried out with great care to ensure consistency and precision. The specimens were meticulously prepared, adhering to strict dimensional requirements. Specifically, they were crafted to possess a height of 60 mm and a diameter of 150 mm. Furthermore, a targeted porosity level of 7.0 $\pm$ 1.0% was maintained during the fabrication process. These meticulous measures were taken to ensure the reliability and validity of the test results, allowing for a comprehensive assessment of the moisture sensitivity of the RAP asphalt mixture. The data obtained from the Hamburg wheel tracking test provided valuable insights into the performance of the RAP asphalt mixture under moisture-induced conditions. By examining the occurrence of stripping points and monitoring settlement, the test results shed light on the mixture's resistance to moisture-induced distress. Ultimately, this information contributed to enhancing the understanding of the RAP asphalt mixture's behavior and aided in the development of more durable and resilient pavement materials.

3. Results and Discussion

3.1. MSCR Test Results

The MSCR test results for the RAP mixture utilizing rejuvenators are depicted in Figure 7. It is evident from the data that the RAP mixture with the newly developed rejuvenators exhibited a significant reduction in cumulative strain rate under the applied creep load, outperforming the conventional rejuvenators. Moreover, the strain recovery slope of the RAP mixture with the new rejuvenators displayed superior characteristics across different creep load levels. For instance, at an MSCR level of 3.2 kPa and 160 s, the strain value for the original RAP mixture was about 670%, whereas for the RAP mixture with the new rejuvenators, it reached around 530%. These results indicate the improved rejuvenation effect achieved by the new rejuvenators in the RAP binder, leading to enhanced stiffness and minimized strain accumulation.

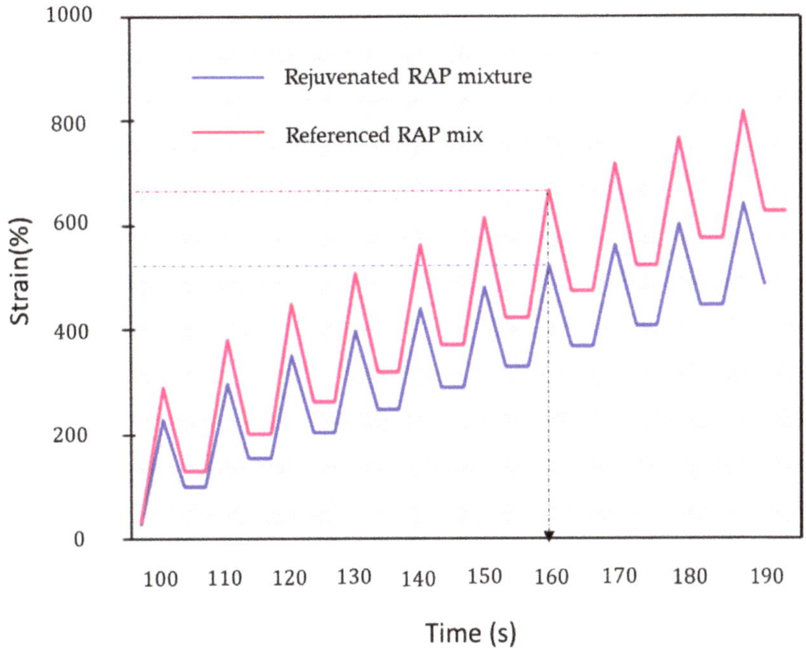

Figure 7. MSCR test results.

Analyzing the non-recoverable compliance under a shear creep load of 3.2 kPa, it was observed that the RAP mixture with the new rejuvenators exhibited behavior similar to that of the conventional rejuvenators in the initial stages (between 100 and 120 s). However, as the testing duration increased, the RAP mixture exhibited significant strain accumulation, resulting in a wider disparity between the two mixtures. This observation highlights the superior performance of the RAP mixture with the new rejuvenators, demonstrating enhanced resistance to deformation and reduced susceptibility to strain accumulation.

These results validate the effectiveness of incorporating Castor wax and Corn oil rejuvenators in enhancing the overall performance and longevity of the RAP mixture. The observed improvements in the MSCR test results can be attributed to the unique properties of Castor wax and Corn oil. This type of rejuvenator properly enhanced the flexibility and resilience of the binder, resulting in improved resistance to the MSCR test. The synergistic effects of these rejuvenators created a modified binder with superior properties, addressing the limitations associated with conventional rejuvenators. In general, the MSCR test results clearly demonstrated the positive impact of a low-viscosity binder incorporating Castor wax and Corn oil as rejuvenators in the RAP mixture. The rejuvenated binder exhibited enhanced properties.

In Figure 8, the dataset presents the percent difference of recovery and non-recovery for two types of RAP binders: the control RAP binder and the modified RAP binder, which incorporated a rejuvenator. In terms of recovery, the control RAP binder exhibited a percent difference of 4.92%, while the modified RAP binder showed a slightly higher percent difference of 5.87%. These values indicate a small variation in the recovery performance between the two binders, with the modified binder showing a slightly higher difference.

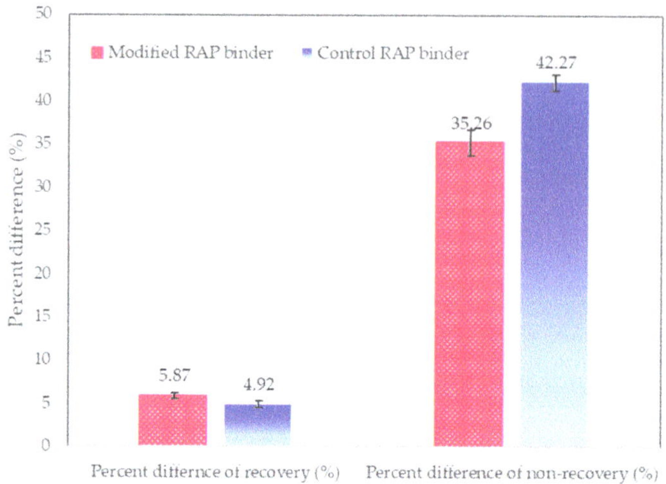

Figure 8. Recovery part (%) and non-recovery part (%).

Regarding non-recovery, the control RAP binder displayed a percent difference of 42.27%, indicating a significant deviation from the expected non-recovery value. Conversely, the modified RAP binder, with the rejuvenator, showed a lower percent difference of 35.26% in non-recovery. These findings suggest that the addition of the rejuvenator in the RAP binder resulted in a reduction in non-recovery variability compared with the control binder.

The observed differences in recovery and non-recovery between the control and modified RAP binders highlight the positive influence of the rejuvenator on the binder's rheological behavior. The lower percent difference in non-recovery for the modified binder indicates a reduced tendency for permanent deformation under applied stress. These findings underscore the effectiveness of the rejuvenator in improving the recovery and non-recovery characteristics of RAP binders, contributing to the development of more resilient asphalt mixtures with enhanced long-term performance and durability.

3.2. TSR Test Results

The TSR test results indicated a significant improvement in the moisture susceptibility of the rejuvenated RAP mixture. The TSR values were calculated as the ratio of the average wet IDT (Indirect Tensile Strength) strength to the average dry IDT strength. Comparing the TSR values of the conventional RAP mixture to the rejuvenated RAP mixture based on Figure 9, there was an observed increase of around 9.47% in the TSR ratio. This suggests that the rejuvenated mixture exhibited enhanced resistance to moisture-induced damage compared with the conventional RAP mixture. The improved TSR ratio can be attributed to the beneficial effects of the low-viscosity binder, Castor wax, and Corn oil rejuvenators. The low-viscosity binder facilitated better coating and adhesion of the binder to the RAP aggregates, resulting in improved bond strength and reduced susceptibility to moisture damage. Additionally, the Castor wax and Corn oil rejuvenators enhanced the binder's flexibility, further contributing to its moisture resistance properties. These findings align with related research studies that have demonstrated the positive influence of rejuvenators in enhancing the TSR values of asphalt mixtures. The observed increase in the TSR ratio is consistent with the desired goal of improving the moisture resistance and durability of asphalt pavements.

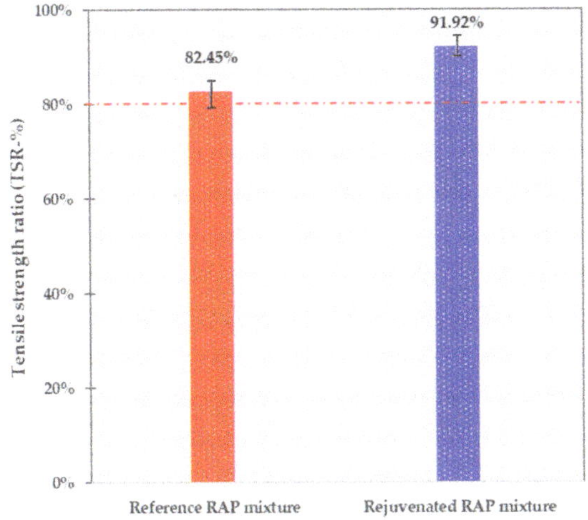

Figure 9. TSR test results.

3.3. Dynamic Modulus Test Results

The evaluation of viscoelastic behaviors for the RAP mixtures aimed to assess their performance characteristics. The dynamic modulus test results provide valuable insights into the mechanical behavior and stiffness characteristics of the RAP mixture. Figure 10 illustrates the dynamic modulus values of the RAP mixture at various testing frequencies and temperatures. For instance, at a testing temperature of 25 °C and a frequency of 10 Hz, the dynamic modulus of the RAP mixture with the rejuvenated binder was found to be 25.5 MPa, whereas the reference RAP binder exhibited a dynamic modulus of 28.7 MPa. This represents a slight reduction of 12.7% in the dynamic modulus for the rejuvenated binder, indicating a softer behavior under cold regions to counter the thermal cracking in winter. Moreover, the percentage reduction in dynamic modulus values was observed across different temperatures and frequencies; however, the gap between the two mixed-design scenarios was negligible.

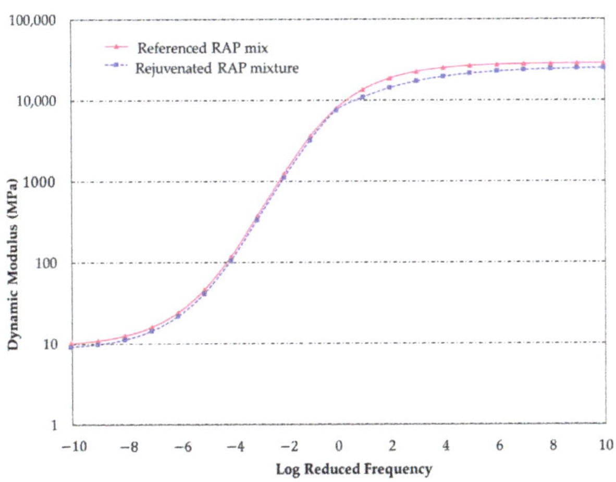

Figure 10. Dynamic modulus test results.

As shown in Figure 11, the incorporation of rejuvenators in the RAP mixture resulted in a notable enhancement in the elastic behavior, as evidenced by the phase angle results.

The modified mixture exhibited a significantly lower phase angle value (approximately 3 degrees) compared with that of the controlled mixture. This reduction in the phase angle confirmed the reinforcement of the ductile behavior of the RAP mixture under service conditions. It demonstrated that the rejuvenated mixture possessed improved resistance to deformation and a better ability to withstand stress and strains, thereby enhancing its overall performance and durability.

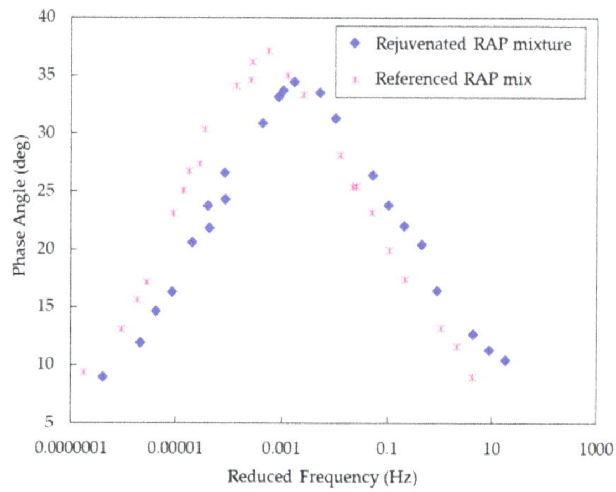

Figure 11. Phase angle analysis.

These findings indicate that the incorporation of the low-viscosity RAP binder and the Castor wax and Corn oil rejuvenators had a significant positive impact on the ductile capacity of the RAP mixture. The comparisons highlight the superiority of the rejuvenated binder in terms of dynamic modulus performance compared with that of the reference RAP binder.

3.4. HWT Test Results

The HWT test was conducted to evaluate the moisture sensitivity and resistance to rutting of the RAP mixture. Figure 12 presents the results of the test, showing the extent of rutting in terms of the rut depth at different numbers of loading cycles. In general, all mixtures displayed a sharp increase in rutting depth in the initial 5000 cycles. At this stage, the referenced RAP mixture exhibited a greater rutting resistance compared with the modified mixes since the brittleness behavior of the aged RAP binder can considerably counter the permanent deformation. However, after the first stripping point at 12,800 HWT cycles, the results indicate that the RAP mixture with the rejuvenated binder exhibited superior resistance to rutting compared with the conventional RAP binder. For instance, the rut depth of the reference RAP binder was measured at 11.77 mm, while the rejuvenated binder showed a significantly lower rut depth of only 10.59 mm and 9.2 mm for the RAP mixture with a high penetration asphalt binder and the RAP mixture containing a low-viscosity binder with rejuvenator, respectively. This represents a reduction of 10.2% and 21.83% in rut depth, demonstrating the improved resistance to deformation and rutting for the rejuvenated binder.

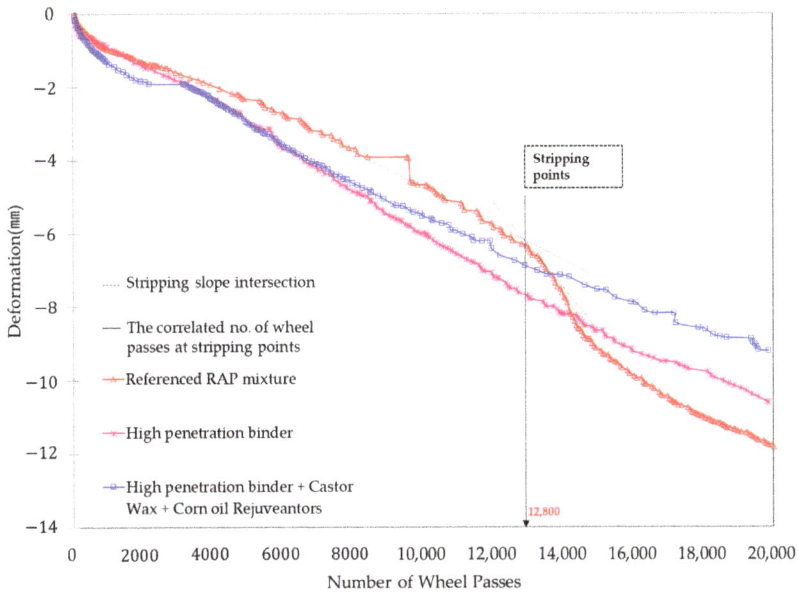

Figure 12. Hamburg wheel tracking test results.

Additionally, it is noteworthy that the stripping phenomenon was absent in the modified mixture, indicating its superior resistance to moisture-induced damage. In contrast, the RAP mixture exhibited signs of stripping at approximately 12,800 cycles. This stark contrast confirms the effectiveness of the rejuvenated mixtures in enhancing their moisture resistance. The incorporation of rejuvenators, such as the low-viscosity binder, Castor wax, and Corn oil, contributed to improved adhesion between the binder and aggregate particles, mitigating the detrimental effects of moisture infiltration. The absence of stripping in the modified mixtures demonstrates their ability to maintain structural integrity and durability even under prolonged exposure to moisture, further validating their suitability for long-lasting pavement applications.

The findings from the Hamburg wheel tracking test support the suitability of the rejuvenated RAP mixture for high traffic and challenging environmental conditions. The reduced rut depth and improved resistance to deformation are indicative of the superior quality and durability of the rejuvenated binder. These results emphasize the potential of using rejuvenation techniques to enhance the performance and longevity of RAP mixtures, making them a promising solution for sustainable and resilient pavement applications.

3.5. Fatigue Test Results

The fatigue test results, as depicted in Figure 13, reveal valuable insights into the performance of the RAP mixture in terms of fatigue crack resistance. The fatigue test was conducted to evaluate the resistance of the RAP mixture to repeated loading and its ability to withstand fatigue-induced cracking. It is important to note that the criterion for fatigue in the fatigue crack test results was set at a 50% reduction in the stiffness of the material compared with its original value. This criterion was used to determine the point of fatigue failure, indicating that the material's stiffness had dropped to half of its initial stiffness after enduring repeated and continuous loading cycles.

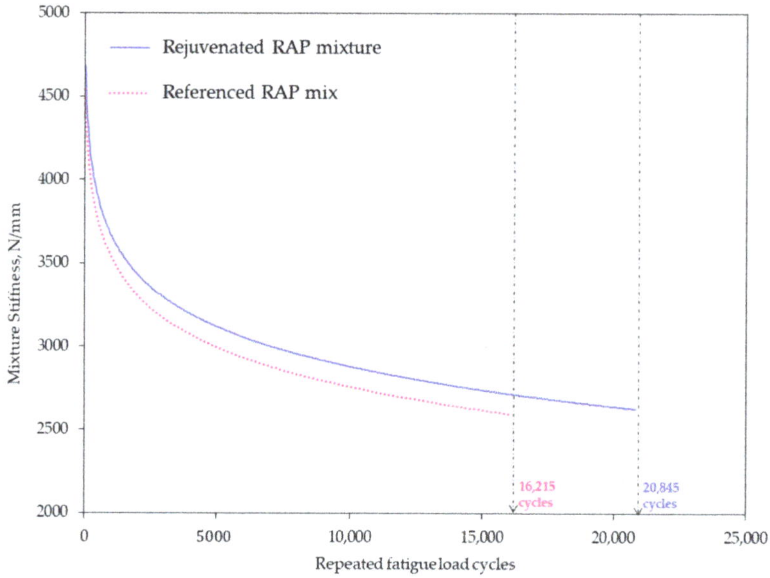

Figure 13. Fatigue test results.

Figure 13 illustrates the results of the fatigue test, showing the number of load cycles to failure for both the reference RAP binder and the rejuvenated binder. The results demonstrate a substantial improvement in fatigue performance with the rejuvenated binder compared with those of the reference RAP binder. The rejuvenated binder exhibited a 28.55% increase in the number of load cycles to failure, indicating a significant enhancement in resistance to fatigue cracking. For instance, the reference RAP binder failed at 16,215 load cycles, while the rejuvenated binder showed no signs of cracking and continued to endure the applied load even after 20,845 load cycles. These findings are consistent with related research studies that have highlighted the positive impact of binder rejuvenation on fatigue resistance.

The improved fatigue performance of the rejuvenated binder can be attributed to the unique properties of the Castor wax and Corn oil rejuvenators. Castor wax enhances the binder's stiffness and resistance to rutting, while Corn oil improves flexibility and resilience, resulting in enhanced resistance to fatigue cracking. The synergistic effects of these rejuvenators contributed to the overall improved fatigue performance of the RAP mixture. The observed 28.55% increase in fatigue life with the rejuvenated binder provides strong evidence of its superior durability and potential for long-lasting pavement applications. These findings align with previous studies that have demonstrated the effectiveness of rejuvenation techniques in enhancing the fatigue resistance of RAP mixtures. The use of rejuvenation techniques presents a promising approach to extending the service life of RAP pavements and improving their overall sustainability.

3.6. General Discussions

The incorporation of a low-viscosity binder and Castor wax with Corn oil rejuvenators played a major role in improving the performance of the RAP binder. These additives addressed key challenges associated with RAP mixtures and enhanced their overall performance. The low-viscosity binder played a crucial role in improving the workability and blending characteristics of the RAP binder. Its high penetration grade facilitated better coating and adhesion between the RAP aggregates and binders, resulting in improved bonding and overall structural integrity. This improvement in bonding contributed to enhanced resistance against fatigue-induced cracking and improved durability of the RAP binder. Castor wax and Corn oil rejuvenators also played a significant role in the improved performance of the RAP binder. The regulator enhanced the flexibility and resilience of the binder, leading to improved resistance against fatigue cracking. The combination of these rejuvenators resulted in a synergistic effect, further enhancing the overall performance of the RAP binder. Comparisons with related research studies have shown consistent findings regarding the positive impact of bio-oil rejuvenators on RAP mixtures. Similar studies have reported improved workability, bonding, rutting resistance, and fatigue performance when these additives were incorporated. This consistency in findings supports the effectiveness and potential of these additives in enhancing RAP mixtures.

However, it is important to note the limitations of the current study. The research focused on laboratory-scale testing, and the findings need to be validated through field trials and long-term performance monitoring. Additionally, the study primarily examined the impact of the additives on specific performance characteristics, and further investigation is needed to explore their effects on other properties of the RAP mixture, such as moisture sensitivity and aging resistance. The results of this study provide a foundation for future research in this field. The next stage of this research could involve conducting field trials to evaluate the performance of the rejuvenated RAP binder under real-world traffic and environmental conditions. Long-term monitoring and performance evaluation would provide valuable insights into the durability and sustainability of RAP mixtures with a low-viscosity binder and Castor wax and Corn oil rejuvenators. Further investigation into the optimal dosage and combination of these additives can also be explored to maximize their benefits and improve the performance of RAP mixtures.

4. Conclusions

In this study, the effectiveness of incorporating a low-viscosity binder and Castor wax with Corn oil rejuvenators in improving the performance of Reclaimed Asphalt Pavement (RAP) mixtures was investigated. Through a comprehensive evaluation of various laboratory tests, the following conclusions can be drawn:

- The MSCR test results revealed that the RAP mixture with the newly developed rejuvenators exhibited a 20% reduction in cumulative strain rate under the applied creep load compared with that of the conventional rejuvenators. At an MSCR level of 3.2 kPa and 160 s, the strain value for the RAP mixture with the new rejuvenators was approximately 20% lower (around 530%) compared with the original RAP mixture (about 670%). The RAP mixture with the new rejuvenators initially exhibited behavior similar to the conventional rejuvenators in the early stages of testing (100–120 s) under a shear creep load of 3.2 kPa. However, as the testing duration increased, the RAP mixture with the new rejuvenators showed significantly reduced strain accumulation compared with the conventional mixture.
- The TSR test results demonstrate the effectiveness of the rejuvenated RAP mixture. The Tensile Strength Ratio (TSR) for the rejuvenated mixture showed a notable 9.47% improvement compared with that of the conventional RAP mixture. This improvement in TSR indicates enhanced resistance to moisture-induced damage, such as stripping and cracking, contributing to the longevity of the pavement structure.
- The evaluation of viscoelastic behaviors for the RAP mixtures revealed that the rejuvenated binder exhibited a slight reduction of 12.7% in dynamic modulus compared with the reference binder, indicating a softer behavior under cold temperatures. Furthermore, the incorporation of rejuvenators resulted in a significant improvement in the elastic behavior, as evidenced by the lower phase angle value (approximately 3 degrees) in the modified mixture. This indicates enhanced resistance to deformation and improved overall performance and durability of the RAP mixture.
- The results demonstrate that the RAP mixture with the rejuvenated binder exhibited superior resistance to rutting compared with the conventional RAP binder. The rut depth for the reference RAP binder was measured at 11.77 mm, while the rejuvenated binder showed significantly lower rut depths of only 10.59 mm and 9.2 mm for the

RAP mixture with a high penetration asphalt binder and the RAP mixture containing a low-viscosity binder with rejuvenator, respectively. This represents a reduction of 10.2% and 21.83% in rut depth, indicating improved resistance to deformation and rutting. Additionally, the absence of the stripping phenomenon in the modified mixture further confirms its superior moisture resistance.

- The results demonstrate a significant improvement in the fatigue performance of the rejuvenated binder compared with the reference RAP binder. The rejuvenated binder exhibited a remarkable 28.55% increase in the number of load cycles to failure, indicating a substantial enhancement in resistance to fatigue cracking. In contrast, the reference RAP binder failed at 16,215 load cycles, while the rejuvenated binder showed no signs of cracking and continued to withstand the applied load even after 20,845 load cycles.
- In general, the incorporation of a low-viscosity binder, Castor wax, and Corn oil rejuvenators in the RAP binder improved its workability, bonding, and overall performance. The low-viscosity binder enhanced coating and adhesion between RAP aggregates and binders, resulting in improved structural integrity and resistance against fatigue-induced cracking. Castor wax and Corn oil may enhance flexibility and resilience, further enhancing the binder's performance. The combined effect of these rejuvenators resulted in a synergistic improvement in the overall performance of the RAP binder.

It is essential to acknowledge the limitations of this study. Laboratory testing provides valuable insights, but real-world conditions and long-term performance may present different challenges. Therefore, further validation through extensive field trials and monitoring is recommended to confirm the findings and assess the rejuvenated RAP mixture's performance in actual pavement applications. Future research should focus on optimizing the dosage and combination of rejuvenators to maximize their benefits and fine-tune the rejuvenated RAP mixture's performance characteristics. Additionally, long-term performance monitoring of the rejuvenated RAP mixture in field applications will provide valuable data on its durability and sustainability.

Author Contributions: Conceptualization, T.H.M.L.; methodology, T.H.M.L.; validation, K.K. and T.H.M.L.; formal analysis, K.K. and T.H.M.L.; investigation, K.K. and T.H.M.L.; resources, K.K. and T.H.M.L.; writing—original draft preparation, K.K. and T.H.M.L.; writing—review and editing, K.K. and T.H.M.L.; visualization, K.K. and T.H.M.L.; supervision, K.K.; project administration, K.K.; funding acquisition, K.K. All authors have read and agreed to the published version of the manuscript.

Funding: Research for this paper was supported by the KICT Research Program (project no. 20230202-001).

Institutional Review Board Statement: Not applicable.

Informed Consent Statement: Not applicable.

Data Availability Statement: Data will be made available on request.

Acknowledgments: Research for this paper was carried out under the KICT Research Program (project no. 20230202-001, Development of high performance, low cost MAST (Multi-Layered Asphalt Surface Treatments) composite pavement system for improving unpaved roads in Vietnam) funded by the Ministry of Science and ICT.

Conflicts of Interest: The authors declare no conflict of interest.

References

1. Dong, R.; Gao, A.; Zhu, Y.; Xu, B.; Du, J.; Ping, S. The Development of a New Thermoplastic Elastomer (TPE)-Modified Asphalt. *Buildings* **2023**, *13*, 1451. [CrossRef]
2. Chen, S.; Yang, Q.; Qiu, X.; Liu, K.; Xiao, S.; Xu, W. Use of MD Simulation for Investigating Diffusion Behaviors between Virgin Asphalt and Recycled Asphalt Mastic. *Buildings* **2023**, *13*, 862. [CrossRef]
3. Lee, S.-Y.; Le, T.H.M. Laboratory and Full-Scale Testbed Study in the Feasibility of Styrene-Butadiene-Styrene Asphalt Pavement Having Epoxy Resin and Crumb Rubber Powder. *Buildings* **2023**, *13*, 652. [CrossRef]

4. Lee, S.Y.; Kim, Y.M.; Le, T.H.M. Laboratory and Field Testbed Evaluation of the Performance of Recycled Asphalt Mixture Using High-Penetration Asphalt. *Buildings* **2023**, *13*, 529. [CrossRef]
5. Jiangsan, H.; Lan, W.; Hui, W.; Zihao, Z.; Zhaoyang, G.; Heng, L. Quantitative Study on Regeneration Mechanism of Warm Mixed Recycled Asphalt. *J. Clean. Prod.* **2023**, *408*, 137104. [CrossRef]
6. Shi, L.; Wang, Y.; Li, H.; Liang, H.; Lin, B.; Wang, D. Recycled Asphalt Mixture's Discrete Element Model-Based Composite Structure and Mesoscale-Mechanical Properties. *Case Stud. Constr. Mater.* **2023**, *18*, e01987. [CrossRef]
7. Wang, L.; Shen, A.; Mou, G.; Guo, Y.; Meiquan, Y. Effect of RAP Gradation Subdivision and Addition of a Rejuvenator on Recycled Asphalt Mixture Engineering Performance. *Case Stud. Constr. Mater.* **2023**, *18*, e02136. [CrossRef]
8. Zhao, S.; Zhang, H.; Gao, M.; Zhang, Q.; Sun, Q.; Dong, Q. Nano-Microscopic Analysis on the Interaction of New and Old Asphalt Mortar in Recycled Asphalt Mixture. *Chem. Phys. Lett.* **2023**, *825*, 140593. [CrossRef]
9. Wang, H.; Liu, X.; Apostolidis, P.; Wang, D.; Leng, Z.; Lu, G.; Erkens, S.; Skarpas, A. Investigating the High- and Low-Temperature Performance of Warm Crumb Rubber–Modified Bituminous Binders Using Rheological Tests. *J. Transp. Eng. Part B Pavements* **2021**, *147*, 4021067. [CrossRef]
10. Wang, D.; Cannone Falchetto, A.; Hugener, M.; Porot, L.; Kawakami, A.; Hofko, B.; Grilli, A.; Pasquini, E.; Pasetto, M.; Tabatabaee, H.; et al. Effect of Aging on the Rheological Properties of Blends of Virgin and Rejuvenated RA Binders. *RILEM Bookseries* **2022**, *27*, 3–10. [CrossRef]
11. Renken, P.; Büchler, S.; Falchetto, A.C.; Wang, D.; Wistuba, M.P. Warm Mix Asphalt-a German Case Study. *Asph. Paving Technol. Assoc. Asph. Paving Technol. Tech. Sess.* **2018**, *87*, 685–714. [CrossRef]
12. Chen, Q.; Li, Y.; Lin, Z.; Yu, H. Surface Activation of Wax-Based Additives to Enhance Asphalt Rheological Properties via Rotating Plasma Treatment. *Buildings* **2023**, *13*, 514. [CrossRef]
13. Barraj, F.; Khatib, J.; Castro, A.; Elkordi, A. Effect of Chemical Warm Mix Additive on the Properties and Mechanical Performance of Recycled Asphalt Mixtures. *Buildings* **2022**, *12*, 874. [CrossRef]
14. Wang, W.; Li, J.; Wang, D.; Liu, P.; Li, X. The Synergistic Effect of Polyphosphates Acid and Different Compounds of Waste Cooking Oil on Conventional and Rheological Properties of Modified Bitumen. *Materials* **2022**, *15*, 8681. [CrossRef]
15. Riccardi, C.; Wang, D.; Wistuba, M.P.; Walther, A. Effects of Polyacrylonitrile Fibres and High Content of RAP on Mechanical Properties of Asphalt Mixtures in Binder and Base Layers. *Road Mater. Pavement Des.* **2022**. [CrossRef]
16. Poulikakos, L.D.; Pasquini, E.; Tusar, M.; Hernando, D.; Wang, D.; Mikhailenko, P.; Pasetto, M.; Baliello, A.; Cannone Falchetto, A.; Miljković, M.; et al. RILEM Interlaboratory Study on the Mechanical Properties of Asphalt Mixtures Modified with Polyethylene Waste. *J. Clean. Prod.* **2022**, *375*, 134124. [CrossRef]
17. Shi, K.; Ma, F.; Liu, J.; Fu, Z.; Song, R.; Yuan, D.; Li, C. Rejuvenation Effect of Aged SBS-Modified Asphalt Utilizing Molecule Analysis. *J. Clean. Prod.* **2023**, *405*, 136964. [CrossRef]
18. Moosom, J.J.; Goh, T.S.; Kong, S.Y. Use of Asphalt Milling Material in Construction of the Roadway. *E3S Web Conf.* **2022**, *347*, 1014. [CrossRef]
19. Yang, T.; Jia, Y.; Pan, Y.; Zhao, Y. Evaluation of the Low-Temperature Cracking Performance of Recycled Asphalt Mixture: A Development of Equivalent Fracture Temperature. *Buildings* **2022**, *12*, 1366. [CrossRef]
20. Xing, C.; Li, M.; Liu, L.; Lu, R.; Liu, N.; Wu, W.; Yuan, D. A Comprehensive Review on the Blending Condition between Virgin and RAP Asphalt Binders in Hot Recycled Asphalt Mixtures: Mechanisms, Evaluation Methods, and Influencing Factors. *J. Clean. Prod.* **2023**, *398*, 136515. [CrossRef]
21. Nguyen, H.L.; Tran, V.Q. Data-Driven Approach for Investigating and Predicting Rutting Depth of Asphalt Concrete Containing Reclaimed Asphalt Pavement. *Constr. Build. Mater.* **2023**, *377*, 131116. [CrossRef]
22. Yan, S.; Dong, Q.; Chen, X.; Zhao, X.; Wang, X. Performance Evaluation of Waste Cooking Oil at Different Stages and Rejuvenation Effect of Aged Asphalt through Molecular Dynamics Simulations and Density Functional Theory Calculations. *Constr. Build. Mater.* **2022**, *350*, 128853. [CrossRef]
23. Dalhat, M.A.; Osman, S.A.; Dalhat Mu'azu, N.; Alagha, O. Utilization of Oil Sludge as Rejuvenator in Hot-Mix-Asphalt Containing Reclaimed Asphalt Concrete. *Constr. Build. Mater.* **2022**, *338*, 127483. [CrossRef]
24. Qiao, Y.; Dave, E.; Parry, T.; Valle, O.; Mi, L.; Ni, G.; Yuan, Z.; Zhu, Y. Life Cycle Costs Analysis of Reclaimed Asphalt Pavement (RAP) Under Future Climate. *Sustainability* **2019**, *11*, 5414. [CrossRef]
25. Polo-Mendoza, R.; Peñabaena-Niebles, R.; Giustozzi, F.; Martinez-Arguelles, G. Eco-Friendly Design of Warm Mix Asphalt (WMA) with Recycled Concrete Aggregate (RCA): A Case Study from a Developing Country. *Constr. Build. Mater.* **2022**, *326*, 126890. [CrossRef]
26. Sihombing, A.V.R.; Subagio, B.S.; Hariyadi, E.S.; Mulyadi, A.M.; Utami, R.; Sihombing, R.P. Effect of Biorejuvenator Types on Microstructure, Mechanistic Performance, and Resilient Modulus Prediction Model of Asphalt Concrete Containing Recycled Materials in Indonesia. *J. King Saud Univ.-Eng. Sci.* **2023**, *in press*. [CrossRef]
27. Lee, S.Y.; Ho Minh Le, T.; Kim, Y.M. Full-Scale and Laboratory Investigations on the Performance of Asphalt Mixture Containing Recycled Aggregate with Low Viscosity Binder. *Constr. Build. Mater.* **2023**, *367*, 130283. [CrossRef]
28. Zhao, Y.; Goulias, D.; Peterson, D. Recycled asphalt pavement materials in transport pavement infrastructure: Sustainability analysis & metrics. *Sustainability* **2021**, *13*, 8071. [CrossRef]
29. Yao, Y.; Yang, J.; Gao, J.; Zheng, M.; Xu, J.; Zhang, W.; Song, L. Strategy for Improving the Effect of Hot In-Place Recycling of Asphalt Pavement. *Constr. Build. Mater.* **2023**, *366*, 130054. [CrossRef]

30. Dinh, B.H.; Park, D.W.; Le, T.H.M. Effect of Rejuvenators on the Crack Healing Performance of Recycled Asphalt Pavement by Induction Heating. *Constr. Build. Mater.* **2018**, *164*, 246–254. [CrossRef]
31. Praticò, F.G.; Vaiana, R.; Iuele, T. Permeable Wearing Courses from Recycling Reclaimed Asphalt Pavement for Low-Volume Roads. *Transp. Res. Rec.* **2015**, *2474*, 65–72. [CrossRef]
32. *ASTM D2172-17*; Standard Test Methods for Quantitative Extraction of Asphalt Binder from Asphalt Mixtures. American Society for Testing and Materials: Conshohocken, PA, USA, 2017; pp. 1–10.
33. *ASTM D5*; Standard Test Method for Penetration of Bituminous Materials. ASTM International: Conshohocken, PA, USA, 2019.
34. *ASTM C117*; Standard Test Method for Materials Finer than 75-Mm (No. 200) Sieve in Mineral Aggregates by Washing. ASTM International: Conshohocken, PA, USA, 2017; pp. 1–3.
35. *ASTM D2216-19*; Standard Test Methods for Laboratory Determination of Water (Moisture) Content of Soil and Rock by Mass. ASTM International: Conshohocken, PA, USA, 2019; pp. 1–7.
36. US Department of Transportation FHWA. *Superpave Gyratory Compactors;* US Federal Highway Administration Publisher: Washington, DC, USA, 2010; pp. 1–35.
37. *AASHTO T350-14*; Standard Method of Test for Multiple Stress Creep Recovery (MSCR) Test of Asphalt Binder Using a Dynamic Shear Rheometer (DSR). ASTM International: Conshohocken, PA, USA, 2014; pp. 3–6.
38. Huang, G.; Zhang, J.; Hui, B.; Zhang, H.; Guan, Y.; Guo, F.; Li, Y.; He, Y.; Wang, D. Analysis of Modulus Properties of High-Modulus Asphalt Mixture and Its New Evaluation Index of Rutting Resistance. *Sustainability* **2023**, *15*, 7574. [CrossRef]
39. Huang, Z.; Ling, X.; Wang, D.; Li, P.; Li, H.; Wang, X.; Wang, Z.; Wei, R.; Zhu, W.; Falchetto, A.C. Research on High- and Low-Temperature Rheological Properties of High-Viscosity Modified Asphalt Binder. *Buildings* **2023**, *13*, 1077. [CrossRef]
40. Wang, D.; Zhu, J.; Porot, L.; Cannone Falchetto, A.; Damen, S. Multiple Stress Creep and Recovery Test for Bituminous Binders–Influence of Several Key Experimental Parameters. *Road Mater. Pavement Des.* **2023**, *24*, 290–308. [CrossRef]
41. *ASTM D6931*; Standard Test Method for Indirect Tensile (IDT) Strength of Bituminous Mixtures. ASTM International: Conshohocken, PA, USA, 2017; pp. 1–5.
42. *AASHTO TP62-2017;* Standard Method of Test for Determining Dynamic Modulus of Hot Mix Asphalt (HMA). American Association of State Highway and Transportation Officials: Washington, DC, USA, 2017.
43. *AASHTO T321-17;* Standard Method of Test for Determining the Fatigue Life of Compacted Asphalt Mixtures Subjected to Repeated Flexural Bending. The American Association of State Highway and Transportation Officials: Washington, DC, USA, 2011; pp. 1–11.
44. *AASHTO T324*; Standard Method of Test for Hamburg Wheel-Track Testing of Compacted Hot Mixtures. American Association of State Highway and Transportation Officials: Washington, DC, USA, 2017; pp. 1–8.

buildings

MDPI

Article

Study on Crystallization Mechanism of Asphalt Mixture in Bridge Deck Pavement

Chuanxi Luo [1,2,*], Duanyi Wang [1], Jian Li [2] and Jun He [1,2]

[1] School of Civil Engineering and Transportation, South China University of Technology, Guangzhou 510000, China; jtcdywang@scut.edu.cn (D.W.); 18825181976@163.com (J.H.)
[2] Xiaoning Institute of Roadway Engineering, Guangzhou 510000, China; m15738523861@163.com
* Correspondence: ctlcx@mail.scut.edu.cn; Tel.: +86-18825181976

Abstract: This study focuses on unknown crystal precipitates from an asphalt mixture used in bridge deck pavement layers. X-ray fluorescence spectroscopy was used to analyze the composition and source of crystals in the asphalt mixture used in bridge deck pavement, and infiltration tests, porosity tests, splitting tests and multi-wheel rutting tests were carried out to determine the precipitation area and non-precipitation area to explain the influence of crystals on the road performance of an asphalt pavement. A nuclear-free densitometer and 3D ground-penetrating radar (3D GPR) were used to detect the porosity and thickness uniformity of the whole section to study the formation mechanism of crystals. The results showed that the main components of crystals were water molecules, while the rest mainly came from machine-made sand, and there was no significant difference in pavement performance in the areas where crystals precipitated. The crystals were mainly caused by rainwater penetrating into the pavement through coarse segregation areas and collecting in the depression of the lower bearing layer. Under high temperature, the solution precipitated out of the pavement and formed crystals.

Keywords: asphalt mixture; pavement; crystals; composition analysis

Citation: Luo, C.; Wang, D.; Li, J.; He, J. Study on Crystallization Mechanism of Asphalt Mixture in Bridge Deck Pavement. *Buildings* **2023**, *13*, 1527. https://doi.org/10.3390/buildings13061527

Academic Editors: Andrea Baliello and Di Wang

Received: 25 April 2023
Revised: 30 May 2023
Accepted: 8 June 2023
Published: 14 June 2023

1. Introduction

After the construction of the structure layer of the asphalt mixture used in a bridge deck pavement is completed, it experiences high temperatures and exposure to rainy weather, causing white or light yellow crystals to precipitate on its surface, as shown in Figure 1. The crystals are distributed in blocks, and their contour is similar to that of liquid flow, with an area of about 0.01–0.2 m^2. The crystals have a certain thickness and can be easily scraped off. The source and influence of these crystals on the performance of existing pavements have been puzzling road workers and have, at one point, caused the project to stop. This paper focuses on the source of these crystals, their influence on the properties of mixtures and their formation mechanism.

(a) (b) (c)

Figure 1. Morphology of crystals. (**a**) Crystal profile; (**b**) crystal side detail; (**c**) crystal front detail.

In the past, the asphalt mixture used in the bridge deck pavement often had crystals appearing after alkaline flooding, which were white or yellow water-insoluble attachments distributed in flakes or scattered spots [1–3]. The crystals have a certain thickness, are powder-like crystals, and most of them are soluble in water. There are the obvious differences in morphology and solubility between the two substances. The main reason for flooding is the coarse segregation of the asphalt mixture, which leads to rainwater infiltrating into the materials dissolved in the bridge deck pavement and precipitating through connected gaps. The flooding of the bridge deck pavement indicates that there are connected gaps in the pavement structure layer, and early pit diseases are prone to occur during pavement maintenance [4–7]. Relevant research shows that when rainwater accumulates between pavement and the cement concrete of a bridge deck, it may cause a hydrodynamic scouring phenomenon when a traffic load is present, which would lead to a failure in the connection between the pavement and bridge deck, and even delamination [8–10]. According to the research results of relevant researchers, a relatively dense asphalt mixture has been applied in the bridge deck pavement [11–14]. In recent years, alkaline flooding of the bridge deck pavement has not occurred as often [15–18]; however, the crystal phenomenon has occurred in the asphalt pavement of newly built subgrade sections after alkaline flooding, though there is no relevant report on the diseases caused by alkaline flooding in subgrade sections [19–21]. With the increasing frequency of alkaline flooding, the diseases caused by flooding are being paid attention to by road workers [22]. Some researchers think that the flooding of the structural layer is only the precipitation of substances in the pavement layer, which has little correlation with the performance of the structural layer, and that the early damage was mainly caused by the connecting gaps [23,24]; therefore, some scholars believe that the phenomenon of alkaline flooding can be solved only by replacing the structural layer with a denser asphalt mixture [25–27]. The fact of the matter is, however, that the phenomenon of alkaline flooding has decreased, but it still exists [28–31]. There is still no complete research result on the formation mechanism of alkaline flooding.

Huang Zhiyong et al. showed that alkaline flooding is different from asphalt, aggregate and dust flooding, cannot be washed away by high-pressure water and has a certain adhesion strength [32,33]. Li Weixiong et al. carried out chemical analyses of alkaline flooding substances and found that the main substances were $CaCO_3$, $Mg(OH)_2$ and $CaCO_3$ [34–36]. Related scholars have carried out a series of studies on the phenomenon of alkaline flooding, but crystallization has not been the focus in pavement engineering [37,38]. There were some detailed research results on the precipitation of crystals in soil. Larsen studied the mechanism of calcium carbonate scaling through laboratory core flow tests based on the theory of crystal growth dynamics [39]. Charpentier studied the crystallization of calcium carbonate and sulfuric acid in subsea downhole equipment through dynamic tests [40]. Sadra studied the kinetics of calcium sulfate crystals through experiments [41]. The regularity of groundwater crystallization and its influence on the stability of a high slope were analyzed by Gao Chunqun [42]. Liu Yuyang carried out research on the occurrence rule and crystallization path of crystals in tunnel drainage systems surrounded by calcium-rich rock [43]. Wu Yuzhe analyzed the component characteristics and sources of the porous crystallites in a loess tunnel and proposed that the crystallites were mainly Ca and Mg ions [44]. Cement concrete, similarly to asphalt concrete, also experienced a crystallization phenomenon. When the phenomenon of bad crystallization was too strong, it directly affected the performance of concrete [45,46]. Asphalt concrete may have the same problem. At the present stage, the chemical composition, mechanism and influence on the performance of the pavement layer of the asphalt-mixture bridge deck are not clear.

X-ray fluorescence spectroscopy was used to analyze and compare the composition differences between raw materials, machine-made sand, bridge deck milling waste and crystals, and to study the composition and source of crystals. An infiltration test, porosity test, splitting test and multi-wheel rutting test were carried out for determining the crystalline area and non-crystalline area, and the differences between them were analyzed to introduce the pavement performance of the crystalline area. Finally, the porosity and thickness uniformity were analyzed using a nuclear-free densitometer and 3D GPR, and the formation mechanism of crystals was studied.

2. Methods

2.1. Composition and Source Analysis of Crystals

Crystals appeared 2–3 days after the completion of the bridge deck pavement construction, during which large-scale rainfall and high temperature weather occurred. After the bridge deck pavement construction was completed, traffic was closed off without interference from external factors. The possibility of foreign substances was eliminated by analyzing the distribution law and adhesion of crystals on site. Combined with the external contour of the bridge deck pavement surface, crystals were formed by liquid crystals and the composition of crystals from the bridge deck pavement structure. The structure consists of an asphalt surface layer, waterproof adhesive layer, adhesive layer and milled bridge deck integral layer, as shown in Table 1. Relevant research shows that the phenomenon of alkaline flooding mainly comes from gravel dust in the waterproof bonding layer and milling dust on the bridge deck. The gravel of the waterproof bonding layer of this project must be dusted by a mixing plant and wrapped with 0.3% asphalt, which can eliminate the possibility of crystallization caused by gravel dust. In order to analyze the composition and source of crystals, X-ray fluorescence spectrometry was used to compare and analyze the composition of raw materials, machine-made sand, bridge deck milling waste and crystals.

Table 1. Bridge deck pavement structure.

Layer	Type
Upper pavement	SMA-13 (SBS modified asphalt)
Lower pavement	GAC-20 (SBS modified asphalt)
Waterproof adhesive layer	Synchronous macadam seal (SBS modified asphalt)
Adhesive layer	Modified emulsified asphalt

2.2. Influence Analysis of Pavement Performance

In order to study the influence of crystal on the performance of the pavement, an infiltration test, core sample porosity test, splitting test and multi-wheel rutting test were carried out for the crystal and non-crystal area, respectively, and the performance difference between the crystal and non-crystal area was obtained. Among them, an infiltration test and core sample porosity test were used to analyze the difference in pavement compactness and study the compactness of the crystal precipitation area [33,34]. The core sample splitting test was used to explain the difference of mechanical properties, and study the influence of crystal precipitation on mechanical properties. A multi-wheel rutting test of core samples was used to introduce the difference in high temperature stability, and study the influence of crystal precipitation on rutting deformation resistance. See Figure 2 for the schematic diagram.

2.3. Analysis of Precipitation Mechanism of Crystals

The uniformity analysis of the whole section was carried out using a non-nuclear density meter and 3D GPR, where the non-nuclear density meter detected the porosity distribution, and studied the corresponding relationship between the porosity segregation area distribution position and the crystal precipitation area distribution position. Three-dimensional GPR uses the principle of electromagnetic wave propagation to detect the thickness distribution of the full section of a pavement, and the calculation model is shown in Equation (1). The corresponding relationship between the distribution position of the thickness segregation area and the distribution position of the crystal precipitation area was studied, and the mechanism of crystal formation was analyzed.

$$h = \frac{c}{\sqrt{\varepsilon}}\frac{t}{2} \quad (1)$$

where c is the speed of light, and the theoretical value is 3×10^8 m/s; ε is the dielectric constant of the structural layer, calculated by Equation (2); t is the two-way propagation time of the electromagnetic wave in the structural layer.

$$\varepsilon = \frac{c^2 {t_H}^2}{4H^2} \quad (2)$$

where H is the height of the core sample and t is the two-way propagation time of the electromagnetic wave in the core sample.

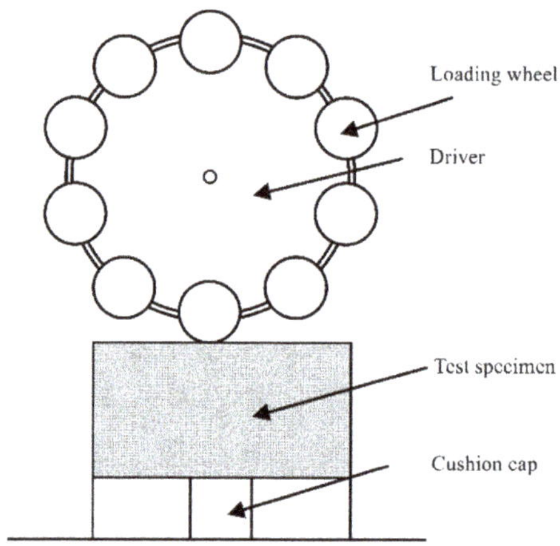

Figure 2. Schematic diagram of multi-wheel rutting test.

3. Experimental Plan

3.1. Raw Materials

The asphalt used in the waterproof bonding layer of the bridge deck pavement and the lower pavement is SBS-modified asphalt, and the test index is shown in Table 2. Limestone was used as aggregate in the lower pavement, and the test index of fine aggregate is shown in Table 3. The lower pavement adopts a skeleton-dense asphalt mixture, and the gradation is shown in Figure 3. The test methods refer to "Test Regulations for Asphalt and Asphalt Mixture of Highway Engineering" (JTG E20-2011).

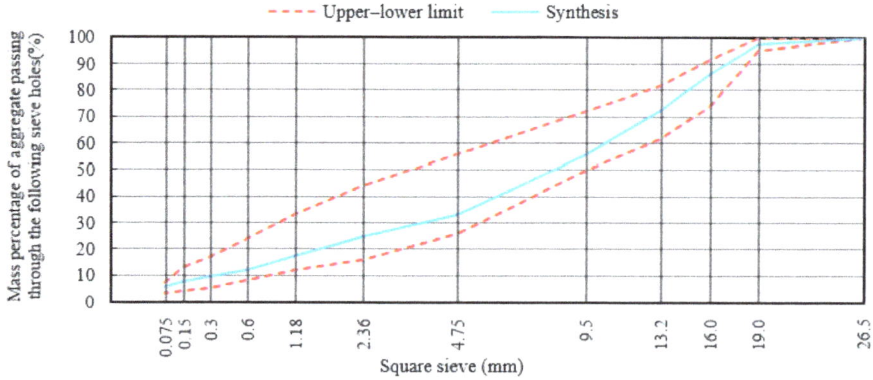

Figure 3. Mineral aggregate gradation of lower pavement's asphalt mixture.

Table 2. Index of SBS-modified asphalt.

Pilot		Test Results
Penetration 25 °C, 100 g, 5 s, (0.1 mm)		49
Ductility 5 °C, 5 cm/min, (cm)		33
Softening point, $T_{R\&B}$ (°C)		88
Solubility (%)		99.8
Storage stability: 163 °C; 48 h; softening point difference, °C		0.8
Elastic recovery 25 °C, %		96
Kinematic viscosity at 135 °C (Pa·s)		2.42
Rolling thin film oven test (RTFOT) Residue (163 °C, 85 min)	Quality change (%)	0.016
	Ductility 5 °C, 5 cm/min, (cm)	22
	Penetration ratio (%)	77.7

Table 3. Fine aggregate index.

Test Index	Unit	Test Results
Apparent relative density	—	2.947
Rigidity (>0.3 mm part)	%	2.7
Sand equivalent	%	69
Methylene blue value	g/kg	0.9
Angularity (flow time)	s	38.7

3.2. Test Plan of X-ray Fluorescence Spectrometry

The sample preparation carried out via X-ray fluorescence spectrometry usually adopts a tableting method or melting method. The tablet pressing method is relatively simple, but its particle size effect is obvious, and the accuracy and repeatability of test results are poor. The melting method can greatly reduce the particle size effect, but it needs a high temperature for melting, so it is impossible to detect H_2O, C and other related elements. According to the characteristics of the samples, the melting method was selected for composition analysis. The machine-made sand, bridge deck milling materials and crystals were analyzed via X-ray fluorescence spectrometry.

First, amounts of 6.0 g of a mixed flux and 0.6 g of crystallites were placed in a platinum crucible and carefully mixed with a glass rod. Then, 1 mL of a lithium nitrate solution was uniformly added, placed on the sample rack of the high-frequency fusion machine, oxidized at 500 °C for 5 min, then fused at 1000 °C for 2 min, and then fused by swinging for 8 min. During the melting process, a small amount of ammonium iodide was added every 2–3 min. It was added for the last time 2 min before the end. After melting, the molten material was poured into a preheated mold carefully. After cooling, the glass sheet was removed and marked. The glass plate containing the crystal was placed in the X-ray fluorescence spectrometer to obtain the chemical elements of the crystal and their proportions.

3.3. Test Plan of Pavement Performance

The morphology of crystals was observed. Combined with the parameters of transverse and longitudinal slopes, the precipitation points of crystals were determined. This point was taken as the center, and a water seepage test was carried out and so was coring; see Figure 4 for the test position. Among them, the four test points were arranged in combination with the combined slope of transverse and longitudinal slopes, so that the test points were at the highest point and the lowest point in the direction of the combined slope to analyze the relationship between the precipitation point and the combined slope.

The other two test points were 90° away from the highest and low points, respectively, and the four test points were 0.3 m away from the central test point. According to the 6 crystal precipitation areas, the water permeability coefficient was detected according to the Code for On-site Test of Highway Subgrades and Pavements (JTG 3450-2019), and the porosity was measured by coring.

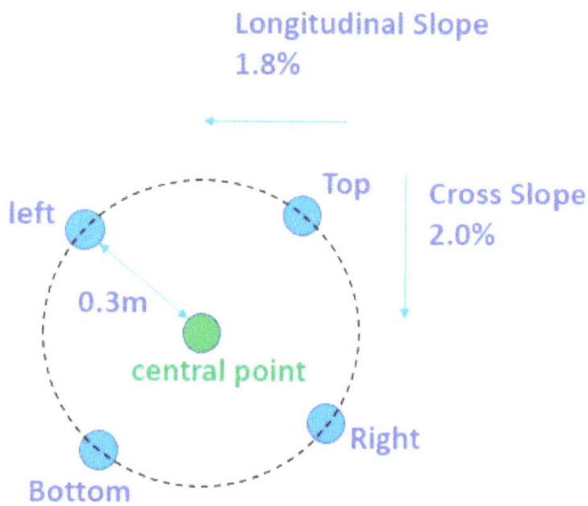

Figure 4. Schematic diagram of test position.

According to the core samples in the above-mentioned three crystal areas, the splitting strength test was carried out according to the Test Code for Asphalt and Asphalt Mixture in Highway Engineering (JTG E20-2011), and the test temperature was 25 °C. Based on the core samples in the other three crystal areas, multi-wheel rutting tests were carried out at 60 °C, and the rutting depth was measured by running the small wheels 16,000 times.

3.4. *Nuclear-Free Density and 3D GPR Test Plan*

A PQI380 nuclear-free densimeter was used to carry out the determination of the porosity distribution of the asphalt structural layer in the whole section. Starting from 0.5 m away from the edge of the bridge deck pavement, the horizontal interval was 0.5 m, the vertical interval was 0.5 m, and the test points were set with a test length of 20 m and a total of 1120 test points. See Figure 5 for the schematic diagram of the test. Three-dimensional GPR was used to detect the thickness of the crystallization area. The longitudinal and transverse interval of the data output was 0.15 m, and 13,266 points of thickness data were the output.

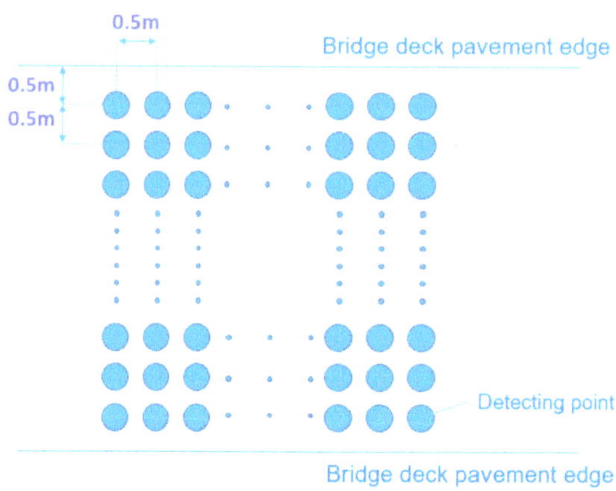

Figure 5. Schematic diagram of test of nuclear-free densimeter.

4. Results and Analysis

4.1. Source Analysis of Crystals

See Figure 6 for the test results of the machine-made sand, bridge deck milling waste and crystals obtained via X-ray fluorescence spectrometry.

As can be seen from Figure 6a, CaO accounts for 54.95% of the composition of the machine-made sand, while the LOI accounts for 42.91%, where the LOI is the ignition loss caused by melting and sample preparation before conducting the X-ray fluorescence spectrometry analysis. The machine-made sand of the asphalt mixture is limestone, and its main component is $CaCO_3$. According to the chemical characteristics of $CaCO_3$, it produces a decomposition reaction at high temperatures, as shown in Equation (3).

$$CaCO_3 \xrightarrow{\text{High Temperature}} CaO + CO_2 \quad (3)$$

$CaCO_3$ decomposed at high temperatures to form CaO and CO_2, in which CO_2 is gaseous and volatilizes during melting and sample preparation, thus producing ignition failure stress. According to the law of conservation of mass, after $CaCO_3$ was decomposed into CaO and CO_2, the mass ratio of CaO to CO_2 was 56:44, while the ratio of CaO to LOI in machine-made sand was 56:44, indicating that CO_2 is the main component in the LOI. As can be seen from Figure 6b, SiO_2 accounts for 46.16%, CaO accounts for 26.58% and the LOI accounts for 17.28% in bridge deck milling waste. As can be seen from Figure 6c, the LOI accounts for 98.95%, Al_2O_3 accounts for 0.23%, SiO_2 accounts for 0.43%, and CaO accounts for 0.23%. A total of 98.95% of crystals evaporated in the gaseous state during sample preparation, and the proportion of combined CaO was only 0.23%, so the volatile gas could not be CO_2-decomposed by $CaCO_3$ at high temperatures. Combined with the composition of the asphalt mixture, considering that H_2O and C cannot be detected via X-ray fluorescence spectrometry under melting and sample preparation, the crystals are dissolved in pure water. Observation shows that most crystals are soluble in water, with only a small amount of precipitates. Combined with the characteristic that the C molecule of organic matter (asphalt) does not melt in pure water, it can be judged that the main component of the LOI is H_2O.

Crystal CaO is mainly caused by $CaCO_3$ forming a Ca $(HCO3)_2$ solution under the action of atmospheric CO_2 and rainwater. The chemical reaction is shown in Equation (4), and the solution precipitates out of the surface by connected gaps. With the increase in temperature, $Ca(HCO_3)_2$ decomposes into $CaCO_3$, CO_2 and H_2O, and the chemical equation is shown in Equation (5). $CaCO_3$ exists in the machine-made sand, bridge deck milling waste and crystals, and it is impossible to analyze the source by comparing CaO quality ($CaCO_3$ can decompose to form CaO and CO_2). As can be seen from Figure 6, the composition of bridge deck milling waste includes P, S, Cl and Ti, while the machine-made sand does not contain the above components, and the crystal composition includes S and Cl, but not P and Ti. The milling material is composed of cement concrete, and the water and rainwater of crystals must contain S and Cl elements, so it is impossible to determine the source via the presence of S and Cl elements. P and Ti elements do not appear in crystals in bridge deck milling waste, so it can be preliminarily speculated that crystals mainly come from the machine-made sand. In addition, SiO_2 and Al_2O_3, which have a high content of components in the melted crystals, should exist as silicate and aluminate in the crystals, and both SiO_2 and Al_2O_3 in the machine-made sand and bridge deck milling waste are silicate and aluminate. The ratio of SiO_2 to Al_2O_3 in crystals is 1.87, and that in machine-made sand is 1.46; the ratio of SiO_2 to Al_2O_3 in bridge deck milling waste is 8.76. The element ratio of crystals is relatively close to that of machine-made sand, so it can be assumed that crystals (after deducting water molecules) mainly come from machine-made sand.

$$CaCO_3 + CO_2 + H_2O \rightarrow Ca(HCO_3)_2 \quad (4)$$

$$Ca(HCO_3)_2 \rightarrow CaCO_3 \downarrow + CO_2 \uparrow + H_2O \quad (5)$$

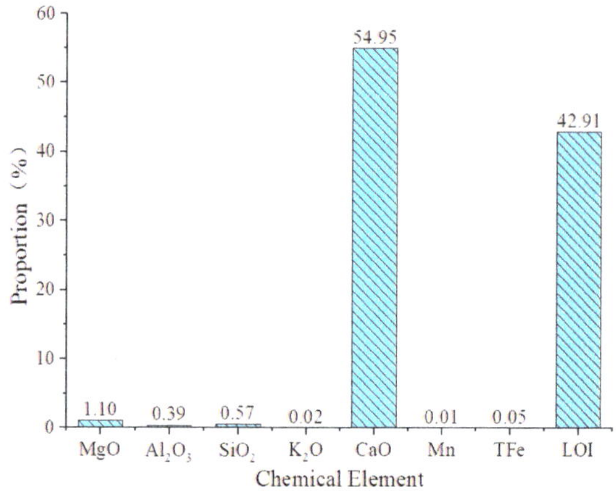

(**a**)

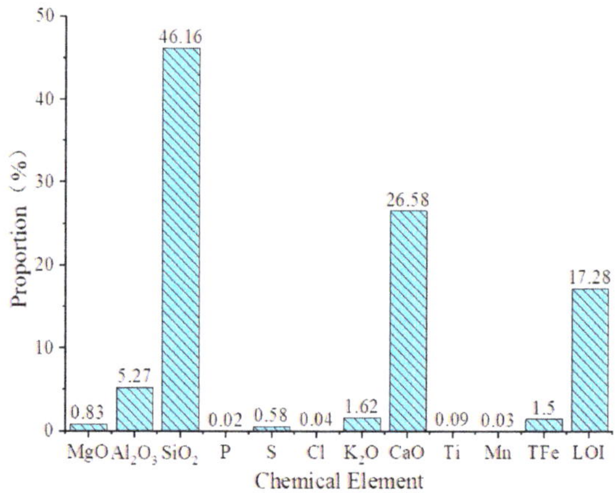

(**b**)

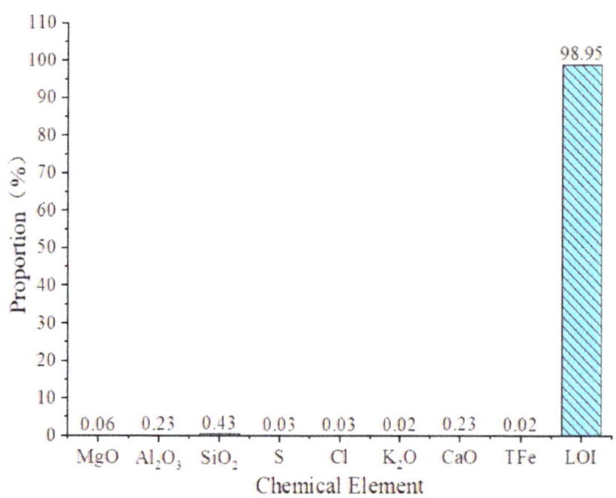

(**c**)

Figure 6. Test results of fluorescence spectrometry. (**a**) Analysis results of machine-made sand; (**b**) analysis results of bridge deck milling materials; (**c**) analysis results of crystals.

4.2. Road Performance Analysis

See Table 4 for the test results of the road performance indexes of crystals and surrounding areas. The average permeability coefficient of the crystalline precipitation area is 39.5 mL/min, and the average permeability coefficient of the peripheral area is 26.5 mL/min. The permeability coefficient of the crystalline precipitation area is slightly higher than that of peripheral area, but it meets the design requirements. The porosity of the crystalline precipitation area is 4.7%, while the porosity of the peripheral area is 4.2%, and the porosity of crystalline precipitation area is slightly higher than that of the peripheral area. The splitting strength and rutting depth of core samples in the crystallization area are similar to those in the surrounding area, but there is no obvious difference.

Table 4. Performance test results.

Indicators	Position	Average	Difference
Water permeability coefficient (mL/min)	Center point	39.5	−13
	Peripheral average	26.5	
Porosity (%)	Center point	4.7	−0.5
	Peripheral average	4.2	
Splitting strength (MPa)	Center point	0.737	0.034
	Peripheral average	0.771	
Rutting depth (mm)	Center point	1.59	−0.05
	Peripheral average	1.54	

In order to further analyze the performance difference between the crystal precipitation area and the surrounding area, data analysis was carried out for different crystal areas, and the test results of the crystal precipitation area were taken as the reference numbers and compared with the test results of the surrounding area; that is, "test results of the surrounding area-test results of the crystal precipitation area". The calculation results are shown in Figure 7.

From Figure 7a, it can be seen that there are obvious differences between the water permeability coefficient of peripheral areas in different directions and the water permeability coefficient of crystal precipitation areas. The water permeability coefficient of the highest point in the peripheral areas of five crystals is greater than that of the crystal precipitation areas, and the water permeability coefficient of five of the other three directions is smaller than that of crystal precipitation areas. The percolation coefficient of the highest point in the surrounding area is too large, and the percolation coefficient of the surrounding area in other directions is too small, so the percolation coefficient is not the key factor affecting the precipitation of crystals.

As can be seen from Figure 7b, the porosity of the highest point in the surrounding area of five crystals is greater than that in the precipitation area, and the porosity of the lowest point of four crystals is smaller than that in the precipitation area. The void fraction of the highest point is too large and the void fraction of the lowest point is too small in the surrounding area, so the void fraction is not the key factor affecting precipitation.

As can be seen from Figure 7c,d, there is no obvious law in the splitting strength and rutting depth. In order to further study the difference between crystals and surrounding areas, different crystals areas are analyzed separately, as shown in Table 5.

It can be seen from Table 5 that there is no obvious difference in splitting strength and rutting depth between the crystal area and the surrounding area. Fitting analysis was carried out using the core sample's porosity, splitting strength and rutting depth, as shown in Figures 8 and 9.

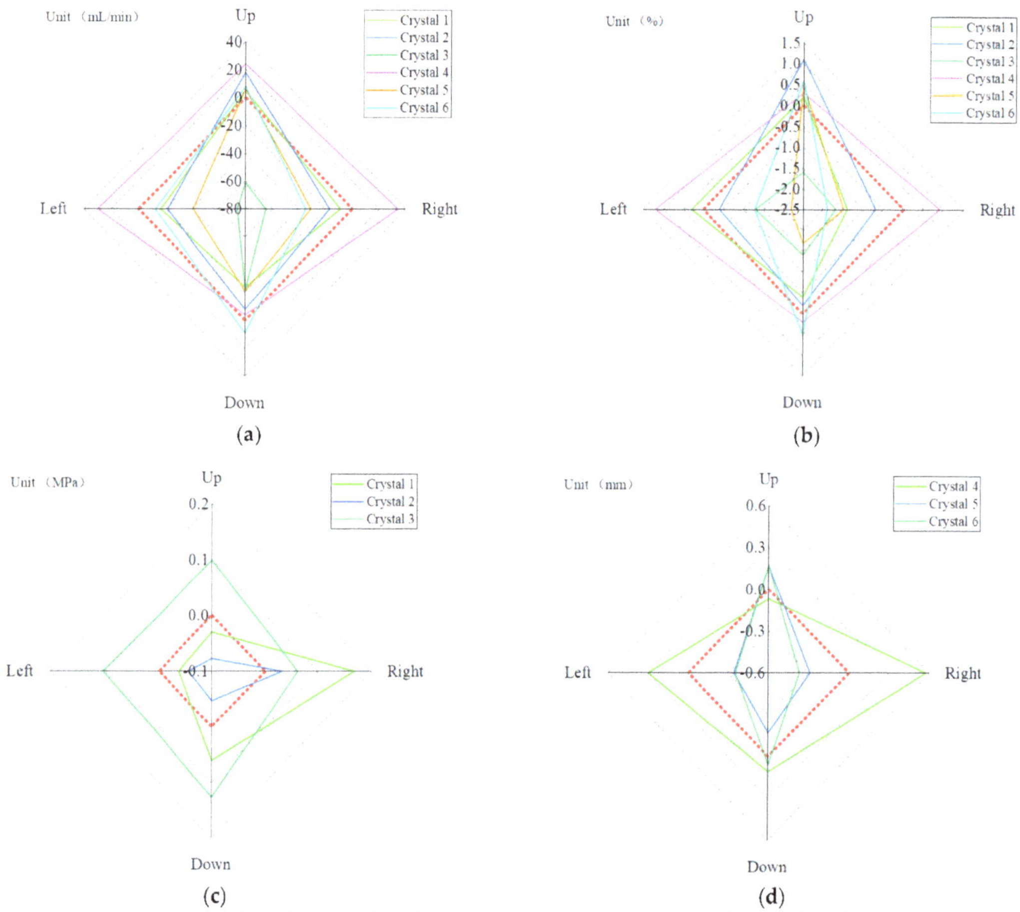

Note: the gray dotted lines are coordinate grid lines; the red dotted line is the reference line with the coordinate value 0.

Figure 7. Test results of regional properties of different crystals. (**a**) Comparison results of water permeability coefficients; (**b**) comparison results of porosity; (**c**) splitting strength comparison result; (**d**) rutting depth comparison result.

Table 5. Analysis results of splitting strength and rutting depth.

Position	Splitting Strength (MPa)			Rutting Depth (mm)		
	Crystal 1	Crystal 2	Crystal 3	Crystal 4	Crystal 5	Crystal 6
Crystal region	0.723	0.803	0.686	1.32	1.86	1.72
Average value of peripheral area	0.763	0.766	0.783	1.55	1.7	1.6
Difference	6%	−5%	14%	17%	−9%	−7%

It can be seen from Figures 8 and 9 that the splitting strength decreases with the increase in the porosity of the core samples, and the rutting depth increases with the increase in porosity. There is a good relationship model between the splitting strength, rutting depth and porosity, as shown in Equations (6) and (7), which shows that porosity is the main factor affecting splitting strength and rutting depth. Therefore, crystals do not have a significant impact on the mechanical properties and high temperature stability of the asphalt mixture.

$$y = 1.1 + 0.81x \qquad R^2 = 0.71 \tag{6}$$

$$y = 0.7 + 0.21x \qquad R^2 = 0.65 \tag{7}$$

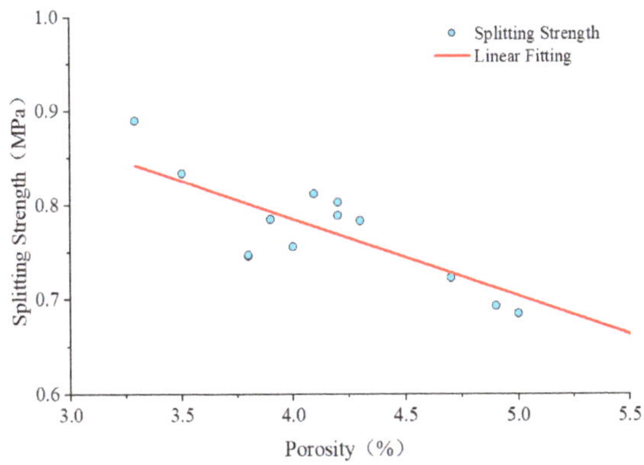

Figure 8. Fitting results of splitting strength and void fraction.

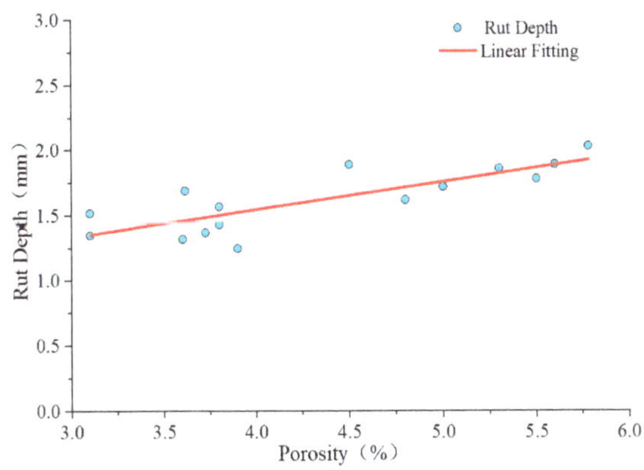

Figure 9. Fitting results of rutting depth and void fraction.

4.3. Nuclear-Free and Radar Test

In order to effectively analyze the spatial distribution of porosity, yellow represents the area with a porosity of less than 3%, green represents that of between 3% and 5%, and red represents an area greater than 5%. See Figure 10 for the porosity test results. It was found that the porosity is too large at the edges of both sides of the road. Because two pavers are used for paving in parallel, there is an obvious segregation phenomenon at the lap joint, which leads to the porosity being too large.

In order to effectively analyze the spatial distribution of thickness, purple is used to represent the area with a thickness greater than 8.8 cm, green represents that between 7.2–8.8 cm, and blue represents a smaller area. See Figure 11 for the thickness test results. It was found that the area with a large thickness is distributed in blocks. Because the interface is a deck pavement, the flatness of the lower bearing layer is not good. When there is a depression in the base surface, the thickness of the pavement will increase abnormally. According to the test results of 3D GPR, there are six basal depression areas in the test range.

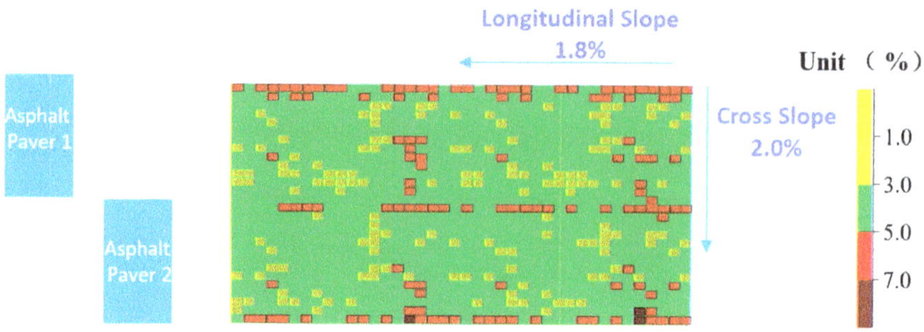

Figure 10. Result of porosity test.

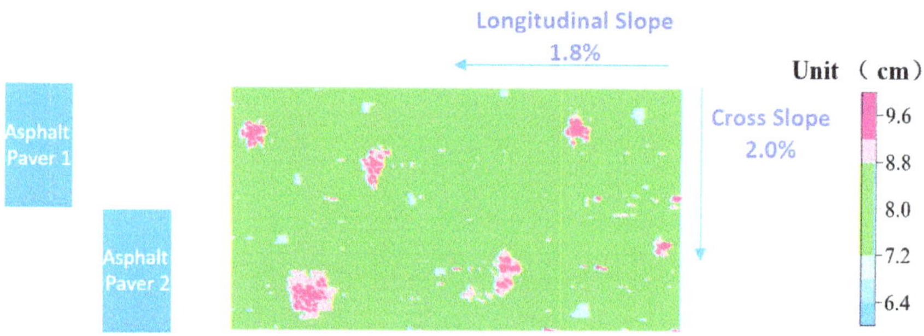

Figure 11. Thickness test results.

In order to further analyze the formation mechanism of crystals, the crystal region, the porosity region and the basal depression region are analyzed. The results are shown in Figure 12. It can be found that crystals precipitate at points A, B, C and F, all of which are located in the basal depression area, and there are areas with large porosity at the high places in the synthesis slope direction. It can be seen that rainwater infiltrates into the pavement structure from the area with large porosity, and then collects to the base depression area through the connecting void, precipitates upward under the action of high temperature, and loses water to form crystals. However, although there are areas with large porosity in the direction of synthetic slope at points D and E, the distance is relatively far, so it is impossible to form connected flow, so it is impossible to collect rainwater.

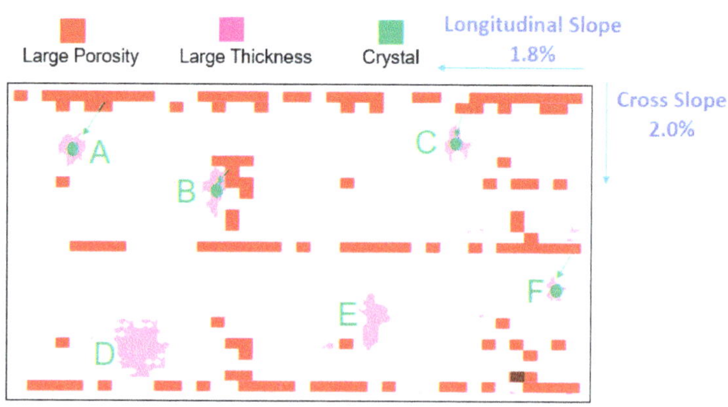

Figure 12. Analysis results of porosity and thickness combination.

In order to avoid crystals in the pavement, the uniformity of pavement construction should be improved as much as possible in the construction process to avoid segregation. For edge segregation and lap segregation areas, rubber wheels can be rolled once more to minimize the connected gap. If a segregation phenomenon is found after construction, emulsified asphalt and other materials can be used to block the connected gap. The flatness of the lower bearing layer of the bridge deck pavement should be improved, and local depressions can be leveled via milling to avoid the accumulation of infiltrated rainwater.

5. Conclusions

In this study, X-ray fluorescence spectroscopy was used to analyze the crystal composition of an asphalt-mixture bridge deck pavement, and the road indexes of the crystal precipitation area and non-precipitation area were detected. Nuclear-free densitometer and 3D GPR were used to detect the porosity and thickness uniformity of the whole section, and the following results were obtained:

(1) In total, 99% of the crystals are composed of H_2O, and the other components with a higher content include Al_2O_3, SiO_2 and CaO.
(2) The crystals (after deducting water molecules) mainly come from machine-made sand.
(3) The water permeability coefficient and porosity in the precipitation area are slightly larger than those in the non-precipitation area, but they all meet the design requirements; for splitting strength and multi-wheel rutting depth, there is no obvious difference between the precipitated area and non-precipitated area.
(4) The formation mechanism of crystals; rainwater penetrates into the pavement through the coarse segregation area, and collects in the depression of the lower bearing layer through the connecting gap. Under the action of high temperature, the solution precipitates out of the pavement surface and undergoes crystallization.
(5) In order to avoid the formation of crystals, the voidage ratio should be properly reduced in the design of the mixture to improve the dense water performance. The construction process should ensure the formation of the bottom bearing layer, to avoid the accumulation of water seepage. At the same time, the uniformity of the pavement should be further improved to ensure the watertight performance of the paving layer.

Through the study in this paper, it was determined that the crystallites come from machine-made sand, but machine-made sand should be wrapped in asphalt in the mixture, and how the infiltration rainwater melts machine-made sand has not been thoroughly analyzed and studied in this paper. In a follow-up study, relevant laboratory experiments can be carried out to reproduce crystallites, and CT technology or standard aggregate can be used to reproduce the whole process of crystallinity formation, so as to further analyze the conditions and requirements of crystallinity.

Author Contributions: Conceptualization, C.L. and D.W.; methodology, C.L.; software, J.L.; validation, C.L., D.W. and J.L.; formal analysis, J.H.; investigation, J.H.; resources, J.H.; data curation, J.L.; writing—original draft preparation, C.L.; writing—review and editing, C.L.; visualization, C.L.; supervision, D.W.; project administration, D.W.; funding acquisition, D.W. All authors have read and agreed to the published version of the manuscript.

Funding: This research was funded by the Natural Science Fund of Guangdong Province, grant number 2019A1515011965, and Natural Science Foundation of China, grant number 51808228.

Data Availability Statement: The data presented in this study are available upon request from the corresponding author.

Acknowledgments: The authors gratefully acknowledge the Natural Science Fund of Guangdong Province, grant number 2019A1515011965, and Natural Science Foundation of China, grant number 51808228.

Conflicts of Interest: The authors declare no conflict of interest.

References

1. Baojun, M. Study on Pavement Quality Control Technology of Asphalt Concrete Bridge Deck of Highway in Mountainous Areas. Master's Thesis, Chang'an University, Xi'an, China, 2020.
2. Hongjun, W.; Yuxia, L.; Jianlin, S. Types and Causes of asphalt pavement damage of cement concrete bridge deck. *Highw. Transp. Sci. Technol. (Appl. Technol. Ed.)* **2014**, *10*, 213–216.
3. Changan, Z. Damage cause analysis and preventive measures of asphalt concrete bridge deck pave-ment. *Gansu Sci. Technol.* **2013**, *29*, 110–111.
4. Yue, F. Study on Color Generation and Decay Mechanism of Asphalt Pavement. Master's Thesis, Fuzhou University, Fuzhou, China, 2019.
5. Cui, P.; Xiao, Y.; Fang, M.; Chen, Z.; Yi, M.; Li, M. Residual Fatigue Properties of Asphalt Pavement after Long-Term Field Service. *Materials* **2018**, *11*, 892. [CrossRef] [PubMed]
6. Karki, P.; Zhou, F. Systematic Method for Quantifying Re-Refined Engine Oil Bottom Content in Binders Using X-Ray Fluorescence Spectroscopy. *Transp. Res. Rec.* **2017**, *2632*, 52–59. [CrossRef]
7. Qian, G.P.; Li, S.J.; Yu, H.N.; Gong, X.B. Interlaminar Bonding Properties on Cement Concrete Deck and Phosphorous Slag Asphalt Pavement. *Materials* **2019**, *12*, 1427. [CrossRef]
8. Muyu, L.; Luy, W.; Chunling, Y.; Juan, T. An experimental study on fading and whitening of asphalt pavement in Guangxi Region. *Smart Cities* **2021**, *7*, 31–32. [CrossRef]
9. Lei, Y.; Hu, X.D.; Wang, H.N.; You, Z.P.; Zhou, Y.L.; Yang, X. Effects of vehicle speeds on the hydrodynamic pressure of pavement surface: Measurement with a designed device. *Measurement* **2017**, *98*, 1–9. [CrossRef]
10. Liang, B.; Jiang, B.L. Study on Composition and Simulation Analysis of Traffic Loads in Submerged Floating Tunnels. In Proceedings of the 2nd International Symposium on Submerged Floating Tunnels and Underwater Tunnel Structures (SUFTUS), Chongqing, China, 16–18 December 2016; pp. 180–189.
11. Bocci, E.; Graziani, A.; Canestrari, F. Mechanical 3D characterization of epoxy asphalt concrete for pavement layers of orthotropic steel decks. *Constr. Build. Mater.* **2015**, *79*, 145–152. [CrossRef]
12. Budzinski, B.; Mieczkowski, P.; Slowik, M.; Mielczarek, M.; Bilski, M.; Fornalczyk, S. Assessment of the low-temperature performance of asphalt mixtures for bridge pavement. *Road Mater. Pavement Des.* **2023**, *24*, 409–423. [CrossRef]
13. Graczyk, M.; Zbiciak, A.; Michalczyk, R.; Kowalewski, L. Numerical modelling of bubbles formation in the bridge asphalt pavement under gas pressure impact. *Transp. Res. Proc.* **2016**, *14*, 3925–3934. [CrossRef]
14. Zou, G.L.; Xu, X.Y.; Li, J.X.; Yu, H.Y.; Wang, C.J.; Sun, J. The Effects of Bituminous Binder on the Performance of Gussasphalt Concrete for Bridge Deck Pavement. *Materials* **2020**, *13*, 364. [CrossRef]
15. Shi, C. Research on Causes and Maintenance Measures of High Speed Bridge Damage in Liaoning Province. Master's Thesis, Chang'an University, Xi'an, China, 2017.
16. Tianyu, C. Analysis of deck pavement disease of hollow slab Bridge in Alpine region. *Smart Cities* **2021**, *7*, 69–70. [CrossRef]
17. Guanhua, H.; Liting, Z.; Yapeng, Y.; Mengrou, Y. Asphaltic Concrete bridge Deck paving mechanism and prevention measures. *Highway* **2019**, *64*, 21–25.
18. Fuxing, R. Causation analysis of asphalt concrete bridge deck whitening disease. *Commun. World* **2018**, *8*, 103–104. [CrossRef]
19. Chao, H. A brief discussion on the whitening pulp disease of asphalt concrete bridge deck pavement. *Heilongjiang Sci. Technol. Inf.* **2013**, *25*, 230.
20. Renjie, Q.; Lin, Z.; Cao, S.; Bin, L.; Ming, S. Causes and solutions of AC-20 whitening in the middle surface layer of Guilai Ex-pressway. *Highw. Automob. Transp.* **2015**, *5*, 119–121.
21. Yang, Y.M.; Qian, Z.D.; Song, X. A pothole patching material for epoxy asphalt pavement on steel bridges: Fatigue test and numerical analysis. *Constr. Build. Mater.* **2015**, *94*, 299–305. [CrossRef]
22. Aiguo, L.; Yang, L.; Xicai, M. Analysis of seepage causes and control technology of asphalt pavement layer of cement concrete bridge deck. *Highw. Transp. Sci. Technol. (Appl. Technol. Ed.)* **2012**, *8*, 59–61.
23. Hao, P.W.; Zhang, M.Y.; Zhang, Q.; Xu, J.Z.; Dong, S. Study on evaluation method of mud-pumping of cement concrete bridge deck pavement. *Constr. Build. Mater.* **2018**, *167*, 106–114. [CrossRef]
24. Tarefder, R.A.; Ahmad, M. Evaluation of pore structure and its influence on permeability and moisture damage in asphalt concrete. *Int. J. Pavement Eng.* **2017**, *18*, 274–283. [CrossRef]
25. Fu, J.C.; Shen, A.Q.; Zhang, H. Study on the Influence and Law of Waterproof System Design Factors on the Typical Stress of Bridge Deck Pavement. *Coatings* **2021**, *11*, 1540. [CrossRef]
26. Haynes, M.A.; Coleri, E.; Obaid, I. Performance of Waterproofing Membranes to Protect Concrete Bridge Decks. *Transp. Res. Rec.* **2021**, *2675*, 1693–1706. [CrossRef]
27. Wang, J.W.; Ji, B.F.; Chen, B.; Chen, S.Q. Application of High-Viscosity Modified Asphalt Mixture in Curved Bridge Pavement. *Sustainability* **2023**, *15*, 3411. [CrossRef]
28. Gao, Y. Causation analysis and treatment suggestion of asphalt surface of a highway bridge in Gansu Province. *Gansu Sci. Technol.* **2022**, *38*, 19–23.
29. Haimin, S. Analysis of the causes of asphalt pavement whitening based on surface energy. *Hei-Longjiang Transp. Sci. Technol.* **2019**, *42*, 91–93. [CrossRef]

30. Chao, Z. Key points and Technical Analysis of Fujian high-speed Asphalt Pavement Maintenance Engineering Design. *Fujian Transp. Sci. Technol.* **2021**, *1*, 6–8.
31. Xiong, Z.; Ningbo, L.; Peiwen, H.; Yujing, W.; Shaohui, L. Evaluation method and influencing factors of anti-whitening performance of asphalt mixture for bridge deck pavement. *Inn. Mong. Highw. Transp.* **2021**, *6*, 9–16. [CrossRef]
32. Zhiyong, H.; Chunlong, X.; Weixiong, L.; Wen, N.; Jiangmiao, Y. Analysis and control of caustic flooding mechanism of newly built asphalt pavement. *Highway* **2019**, *64*, 40–47.
33. Hu, J.; Qian, Z.D.; Liu, P.F.; Wang, D.W.; Oeser, M. Investigation on the permeability of porous asphalt concrete based on microstructure analysis. *Int. J. Pavement Eng.* **2020**, *21*, 1683–1693. [CrossRef]
34. Li, W.X.; Wang, D.Y.; Xiong, C.L.; Yu, J.M.; Chen, B.; Yu, X.S. Mechanism, prevention and remedy of alkali-pumping in new constructed asphalt pavement. *Int. J. Pavement Eng.* **2022**, *23*, 2689–2702. [CrossRef]
35. Sasaki, I.; Moriyoshi, A.; Tsunekawa, M. Water accumulation and behavior of surfactant associated with moisture permeation in bituminous pavement on concrete deck bridge. *J. Jpn. Pet. Inst.* **2006**, *49*, 315–320. [CrossRef]
36. Xu, S.F.; Lu, G.Y.; Hong, B.; Jiang, X.Y.; Peng, G.; Wang, D.W.; Oeser, M. Experimental investigation on the development of pore clogging in novel porous pavement based on polyurethane. *Constr. Build. Mater.* **2020**, *258*, 12–37. [CrossRef]
37. Jahandari, S.; Tao, Z.; Saberian, M.; Shariati, M.; Li, J.; Abolhasani, M.; Kazemi, M.; Rahmani, A.; Rashidi, M. Geotechnical properties of lime-geogrid improved clayey subgrade under various moisture conditions. *Road. Mater. Pavement* **2022**, *23*, 2057–2075. [CrossRef]
38. Toghroli, A.; Shariati, M.; Sajedi, F.; Ibrahim, Z.; Koting, S.; Mohamad, E.T.; Khorami, M. A review on pavement porous concrete using recycled waste materials. *Smart Struct. Syst.* **2018**, *22*, 433–440. [CrossRef]
39. Larsen, T.; Randhol, P.; Lioliou, M.; Josang, L.O.; Ostvold, T. Kinetics of $CaCO_3$ scale formation during core flooding. In Proceedings of the 9th International Conference on Oilfield Scale 2008—"Managing Scale Through the Field Lifetime", Aberdeen, UK, 28–29 May 2008; pp. 162–173.
40. Charpentier, T.V.J.; Neville, A.; Baraka-Lokmane, S.; Hurtevent, C.; Ordonez-Varela, J.R.; Nielsen, F.M.; Eroini, V.; Olsen, J.H.; Ellingsen, J.A.; Bache, O. Evaluation of anti-fouling surfaces for prevention of mineral scaling in sub-surface safety valves. In Proceedings of the SPE International Conference and Exhibition on Oilfield Scale 2014, Aberdeen, UK, 14–15 May 2014; pp. 31–45.
41. Foroutan, S.; Moghadasi, J. A neural network approach to predict formation damage due to calcium sulphate precipitation. In Proceedings of the SPE European Formation Damage Conference and Exhibition 2013: Unconventional and Conventional Solutions to Challenging Reservoirs, Noordwijk, The Netherlands, 5–7 June 2013; pp. 714–725.
42. Gao, C.Q. Study on the Regularity of Groundwater Crystallization Clogging Drainage Pipe and Its Influence on the Stability of High Slope. Ph.D. Thesis, Chongqing Jiaotong University, Chongqing, China, 2020.
43. Liu, Y.Y.; Lai, H.P.; Zhao, Y.B.; Sun, W.X. Spatio-Temporal Variation of Crystallites and Water Environment and Crystallization Path Analysis in Tunnel Drainage System of Calcium-Rich Surrounding Rock. *Mod. Tunn. Technol.* **2023**, 1–10. Available online: http://kns.cnki.net/kcms/detail/51.1600.U.20230331.1343.006.html (accessed on 14 May 2023).
44. Wu, Y.Z.; Li, G.P.; Jia, Q.Q.; Qi, H.Y.; Li, D.W. Study on the component characteristics and sources of porous crystallites in loess tunnel. *Mod. Tunn. Technol.* **2020**, *57*, 154–159. [CrossRef]
45. Jiang, X.; Mu, S.; Liu, J.P. Influence of chlorides and salt concentration on salt crystallization damage of cement-based materials. *J. Build. Eng.* **2022**, *61*, 105260. [CrossRef]
46. Yang, S.Y.; Han, M.; Chen, X.L.; Song, J.J.; Yang, J.S. Influence of sulfate crystallization on bond-slip behavior between deformed rebar and concrete subjected to combined actions of dry-wet cycle and freeze-thaw cycle. *Constr. Build. Mater.* **2022**, *345*, 128368. [CrossRef]

buildings

MDPI

Article

Research on High- and Low-Temperature Rheological Properties of High-Viscosity Modified Asphalt Binder

Zhongcai Huang [1], Xianwu Ling [2,*], Di Wang [3], Pengfei Li [2], Huaquan Li [1], Xinyu Wang [1], Zujian Wang [1], Rong Wei [1], Weining Zhu [1] and Augusto Cannone Falchetto [3,*]

1 Guangxi Communications Investment Group Corporation Ltd., Nanning 530022, China
2 School of Highway, Chang'an University, Xi'an 710064, China
3 Department of Civil Engineering, Aalto University, 02150 Espoo, Finland
* Correspondence: 2021121186@chd.edu.cn (X.L.); augusto.cannonefalchetto@aalto.fi (A.C.F.)

Abstract: This study evaluates the critical high- and low-temperature rheological properties of a high-viscosity modified asphalt (HVMA) binder by analyzing one neat and three high-viscosity modified binders (B-type, Y-type, and H-type) using temperature sweep tests and multi-stress creep recovery tests (MSCR) through the dynamic shear rheometer (DSR), and low-temperature creep stiffness properties by the bending beam rheometer (BBR). Technical indexes such as the softening point temperature, dynamic viscosity, rutting factor, unrecoverable creep compliance, and the creep recovery rate are measured and calculated for high-temperature properties, while the m/S value, dissipation energy ratio, relaxation time, elongation, creep stiffness, and creep speed are used as technical indexes for low-temperature properties. The results show that the incorporation of high-viscosity modifiers reduces the unrecoverable creep compliance and increases the creep recovery rate of the asphalt binder. Non-recoverable creep compliance is found to be a reliable indicator for high-temperature performance, while at low temperatures, the relaxation time decreases, the dissipation energy increases, and the stress relaxation ability improves. The dissipation energy ratio and m/S value are suggested to evaluate the low-temperature performance of HVMA binders using the Burgers model based on the BBR bending creep stiffness test. Therefore, this study recommends using the unrecoverable creep compliance via MSCR to evaluate high-temperature properties and dissipation energy ratio and m/S value for low-temperature properties in the evaluation of HVMA binders.

Keywords: HVMA binder; rheological theory; unrecoverable creep compliance; dissipation energy ratio; m/S value

Citation: Huang, Z.; Ling, X.; Wang, D.; Li, P.; Li, H.; Wang, X.; Wang, Z.; Wei, R.; Zhu, W.; Falchetto, A.C. Research on High- and Low-Temperature Rheological Properties of High-Viscosity Modified Asphalt Binder. *Buildings* **2023**, *13*, 1077. https://doi.org/10.3390/buildings13041077

Academic Editor: Salvatore Antonio Biancardo

Received: 19 March 2023
Revised: 11 April 2023
Accepted: 18 April 2023
Published: 19 April 2023

1. Introduction

With the advancement of pavement materials and mixtures capable of meeting specific performance under different temperatures, deformations, and functions, higher and higher requirements are put forward for the creation of modified asphalt materials such as SBS, rubber, and high viscosity modified binders by increasing the amount of asphalt binder in the mix [1–4]. Because modified asphalt material can improve the road performance requirements of asphalt binders, it is widely used in road engineering. Among them, high-viscosity modified asphalt (HVMA) binders refer to a modified material with a viscosity higher than 20,000 Pa·s at 60 °C [5–7]. Because of its excellent bonding force and strong adhesion to aggregates, it can be used as one of the future recycling mixtures [8–10] and has been widely adopted in porous asphalt mixtures [11], stress absorption layers [12], bridge deck pavement layers [13] and other domains [14–16].

The main areas of research on HVMA in recent years have been the making of modified asphalt, the performance of modified asphalt at high and low temperatures, and its use as a binder in pavement materials. More investigation has conducted additional research on the rheological characteristics of HVMA binders, particularly high-temperature responses.

Chen et al., investigated the effects of high-viscosity modifier (HVM) content on the characteristics of modified binders and contrasted it with neat asphalt using 60 °C viscosity as the key indicator. The findings demonstrated that high-viscosity modified materials were quite compatible with a neat bituminous binder at 60 °C [17]. Li et al., used vacuum decompression capillary and dynamic shear rate scanning experiments to examine the capillary viscosity and zero shear viscosity of a HVM binder. The findings demonstrate that the viscosity characteristics of high-viscosity asphalt binders vary depending on the state they are in, and for those with viscosities greater than 30,000 Pa·s, zero shear viscosity can be used to characterize the material's bonding properties [18]. Tan et al., created HVMA by modifying matrix asphalt with thermoplastic elastomer and SBS particles. The effectiveness was then compared to engineering high-viscosity asphalt, SBS-modified asphalt, and 70 # asphalt. The findings revealed that the modified materials dispersed well in asphalt and that the TPE particles, which function as high-elastic interlocking units, were spread uniformly throughout the network structure. The T-HVA also displayed better high-temperature stability and low-temperature flexibility [19]. PG high temperature grading, multi-stress repeated creep, accelerated fatigue, temperature scanning, and other procedures were used by Li et al., to set up HVMA for use in high-temperature environments. They compared and analyzed the alterations of the high temperature, fatigue resistance, and shear resistance indexes. The study's findings are beneficial for the development of porous asphalt mixtures and their use in high-temperature regions [20]. Yu et al., developed waste-LDPE/SBS composites to produce environmentally beneficial HVMA. The rheological properties and microstructure of asphalt were examined using a DSR, BBR, and fluorescence imaging. The findings demonstrate that waste polymers can significantly increase the performance of asphalt at high temperatures and lengthen the fatigue life of asphalt materials [21]. Using a DSR, Guo et al., examined the dynamic viscoelastic characteristics of a high viscosity, high elasticity, high strength, and SBS-modified glue. The results showed that different kinds of modified asphalt binders performed differently at high temperatures depending on their equilibrium complex shear modulus [22]. The rheological characteristics of three different types of high-viscosity changed materials were examined using DSR by Qin et al., who also compared them to the neat binder. It was discovered that HVMA binders had superior thermal stability and deformation recovery ability at high temps compared to the neat binder [23]. By using a BBR, Zhang et al., investigated the low-temperature rheological characteristics of neat asphalt and high-viscosity modified binders (BBR). The findings show that the production of the HVMA binder with a plasticizer and crosslinking agent can enhance its low-temperature performance [24]. Zhang et al., used macro and micro tests, including BBR and FTIR, to examine the efficacy of HVMA. The findings demonstrate a relationship between the structural properties of the compounds and the thermal properties of asphalt for the HVMA binder's low-temperature rheological properties [25].

Researchers have studied the high- and low-temperature rheological properties of high-viscosity modified binders over time, and on the basis of their findings [26–28], they have proposed the assessment index of high-temperature rheological properties of HVMA binders. Among them, most of the rheological properties of high-viscosity modified asphalt are similar to those of neat asphalt, such as complex shear modulus, m/S value and rutting factor. Still, the research needs to reach a deeper level, considering specific indicators and expanding the limited analysis of low-temperature rheological properties of high-viscosity modified materials. As a result, it is essential to research the crucial high- and low-temperature rheological characteristics of HVMA binders. In order to methodically investigate these characteristics, the DSR and the BBR experiments were performed on three different types of high-viscosity modified binders. To find more useful high- and low-temperature performance key evaluation indexes for HVMA binders and to establish a foundation for judging the caliber of high-viscosity modified binders, seven high-temperature performance evaluation indexes and six low-temperature performance evaluation indexes were chosen for a differentiation analysis.

2. Raw Materials and Test Methods

2.1. Asphalt Material

The neat material was the SK-90 road asphalt binder, and two commercially available high-viscosity modifiers (B-type and Y-type) were selected. The thermoplastic rubber that makes up the high-viscosity modifier is mixed with granular modifiers such as adhesive resin, plasticizer, and anti-aging substance. Compared with the Y-type high viscosity modifier, the B-type high viscosity modifier also added a small amount of carbon black. The main physical indexes are depicted in Table 1. The HVMA binders were made by adding HVM to the neat material. The neat asphalt binder was melted into a flowing state during the preparation process, and the high-viscosity modifier was then applied at a temperature of 170–180 °C (the content was 12% of the mass of the neat binder [23]). After operating at moderate speed (2000 r/min) for 10 min, the high speed (5000 r/min) was maintained for 30 min. To remove the bubbles created during the shearing procedure, it was put in an oven at 175 °C for 10 min after shearing was finished. For comparison, the study also selected a completed high-viscosity modified binder (H-type). Table 2 displays the key characteristics for the four various bitumen binders.

Table 1. Physical indexes of the high-viscosity modifier.

Test Item	Test Value	Index
Particle Size/mm	4.5	≤5
Density/(g/cm^3)	0.8	0.7~1.0
Water Absorption/%	0.4	<1%

Table 2. Characteristics of asphalt binders.

Type	Penetration (25 °C, 5 s, 100 g)/(0.1 mm)	Softening Point/°C	Ductility (5 °C)/cm	Dynamic Viscosity (60 °C)/(Pa·s)
SK-90	97.1	47.4	9.7	140.3
B-type	51.1	84.3	64.3	38,696.9
Y-type	54.6	85.9	59.9	20,425.1
H-type	49.3	94.2	48.6	99,635.4

2.2. Test Method

The DSR test can be used to gauge the viscosity and elastic characteristics of asphalt binders. In the study, temperature sweep test and MSCR test of HVMA binders were used to investigate the high temperature rheological properties. HVMA binders' low temperature creep stiffness properties were investigated using a BBR test.

2.2.1. Temperature Sweep Test

Consequently, the AASHTO T315 [29] temperature scanning tests are performed using the DSR. The measurement is conducted in the 3 °C to 80 °C temperature range using a 25 mm parallel plate with a 1 mm gap. Then, the step size is 5 °C, and the angular frequency is 10 rad/s.

2.2.2. MSCR Test [30]

A DSR was used to conduct the test, which involved shear creep and recovery at two different stress levels at a predetermined temperature. First, 0.1 kPa and 3.2 kPa were the two creep stress values that were applied. At the 0.1 kPa stress level, twenty cycles were run. Then, for a total runtime of 300 s, ten cycles at the 3.2 kPa stress level were performed. Each cycle was further split into a 1 s stress loading stage and a 9 s zero stress recovery stage [31,32]. Considering the actual reference pavement conditions [33], the test temperature was imposed to be 64 °C. The asphalt binder specimens were short-term aged and tested with the 8 mm plate and 2 mm gap. MSCR test uses the non-recoverable creep compliance J_{nr} and the creep recovery rate R as the evaluation index, and the calculation formula is shown in Equations (1) and (2).

$$J_{nr} = \frac{\varepsilon_u}{\sigma} \quad (1)$$

$$R = \frac{\varepsilon_p - \varepsilon_u}{\varepsilon_p} \times 100\% \quad (2)$$

above ε_p is the peak strain; ε_u is the unrecovered strain; and σ is the stress.

2.2.3. BBR Test

The BBR test measures the stiffness and relaxation performance of the asphalt binder at low temperatures. The BBR test was employed in this study to assess the asphalt's low temperature rheological characteristics. The measurement temperature was set to −12 °C, −18 °C, and −24 °C in accordance with AASHTO T313 [34] requirements.

3. Test Results and Analysis

3.1. High-Temperature Rheological Properties

3.1.1. Rutting Factor

The variation of the rutting factor ($G^*/\sin(\delta)$) with the temperature for SK-90, B-type, Y-type, and H-type is shown in Figure 1. The $G^*/\sin(\delta)$ of the four different types of asphalt binders diminishes as the temperature rises, suggesting that the resistance to rutting will decrease as well. Then, the $G^*/\sin(\delta)$ of B-type, Y-type and H-type HVMA binders were greater than those of the neat binder, indicating that adding a high-viscosity modifier greatly improved the rutting resistance of the neat binder. In the process of high temperature to low temperature, the anti-rutting ability of HVMA binders is better and better than that of the neat binder, indicating that the high temperature rheological properties of high viscosity asphalt should be selected under high temperature conditions. Based on field data [33], it was noted that asphalt pavement could become as hot as 65 °C in the summer, so the $G^*/\sin(\delta)$ at 65 °C was chosen as the evaluation index of high-viscosity modified binders. The $G^*/\sin(\delta)$ of the three HVMA binders at 65 °C were 15,519 Pa (B-type), 10,964 Pa (Y-type) and 19,921 Pa (H-type), respectively. The order from large to small is H-type > B-type > Y-type, indicating that an H-type high-viscosity modified binder has the best rutting resistance.

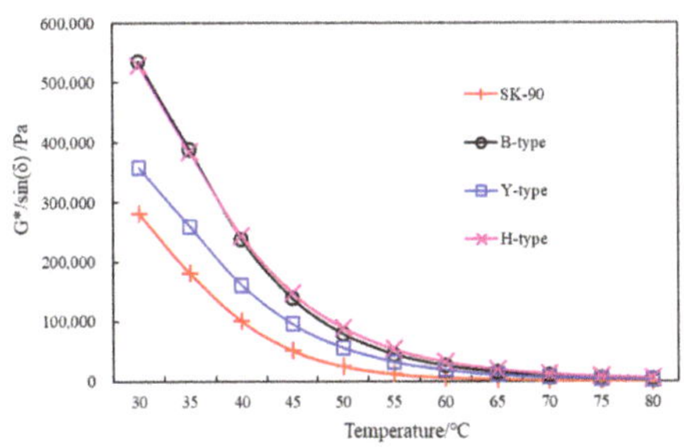

Figure 1. Rutting factor evolution with temperature.

3.1.2. Non-Recoverable Creep Compliance and Creep Recovery Rate

Figure 2 shows the above MSCR test results at 64 °C. Figure 2 shows that under multiple stress loads, the time–strain curve of SK-90 is greater than those of the three HVMA binders. This pattern shows that the neat binder's high-temperature rheological properties can be greatly enhanced by high-viscosity modifiers. Comparing the time–strain curves of the three high-viscosity modified binders, it is evident that the time–strain curve of H-type is the lowest (the best high-temperature rheological properties). In contrast, the time–strain curves of B-type and Y-type are relatively similar, indicating that the difference in the high-temperature rheological properties of B-type and Y-type is insignificant. It is found that the time–strain curves of SK-90 under two stress levels are one order of magnitude smaller than those of HVMA binders under the same stress and temperature conditions, which indicates that the addition of a high-viscosity modifier greatly enhances the high-temperature deformation resistance of asphalt.

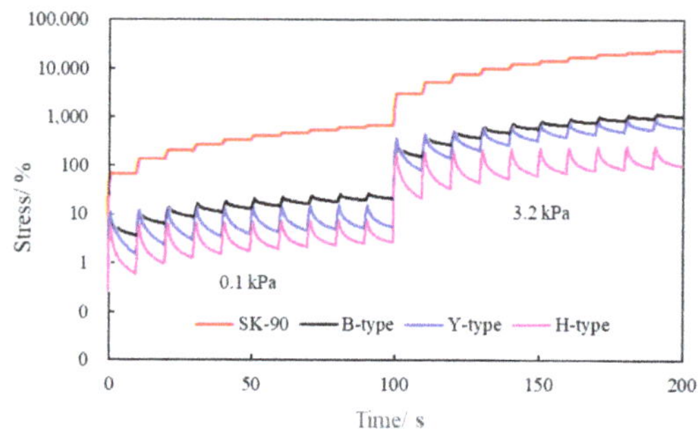

Figure 2. The time–strain relationship of different asphalt's MSCR test.

The J_{nr} and R were calculated according to Equations (1) and (2). The outcomes are displayed in Figures 3 and 4. In the figure, $J_{nr}0.1$ and $J_{nr}3.2$ represent the average values of non-recoverable creep compliance of the binder in ten cycles under 0.1 kPa and 3.2 kPa stress, respectively. $R0.1$ and $R3.2$ represent the average values of the R of the binder in ten cycles under 0.1 kPa and 3.2 kPa stress, respectively. Figure 3 shows that the SK-90 neat, B-type, and Y-type HVMA binders all have lower levels of resistance to irreversible deformation than the H-type HVMA binder. The J_{nr} of the four binders in a high-stress level (3.2 kPa) is higher than that in a low-stress level (0.1 kPa), demonstrating that under higher stress conditions, an asphalt binder is more susceptible to permanent deformation. Figure 4 shows that under the low-stress level, the R values of H-type HVMA and Y-type high-viscosity modified binders are not much different and greater than that of B-type high-viscosity modified and SK-90 neat asphalt binders. Under the high-stress level, the R of the four types of asphalt binders from the largest to the smallest is H-type, Y-type, B-type, and SK-90. Compared to the R value under the low-stress level, R under the high-stress level has different degrees of reduction. Among them, the B-type high-viscosity modified material has the most significant decrease, reaching 15%, and a minimal 0.9% decrease in the H-type HVMA binder.

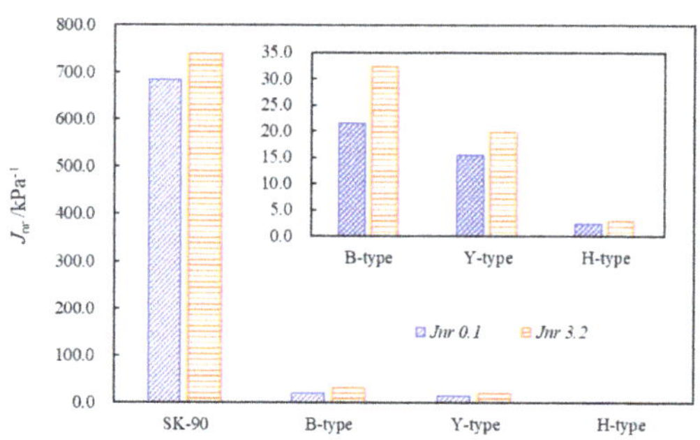

Figure 3. Unrecoverable creep compliance of different asphalt binders.

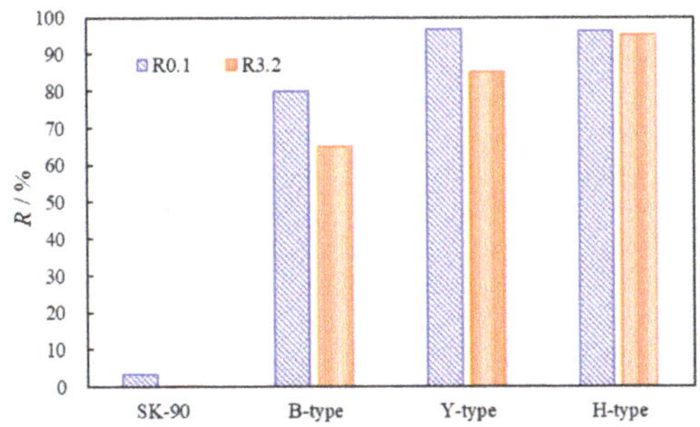

Figure 4. Creep recovery rate of different asphalt binders.

3.1.3. Analysis of High-Temperature Performance Evaluation Index

For the high-temperature evaluation index of HVMA binders, relevant researchers have performed considerable work, still demanding a unified view. In order to assess the high-temperature performance of HVMA binders, it is essential to investigate an evaluation index. Table 3 lists seven different high-temperature evaluation indexes of high-viscosity modified binders, separately for Softening Point, Dynamic Viscosity, $G^*/\sin(\delta)$, the $J_{nr}0.1$ and $J_{nr}3.2$, and the $R0.1$ and $R3.2$. In this paper, the discriminating capability of high-temperature performance evaluation indexes of high-viscosity modified binders is studied using the discrimination analysis method. Discrimination analysis is a method of using an independent sample t-test to study whether there are differences in the collection of test data. Discrimination is the quantitative power of an evaluation index in distinguishing among specific characteristics; the greater the value of discrimination D, the more sensitive the index is [35]. This paper used SPSS software to calculate the discrimination of high-temperature evaluation indexes, shown in Table 4.

Table 3. Different evaluation indexes of high-temperature property.

Type	Softening Point/°C	Dynamic Viscosity (60 °C)/(Pa·s)	$G^*/sin(\delta)$ (65 °C)/Pa	$J_{nr}0.1$ (64 °C)/ kPa^{-1}	$J_{nr}3.2$ (64 °C)/ kPa^{-1}	R0.1 (64 °C)	R3.2 (64 °C)
B-type	84.3	38,696.9	15,518.6	21.54	32.39	80.1	64.9
Y-type	85.9	20,425.1	10,964.2	15.47	19.9	96.9	85.1
H-type	94.2	99,635.4	19,921.1	2.41	2.9	96.2	95.2

Table 4. Discrimination—calculation result of high-temperature property.

Index	Softening Point/°C	Dynamic Viscosity (60 °C)/(Pa·s)	$G^*/sin(\delta)$ (65 °C)/Pa	$J_{nr}0.1$ (64 °C)/ kPa^{-1}	$J_{nr}3.2$ (64 °C)/ kPa^{-1}	R0.1 (64 °C)	R3.2 (64 °C)
D	0.000061	0.134179	0.007432	0.151634	0.190717	0.000309	0.002042

Table 4 shows the ranking of the seven different kinds of high-temperature performance evaluation indexes for HVMA in terms of discrimination: $J_{nr}3.2$ (64 °C) > $J_{nr}0.1$ (64 °C) > Dynamic Viscosity (60 °C) > $G^*/\sin(\delta)$ (65 °C) > $R3.2$ (64 °C) > $R0.1$ (64 °C) > Softening Point. This shows that the MSCR test can better discriminate HVMA binders. The distinguishing ability of unrecoverable creep compliance is the best. Because the softening point test is empirical, it cannot be used to assess the high-temperature performance of HVMA because it has the lowest softening point discrimination. The performance of HVMA binders at high temperatures is recommended to be assessed using the unrecoverable creep compliance of the MSCR test.

3.2. *Low-Temperature Rheological Properties*

3.2.1. Burgers Model

A typical viscoelastic material is an asphalt binder. The viscoelastic properties of modified asphalt binders can be adequately described by the Burgers model. It is a four-element viscoelastic constitutive model with a series-parallel connection made of two spring elements and two dashpot elements. The mathematical equation is [36]:

$$\varepsilon(t)= \sigma_0\left[\frac{1}{E_1}+\frac{1}{\eta_1}t+\frac{1}{E_2}\left(1-e^{-\frac{E_2}{\eta_2}t}\right)\right] \quad (3)$$

In the formula, ε is strain; σ_0 is the applied stress, MPa; E_1 is instantaneous elastic modulus; η_1 is the instantaneous viscosity coefficient; E_2 is the slow deformation after applying stress; and η_2 is a viscosity index whose deformation does not disappear immediately after the applied stress is removed.

The BBR test data of the four asphalt binders were nonlinearly fitted (by 1stOpt—First Optimization software) to identify the Burgers model parameters. The relaxation time (λ) and dissipation energy ratio ($W_d(t)/W_s(t)$) of the four materials were calculated by fitting parameters E_1, E_2, η_1, and η_2. The calculation formula is as follows [36]:

$$\lambda = \eta_1/E_1 \quad (4)$$

$$W_d(t)/W_S(t) = \left[\frac{t}{\eta_1}+\frac{1}{2E_2}(1-e^{-\frac{2E_2}{\eta_2}t})\right]/\left[\frac{1}{E_1}+\frac{1}{2E_2}(1-2e^{-\frac{E_2}{\eta_2}t}+e^{-\frac{2E_2}{\eta_2}t})\right] \quad (5)$$

where t is the stress action time, and s is the stress level.

Relaxation time reflects the measurement of the stress of the asphalt binder with time. The longer the relaxation period, the more unfavorable (slow) the stress dissipation in the asphalt binder is [37]. The evolution of the relaxation time over temperature for the four binders can be ordered using the formulas and as shown in Figure 5: SK-90, Y-type, B-type, and H-type. The H-type high-viscosity modified binder presents the shortest relaxation time, suggesting the best low-temperature performance. As the temperature decreases, the relaxation time increases; this is because at a lower temperature, the binder elasticity increases, then the viscosity effect reduces, the energy consumption rate becomes slower, and the time allowing the stress change becomes longer, resulting in longer relaxation times.

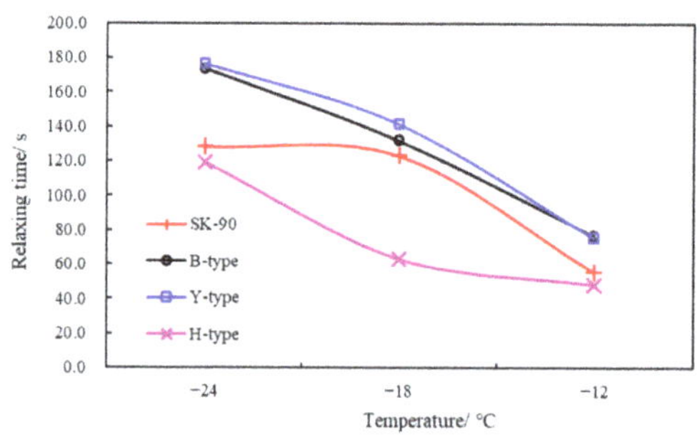

Figure 5. The relaxation time of asphalt binders under different temperatures.

The dissipation energy ratio shows the asphalt binder's ability to relax; the higher the dissipation energy ratio, the greater the material's low-temperature crack resistance [37]. Figure 6 presents the dissipation energy ratio for the four binders and the following trend: H-type, B-type, Y-type, and SK-90. The low-temperature performance of the H-type is the best in terms of dissipation energy, which is consistent with the relaxation time evaluation findings. The dissipation energy ratio of asphalt binders reduces as the temperature drops, demonstrating that the dissipation energy in the binders decreases, the stored energy rises, and low-temperature crack resistance worsens. In the temperature range of −18 °C~−12 °C, the dissipation energy ratio of the four binders decreases significantly, while in the temperature range of−24 °C~−18 °C, the dissipation energy ratio of the four materials exhibits a minor reduction, indicating that as the temperature drops, the elastic ratio of the binders rises, but as the temperature drops further, the binder approaches a glassy state.

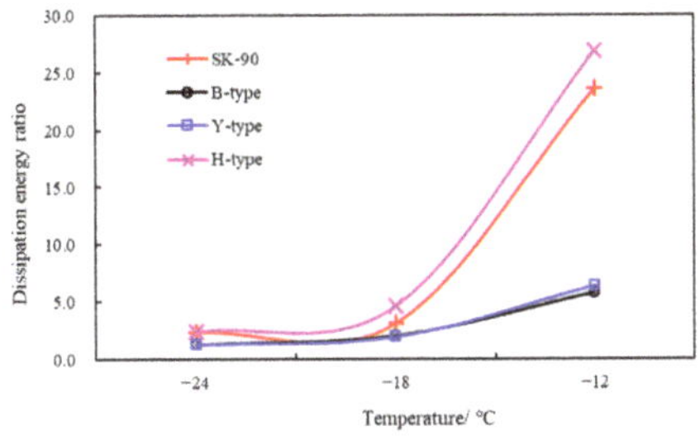

Figure 6. Dissipation energy ratio of the binders under different temperatures.

3.2.2. Creep Stiffness and Creep Speed

Figures 7 and 8 display the S and m numbers of the four binders. Low-temperature crack resistance becomes worse as the temperature lowers due to an increase in asphalt's S value. Under various low-temperature circumstances, the order of *S* for the four binders from large to small is SK-90, B-type, Y-type, and H-type. From the stiffness perspective, the H-type HVM binder has a better low-temperature crack resistance than SK-90, B-type, and Y-type asphalt. While the temperature decreases, the m value of asphalt increases, but there is no apparent consistency in the variation of the m value of the four binders under various temperature conditions. It can be seen that different asphalts show the same law: under low temperature conditions, the creep stiffness modulus of asphalt increases while the creep rate decreases; the creep stiffness modulus can better reflect the modulus difference between different asphalts, but, all in all, using S or m alone to assess the low-temperature performance of HVMA has obvious limitations.

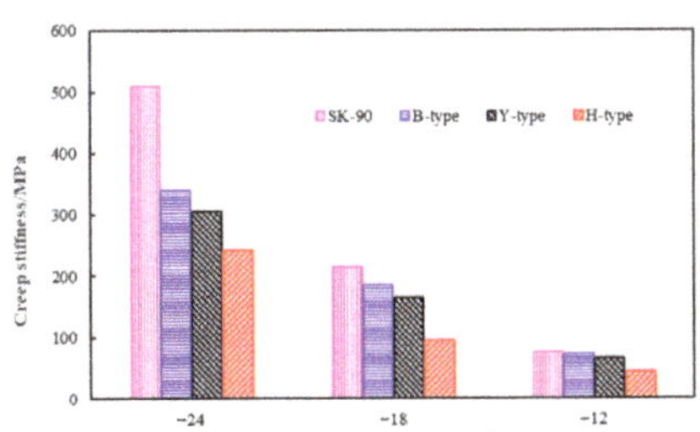

Figure 7. Creep stiffness of different asphalt binders.

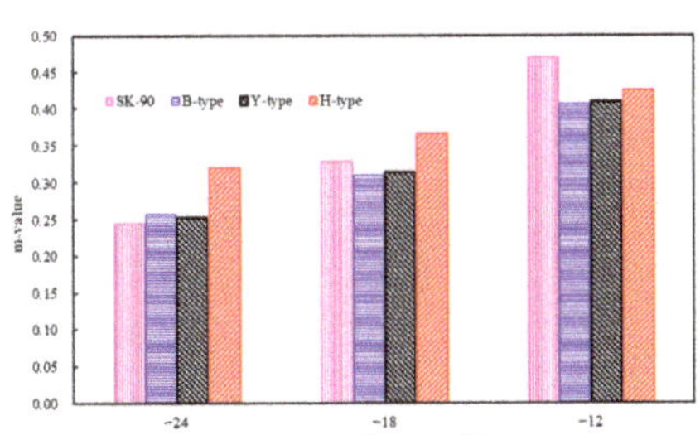

Figure 8. m-value of different asphalt binders.

3.2.3. m/S Value

According to other studies [38], the low-temperature characteristics of asphalt can be described by the m/S ratio; the higher the ratio, the better the low-temperature properties of the asphalt material. The m/S values of the four materials at various temperatures are depicted in Figure 9. The fact that the m/S values of B-type, Y-type, and H-type high-viscosity modified binders are higher than those of neat asphalt binder (SK-90) under low-temperature settings shows that high-viscosity modifiers can greatly enhance binders' low-temperature performance. Among them, the m/S value of H-type high-viscosity modified material is the highest, indicating that H-type HVMA has the greatest low-temperature performance. The increase rate of m/S value of the four materials evaluated in the temperature range of −24 °C~−18 °C is less than that in the temperature range of −18 °C~−12 °C. This trend is due to the increasing temperature leading to the higher energy of molecular motion in the binder, facilitating the activity of the molecular chain segments and structure in the material.

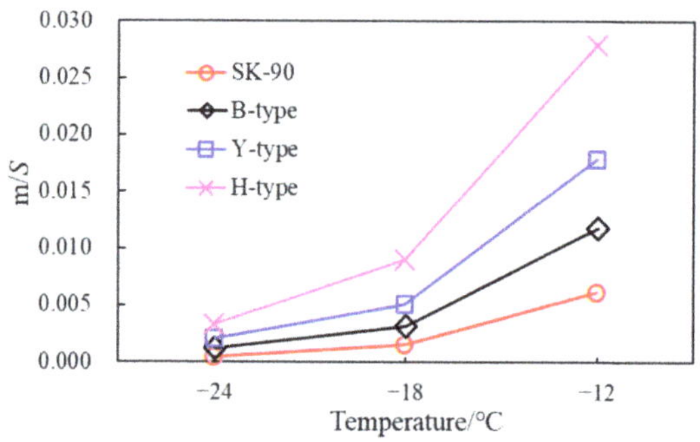

Figure 9. m/S for the different asphalt binders.

3.2.4. Analysis of Low-Temperature Performance Evaluation Index

Over the years, there has been limited research in identifying suitable evaluation indexes for the low-temperature performance of HVMA binders. Consequently, it is essential to investigate an evaluation index suitable for evaluating the low-temperature performance of HVMA binders. Section 3.1.3's discrimination analysis method is also used to investigate the low-temperature performance evaluation index of HVMA binders. In this paper, m/S value (−12 °C), ductility (5 °C), creep stiffness (−12 °C), creep speed (−12 °C), relaxation time (−12 °C), and dissipation energy ratio (−12 °C) are selected as different low-temperature performance evaluation indexes of HVMA binders (see Table 5). The discrimination capability of low-temperature evaluation indexes is presented in Table 6.

Table 5. Different evaluation indexes for low-temperature properties.

Type	m/S (−12 °C)	Density/ (g/cm^3)	S (−12 °C)/ MPa	m (−12 °C)	λ (−12 °C)	$W_d(t)/W_s(t)$ (−12 °C)
B-type	0.00556	64.3	73	0.406	76.87	5.75
Y-type	0.00617	59.9	66.2	0.409	75.30	6.34
H-type	0.01	48.6	42.4	0.424	48.30	26.86

Table 6. Discrimination capability of low-temperature evaluation indexes.

Index	m/S (−12 °C)	Density/ (g/cm^3)	S (−12 °C)/ MPa	m (−12 °C)	λ (−12 °C)	$W_d(t)/W_s(t)$ (−12 °C)
D	0.020454	0.001740	0.011869	0.000007	0.008249	0.407520

Table 6 demonstrates that the sequence of discrimination of low-temperature performance evaluation indexes for high-viscosity modified binders is as follows: $W_d(t)/W_s(t)$ (−12 °C) > m/S (−12 °C) > S (−12 °C) > λ (−12 °C) > elongation (5 °C, 5 cm·min^{-1}) > m (−12 °C). The dissipation energy ratio and m/S value have the best discrimination power. This paper recommends that the dissipation energy ratio and m/S value should be utilized to assess the low-temperature performance of a HVM binder, rather than creep stiffness or creep rate alone.

4. Conclusions

Based on the results of the analysis, the following inferences can be drawn:

(1) At 65 °C, the rut factor values of the three HVMA binders are all greater than 10,000 Pa, and the unrecoverable creep compliance under different stress levels is much smaller than that of neat asphalt, and the creep recovery rate is much larger than that of matrix asphalt. At a low temperature, the dissipation energy ratio and m/S value of the three kinds of HVMA binders are smaller than those of neat asphalt. These show that the use of high viscosity modifiers is positive to improve the performance of asphalt and high-viscosity modifiers can significantly enhance asphalt's high-temperature deformation resistance, significantly reduce the possibility of low-temperature crack resistance, and have good deformation recovery ability.

(2) Considering seven high-temperature performance evaluation indexes of HVMA binders, such as rutting factor, dynamic viscosity, softening point, non-recoverable creep compliance, and creep recovery rate, it is recommended to use the MSCR test's non-recoverable creep compliance to evaluate the high-temperature performance of HVMA.

(3) Elongation, creep stiffness, creep speed, dissipation energy ratio, and relaxation time were chosen as low-temperature performance evaluation indexes of high-viscosity modified binders. The Burgers model's dissipation energy ratio and m/S value are recommended for evaluating the low-temperature performance of high-viscosity modified materials.

(4) This study investigates the high- and low-temperature rheological performance index of a HVMA binder and provides the recommended high and low temperature performance evaluation index. Due to the small number of asphalt samples used in the test, we will test more samples in the follow-up study to study and analyze the more accurate rheological performance index of HVMA binders, and will provide the basis for determining the quality of high-viscosity modified binders.

Author Contributions: Conceptualization, X.L. and Z.H.; data curation, H.L. and R.W.; formal analysis, X.W. and Z.W.; writing—review and editing, D.W.; writing—original draft, P.L.; methodology, P.L. and A.C.F.; software, D.W. and W.Z.; validation, H.L. and X.W.; investigation, Z.W.; resources, R.W. and W.Z.; visualization, Z.H.; supervision, A.C.F.; project administration, X.L. All authors have read and agreed to the published version of the manuscript.

Funding: This research was funded by the Scientific and Technological Development Project of Guangxi Communications Investment Group Corporation Ltd., grant number 2021-001. We would like to thank Dongdong Yuan (Chang'an University, China) for his assistance on laboratory experiments.

Conflicts of Interest: The authors declare no conflict of interest.

References

1. Wang, D.; Baliello, A.; Poulikakos, L.; Vasconcelos, K.; Kakar, M.R.; Giancontieri, G.; Pasquiniet, E.; Porot, L.; Tušar, M.; Riccardi, C.; et al. Rheological properties of asphalt binder modified with waste polyethylene: An interlaboratory research from the RILEM TC WMR. *Resour. Conserv. Recycl.* **2022**, *186*, 106564. [CrossRef]
2. Yuan, D.; Jiang, W.; Xiao, J.; Tong, Z.; Jia, M.; Shan, J.; Ogbon, A.W. Assessment of the aging process of finished product–modified asphalt binder and its aging mechanism. *J. Mater. Civ. Eng.* **2022**, *34*, 04022174. [CrossRef]
3. Tong, Z.; Guo, H.; Gao, J.; Wang, Z. A novel method for multi-scale carbon fiber distribution characterization in cement-based composites. *Constr. Build. Mater.* **2019**, *218*, 40–52. [CrossRef]

4. Jia, M.; Sha, A.; Jiang, W.; Li, X.; Jiao, W. Developing a solid–solid phase change heat storage asphalt pavement material and its application as functional filler for cooling asphalt pavement. *Energy Build.* **2023**, *285*, 112935. [CrossRef]
5. Sha, A.; Liu, Z.; Jiang, W.; Qi, L.; Hu, L.; Jiao, W.; Barbieri, D.M. Advances and development trends in eco-friendly pavements. *J. Road Eng.* **2021**, *1*, 1–42. [CrossRef]
6. Yuan, D.; Jiang, W.; Hou, Y.; Xiao, J.; Ling, X.; Xing, C. Fractional derivative viscoelastic response of high-viscosity modified asphalt. *Constr. Build. Mater.* **2022**, *350*, 128915. [CrossRef]
7. Cai, J.; Wen, Y.; Wang, D.; Li, R.; Zhang, J.; Pei, J.; Xie, J. Investigation on the cohesion and adhesion behavior of high-viscosity asphalt binders by bonding tensile testing apparatus. *Constr. Build. Mater.* **2020**, *261*, 120011. [CrossRef]
8. Cai, J.; Song, C.; Zhou, B.; Tian, Y.; Li, R.; Zhang, J.; Pei, J. Investigation on high-viscosity asphalt binder for permeable asphalt concrete with waste materials. *J. Clean. Prod.* **2019**, *228*, 40–51. [CrossRef]
9. Büchler, S.; Falchetto, A.C.; Walther, A.; Riccardi, C.; Wang, D.; Wistuba, M.P. Wearing Course Mixtures Prepared with High Reclaimed Asphalt Pavement Content Modified by Rejuvenators. *Transp. Res. Rec.* **2018**, *2672*, 96–106. [CrossRef]
10. Hugener, M.; Wang, D.; Cannone Falchetto, A.; Porot, L.; Kara De Maeijer, P.; Tabatabaee, E.; Kawakami, A.; Hafko, B.; Grilli, A.; Pasquini, E.; et al. Recommendation of RILEM TC 264 RAP on the evaluation of asphalt recycling agents for hot mix asphalt. *Mater. Struct.* **2022**, *55*, 31. [CrossRef]
11. Gao, J.; Yao, Y.; Yang, J.; Song, L.; Xu, J.; He, L.; Tao, W. Migration behavior of reclaimed asphalt pavement mastic during hot mixing. *J. Clean. Prod.* **2022**, *376*, 134123. [CrossRef]
12. Jiang, W.; Yuan, D.; Shan, J.; Ye, W.; Lu, H.; Sha, A. Experimental study of the performance of porous ultra-thin asphalt overlay. *Int. J. Pavement Eng.* **2022**, *23*, 2049–2061. [CrossRef]
13. Jia, M.; Sha, A.; Lin, J.; Zhang, Z.; Qi, B.; Yuan, D. Polyurethane asphalt binder: A promising candidate for steel bridge deck-paving material. *Int. J. Pavement Eng.* **2022**, *23*, 3920–3929. [CrossRef]
14. Cao, Y.; Li, J.; Sha, A.; Liu, Z.; Zhang, F.; Li, X. A power-intensive piezoelectric energy harvester with efficient load utilization for road energy collection: Design, testing, and application. *J. Clean. Prod.* **2022**, *369*, 133287. [CrossRef]
15. Yuan, D.; Jiang, W.; Sha, A.; Xiao, J.; Wu, W.; Wang, T. Technology method and functional characteristics of road thermoelectric generator system based on Seebeck effect. *Appl. Energy* **2023**, *331*, 120459. [CrossRef]
16. Xinxin, G.; Zhang, C.; Cui, B.-X.; Wang, D.; Tsai, J. Analysis of Impact of Transverse Slope on Hydroplaning Risk Level. *Procedia Soc. Behav. Sci.* **2013**, *96*, 2310–2319.
17. Chen, Y.; Tan, Y.; Chen, K. Effect of TPS modifier on the properties of high viscosity asphalt. *J. Harbin Inst. Technol.* **2012**, *44*, 82–85.
18. Li, L.; Geng, H.; Sun, Y. Evaluation method and indicator for viscosity of high-viscosity asphalt. *J. Build. Mater.* **2010**, *13*, 352–356.
19. Tan, Y.; Haiyan, Z.; Dongwei, C.; Lei, X.; Rongjie, D.; Zhaoqiang, S.; Rui, D.; Xianhe, W. Study on cohesion and adhesion of high-viscosity modified asphalt. *Int. J. Transp. Sci. Technol.* **2019**, *8*, 394–402.
20. Li, M.; Zeng, F.; Xu, R.; Cao, D.; Li, J. Study on Compatibility and Rheological Properties of High-Viscosity Modified Asphalt Prepared from Low-Grade Asphalt. *Materials* **2019**, *12*, 3776. [CrossRef]
21. Yu, H.; Jin, Y.; Liang, X.; Dong, F. Preparation of Waste-LDPE/SBS Composite High-Viscosity Modifier and Its Effect on the Rheological Properties and Microstructure of Asphalt. *Polymers* **2022**, *14*, 3848. [CrossRef] [PubMed]
22. Guo, Y.; Ni, F. Dynamic viscoelastic properties of modified asphalt and mixture based on DSR. *J. Southeast Univ.* **2014**, *44*, 386–390.
23. Qin, X.; Zhu, S.; He, X.; Jiang, Y. High temperature properties of high viscosity asphalt based on rheological methods. *Constr. Build. Mater.* **2018**, *186*, 476–483. [CrossRef]
24. Zhang, F.; Hu, C. Preparation and properties of high viscosity modified asphalt. *Polym. Compos.* **2017**, *38*, 936–946. [CrossRef]
25. Zhang, F.; Hu, C.; Zhuang, W. The research for low-temperature rheological properties and structural characteristics of high-viscosity modified asphalt. *J. Therm. Anal. Calorim.* **2018**, *131*, 1025–1034. [CrossRef]
26. Yuan, D.; Jiang, W.; Xiao, J.; Lu, H.; Wu, W. Thermal-oxygen aging effects on viscoelastic properties of high viscosity modified asphalt. *J. Chang. Univ.* **2020**, *40*, 1–11.
27. Jiang, W.; Li, P.; Sha, A.; Li, Y.; Xiao, J.; Xing, C. Research on Pavement Traffic Load State Perception Based on the Piezoelectric Effect. *IEEE Trans. Intell. Transp. Syst.* **2023**, 1–15. [CrossRef]
28. Yuan, D.; Jiang, W.; Xiao, J. Comparison of rheological properties between SBS, rubber and high-viscosity modified asphalt binders. *J. Chang. Univ.* **2020**, *40*, 135–142.
29. *AASHTO T315*; Determining the Rheological Properties of Asphalt Binder Using a Dynamic Shear Rheometer (DSR). AASHTO: Washington, DC, USA, 2010.
30. *ASTM D7405-20*; Standard Test Method for Multiple Stress Creep Recovery (MSCR) Test of Asphalt Binder Using a Dynamic Shear Rheometer. American Association of State Transport and Officials: Washington, DC, USA, 2012.
31. Tong, Z.; Gao, J.; Zhang, H. Innovation for evaluating aggregate angularity based upon 3D convolutional neural network. *Constr. Build. Mater.* **2017**, *155*, 919–929. [CrossRef]
32. Sun, Y.; Wang, W.; Chen, J. Investigating impacts of warm-mix asphalt technologies and high reclaimed asphalt pavement binder content on rutting and fatigue performance of asphalt binder through MSCR and LAS tests. *J. Clean. Prod.* **2019**, *219*, 879–893. [CrossRef]
33. Jiang, W.; Yuan, D.; Xu, S.; Hu, H.; Xiao, J.; Sha, A.; Huang, Y. Energy harvesting from asphalt pavement using thermoelectric technology. *Appl. Energy* **2017**, *205*, 941–950. [CrossRef]

34. *AASHTO T313*; Determining the Flexural Creep Stiffness of Asphalt Binder Using the Bending Beam Rheometer (BBR). AASHTO: Washington, DC, USA, 2008.
35. Wang, W.; Jia, M.; Jiang, W.; Lou, B.; Jiao, W.; Yuan, D.; Li, X.; Liu, Z. High temperature property and modification mechanism of asphalt containing waste engine oil bottom. *Constr. Build. Mater.* **2020**, *261*, 119977. [CrossRef]
36. Lin, P.; Huang, W.; Tang, N.; Xiao, F.; Li, Y. Understanding the low temperature properties of Terminal Blend hybrid asphalt through chemical and thermal analysis methods. *Constr. Build. Mater.* **2018**, *169*, 543–552. [CrossRef]
37. Wu, W.; Jiang, W.; Yuan, D.; Lu, R.; Shan, J.; Xiao, J.; Ogbon, A.W. A review of asphalt-filler interaction: Mechanisms, evaluation methods, and influencing factors. *Constr. Build. Mater.* **2021**, *299*, 124279. [CrossRef]
38. Xing, C.; Jiang, W.; Li, M.; Wang, M.; Xiao, J.; Xu, Z. Application of atomic force microscopy in bitumen materials at the nanoscale: A review. *Constr. Build. Mater.* **2022**, *342*, 128059. [CrossRef]

buildings

MDPI

Article

Laboratory and Full-Scale Testbed Study in the Feasibility of Styrene-Butadiene-Styrene Asphalt Pavement Having Epoxy Resin and Crumb Rubber Powder

Sang-Yum Lee [1] and Tri Ho Minh Le [2,*]

[1] Faculty of Civil Engineering, Induk University, 12 Choansan-ro, Nowon-gu, Seoul 01878, Republic of Korea
[2] Faculty of Civil Engineering, Nguyen Tat Thanh University, 300A Nguyen Tat Thanh Street, District 4, Ho Chi Minh City 70000, Vietnam
* Correspondence: lhmtri@ntt.edu.vn

Citation: Lee, S.-Y.; Le, T.H.M. Laboratory and Full-Scale Testbed Study in the Feasibility of Styrene-Butadiene-Styrene Asphalt Pavement Having Epoxy Resin and Crumb Rubber Powder. *Buildings* **2023**, *13*, 652. https://doi.org/10.3390/buildings13030652

Academic Editors: Andrea Baliello and Di Wang

Received: 8 February 2023
Revised: 27 February 2023
Accepted: 27 February 2023
Published: 28 February 2023

Abstract: Conventional asphalt concrete pavements have deteriorated rapidly due to the current increased traffic and extreme climate impacts. In addition to the upgrading in the construction quality, there is an urgent need to expand the utilization of modified asphalt binders to improve road capacity and traffic safety. The proposed research aims to combine epoxy resin (ER) and crumb rubber powder (CRP) contents into conventional Styrene-butadiene-styrene (SBS)-modified asphalt binder to not only reduce the consumption of normal asphalt binder but also promote the usage of recycled waste material in practice. To cope with this research objective, the ER and CRP were designed at 3% and 5% by weight of asphalt binder, respectively. Various laboratory tests were performed to evaluate the performance of modified mixtures (ERCRP), including the Frequency Sweep Test, Multiple Stressed Creep and Recovery, Dynamic Modulus, Semi-Circular Bending (SCB), and Cantabro Durability Tests. Additionally, an assessment of the modified asphalt concrete pavement via field testbed was conducted through Falling Weight Deflectometer and Ground Penetrating Radar. Overall, by adding the ER and CRP, the strain value of the control reference mix can be reduced up to 31.8% and 28.3% at MSCR 0.1 and 3.1 kPa, respectively. Additionally, the dynamic modulus of the ERCRP-modified samples was approximately 32,267 and 189 MPa, while the value of the reference mixture was 28,730 and 105 MPa at the highest and lowest frequency, respectively, indicating an enhancement under repeated loads. Regarding the SCB test results at 0 °C, the peak stress of the ERCRP-modified mixture was 4.75 MPa, while the value of the reference specimens was only 4.2 MPa, noticing the improved stress-bearing capacity. Based on a full-scale testbed, the FLWD elastic modulus of reinforced pavement shows a novel improvement (6.75%) compared with the control pavement, suggesting a potential application of ERCRP-modified asphalt binder for sustainable development purposes.

Keywords: sustainable development; full-scale testbed; modifier asphalt binder; crumb rubber powder; epoxy resin; SBS asphalt pavement

1. Introduction

Over the past decades, the physical properties of road pavement have rapidly deteriorated due to increased traffic and environmental impacts [1,2]. Multiple strategies have been proposed to increase road capacity as well as traffic safety. In addition to the enhancement in construction technologies, the modification of asphalt binders has gained attention recently since this coating material can greatly contribute to the durability and workability of the binder [3]. To cope with this objective, synthetic polymers have been developed to represent the efficiency modifiers in asphalt binders [4]. These materials are often mixed into the asphalt at a weight ratio of 3–8% to improve the physical properties of the asphalt that is used in road pavement [5]. Modified asphalt properties have been altered to improve pavement resistance to damage, such as plastic deformation and thermal cracking [6]. Asphalt modifiers can provide improved resistance performance to the aforementioned damage by mixing the modifier into the asphalt during production (wet-mix) and dry mix type, which introduces the modifier together with the asphalt when producing the HMA (Hot Mix Asphalt) [5].

Asphalt modification that uses polymers is intended to reduce pavement damage, improve road durability, and reduce road management costs. However, not all polymers can be used as asphalt modifiers [7]. The most important requirement is that the polymer has compatibility with asphalt [6]. Therefore, when the polymer is mixed with the asphalt, it should exhibit a single phase without phase separation. In general, if two materials have similar structures or similar polarity, they mix well with each other and become a single-phase mixture [8]. Compatibility equates to storage ability, and it indicates stability. Here, the polymer becomes swollen as it is partially surrounded by maltene, an oily aromatic compound of asphalt.

Polymers that are used as modifiers are divided into three types. They are thermoplastic elastomers, thermoplastics, and reactive polymers. The most typical thermoplastic elastomer (TPE) is the styrene-butadiene-styrene triblock copolymer (SBS), and another typical TPE is the styrene-isoprene (2-methyl butadiene)-styrene triblock copolymer (SIS) which is similar to SBS [9]. These have a biphasic structure in which a crystallized styrene portion is connected to an amorphous butadiene region and overall, it forms a network structure [10]. When these polymers are mixed with asphalt by heat and shear force, the butadiene portion swells due to the asphalt's maltene, and the modified asphalt retains this network structure even when it cools [10]. Therefore, the characteristic elasticity of these polymers is transferred to the modified asphalt, and the asphalt exhibits excellent elastic recovery and permanent deformation resistance [11]. However, these modification methods have two main disadvantages. One is that solid rubber must be dissolved by heat in order to mix the rubber polymers with the asphalt. To do this, high heat is required for a long period of time, and a strong shear force is required during mixing because the dissolved rubber is highly viscous, and this leads to high costs [10,12]. The other disadvantage is the C=C double bond that exists in butadiene. The double bond is vulnerable to heat and oxygen in the air and is easily broken, which reduces the durability of the modified asphalt.

The second type of polymer that is used as a modifier is thermoplastic. PE (polyethylene) and EVA (ethylene-vinyl acetate copolymer) are typical thermoplastics [13]. The advantage of these polymers is that they are easy to purchase and low-cost, and they are particularly environmentally friendly because recycled products can be used. In the case of these modifiers, the high stiffness and deformation resistance that is characteristic of plastic polymers is transferred to the asphalt [3]. However, because PE is highly crystalline, non-polar, and amorphous, it does not mix well with asphalt, which is polar. Therefore, it will float to the top of the asphalt if left alone due to its difference in density (0.94 g/cm^3) [13]. This reduces the storage stability of the modified asphalt [14]. To make it mix well with asphalt, an acetate group (CH3COO) is used to reduce crystallinity, an ester group is used to increase polarity slightly, and the result is EVA [15].

The third type of polymer that is used as a modifier is a reactive polymer. Elvaloy and Lotader are typical reactive polymers. These are ethylene–butyl acrylate–glycidyl methacrylate terpolymers, and their polarity is increased by the acrylic group [16]. Furthermore, the glycidyl group (CH2OCH2-CH2-) reacts with the carboxyl group to form a covalent bond so that they mix well with asphalt [17]. However, this reaction creates an asphalt gel and prevents the formation of a network structure in the modified asphalt. Therefore, these polymers are not suitable as asphalt modifiers, and in practice, they are mainly used as adhesives in the plastics industry [18].

Recent studies have estimated that 2 billion tires are disposed of worldwide each year [19]. This accounts for two to three percent of the entire volume of industrial waste. Sales of vehicles are rising along with the world's population, especially in developing countries with more accessibility to vehicles [8]. Further, additional tires must be replaced after prolonged exposure to commuting increases the volume of discarded tires [20]. As a result, the rapid increase in the number of discarded tires imposes a significant hazard to the environment. Despite resource conservation efforts and government controls, tires continue to be treated improperly. Once they are stored in landfills, tires tend to leach contaminants into the atmosphere, ground, and water, which can cause environmental disruption. Carbon dioxide can be released into the air when used tires are subjected to the sun. This carbon dioxide causes an increase in air pollution, which could cause climate change as well [20].

The utilization of used tires in practical application is proven to be an increasingly viable remedy to the problem [21]. Although only a minimal amount of waste tires is reclaimed each year, recycling waste tires has emerged as the primary method for the handling of this material in the EU, where it represents about 40 percent of the total of all waste tires. As a result, the consumption of recycled tires would considerably enhance the goals of green construction in the world. As a viable reinforced component in the asphalt pavement industry, discarded tire rubber powder has been utilized as a vital additive for hot asphalt mixtures [21]. The benefits of Crumb Rubber Powder (CRP) asphalt include decreased contamination issues, stronger bearing capacity, and resilience to permanent distortion. In contrast to conventional bitumen, CRP asphalt has increased viscosity in hot regions and reduced rigidity in cold areas [22]. Reduced rigidity at low temperatures can contribute to the high capacity to prevent thermal cracking, while the high viscosity property can provide excellent resilience to rutting imposed by transportation volume.

Although various researchers have examined the application of modifiers in asphalt binder, there is a lack of research investigating the combined effect of epoxy resin and crumb rubber powder in SBS-modified asphalt. The full-scale applicability of the SBS-modified asphalt using both ER and CRP also shows limited findings. Therefore, the proposed research aims to determine the adaptability of mixing ER and CRP into polymer-modified asphalt in both laboratory and full-scale testbed experiments to promote the consumption of by-product material in practice as well as ensure the durability of the conventional pavement structure. Regarding the mix design overview, the CRP and ER were substituted 5% and 3% by weight of SBS-modified asphalt binder, respectively. Afterward, the proposed research employed a variety of testing methods to evaluate the physical properties of modified asphalt containing the Frequency Sweep Test, Multiple Stressed Creep and Recovery, Dynamic Modulus, Semi Circular Bending, and Cantabro Durability Tests. Further, a long-term performance investigation in the field testbed was also conducted by using Falling Weight Deflectometer and Ground Penetrating Radar. In the evaluations, conventional modified asphalts were used for a comparative evaluation of the developed material. The general summary of the research is presented in Figure 1.

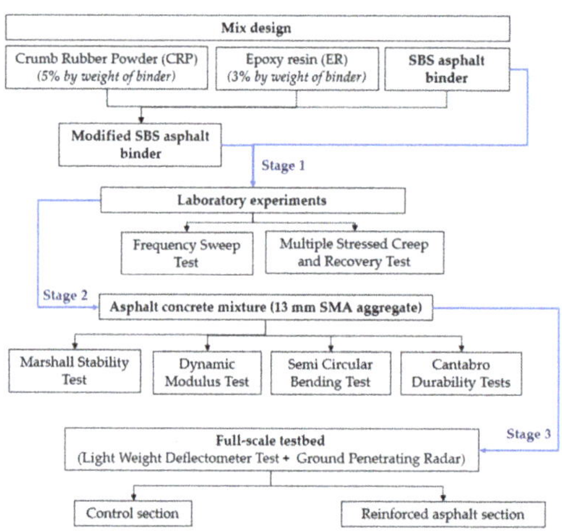

Figure 1. Research flowchart.

2. Materials and Methods

2.1. Aggregates and Asphalt Binder

2.1.1. Aggregates

The aggregate and fillers for this research were provided by a Korean Company, and the basic properties of the material are shown in Table 1. Following previous studies and suggestions of relevant studies, the ideal asphalt binder constitutes 5–6% of the total weight [17,18], and this work developed samples using the Superpave technique. The goal of Superpave was to propose a technique that would maximize the asphalt mixture's durability against rutting, fatigue cracking, and thermal cracking while making the optimum utilization of bitumen pavement methods. The Superpave Gyratory Compactor (SGC) determines the volumetric and compression qualities of the materials, and the data collected by the SGC throughout compression can be used to generate insight into the workability of the specific combination [23].

Table 1. Aggregate and mineral filler properties.

Materials	Properties	Value
Aggregate	Relative apparent density [24]	2.73
	Water absorption [24]	0.182%
	Aggregate crushed value [25]	17.7%
	Los Angeles abrasion value [26]	25.8%
	Flakiness and elongation index [27]	12.8%
Mineral Filler	Relative apparent density [28]	2.31
	Moisture content [28]	0.08%

The development of a polymer-modified asphalt binder mixture (PG 76-22) was performed using a 13 mm SMA aggregate type. Since the stress from the vehicle load is borne entirely by the aggregate in the stone mastic asphalt (SMA), the usage of this aggregate type can counter the binder delamination (desorption) based on the adhesive force between the aggregate [29] in the conventional asphalt mixture. The synthetic particle sizes distribution of the aggregate that was created in this test is summarized in the following Table 2.

Table 2. Mix design for 13 mm SMA mixture design particle size table.

	Corrected Mixing Ratio (%)				
	Synthesis Rate (%)				
Sieve size	14–20 mm	11–14 mm	7–11 mm	Fine Aggregate	Filler
	-	59.0	17.0	13.0	11.0
25 mm	-	-	-	-	-
20 mm	-	100.0	100.0	100.0	100.0
13 mm	-	96.4	100.0	100.0	100.0
10 mm	-	18.3	97.6	100.0	100.0
5 mm	-	1.3	1.7	95.1	100.0
2.5 mm	-	0.6	0.7	69.5	100.0
0.6 mm	-	0.1	0.1	37.5	100.0
>0.3 mm	-	0.0	0.0	27.7	97.2
0.15 mm				18.8	92.2
0.08 mm				10.0	73.0

2.1.2. Asphalt Binder

Styrene-butadiene-styrene (SBS) was employed to strengthen the primary asphalt binders utilized in the proposed approach. SBS is well recognized for being flexible in ambient conditions and having the capability to boost a binder's elasticity. This SBS compound can increase the performance of a mixture and the applicability of bitumen by lowering the mixing energy and the stiffness of asphalt.

Regarding the epoxy resin, bisphenol A diglycidyl ether was employed as the epoxy resin in this research. The accompanied curing agents were also provided by the equivalent Korean chemical company. The ER has the Epoxy Equivalent (g/mol) of 193, while the unhydrolytic chlorine (%) and hydrolyzable chlorine (%) were 0.011 and 0.1, respectively. The epoxy solution to curing ingredient mass distribution is 100:65 based on [30].

As for the CRP, it was produced via a mechanochemical process with enabled CRP having a particle diameter of 30 mesh. A synthetic ingredient (organic disulfide) was applied using mechanical power to speed up the reaction process [31]. Regarding the previous investigation, the activated technique conditions of the mechanochemical method that uses the organic disulfide ingredient (5% of CRP by wt.) were modified [32]. According to the preliminary experimental results, blending SBS and CRP produces a substance with very low workability. Accordingly, in order to produce a uniform solution and achieve the desired air void ratio, the heat of the production processes was tightly regulated. Based on the trial works, the combining procedure was controlled at 165 °C for 40 min using a 3% organic disulfide content.

2.1.3. Mixture Design

Following the mixing process, the normal SBS asphalt binder was conditioned at 165 °C for one hour to a melted condition before adding ER and CRP. SBS asphalt binder was blended with CRP and ER; then the mixture was stirred for 5 min at a speed of 1000 rpm. To achieve a homogeneous combination of the SBS-ER-CRP, the curing agent was introduced to the aforementioned combination and swirled for an additional 5 min at a rate of 1500 rpm. The fundamental characteristics of asphalt binders treated with ERCRP are presented in Table 3, and the general test results of the ERCRP mixture are shown in Table 4.

Table 3. Properties of asphalt binder.

Properties	Value
Penetration (1/10 mm) 25 °C [33]	85.8
Softening point (°C) [34]	68.7
Ductility at 5 °C (cm/min) [35]	99.3
$G^*/\sin\delta$; (Original) [36]	1.71 kPa
$G^*/\sin\delta$ (after RTFO) [36]	2.43 kPa
$G^* \times \sin\delta$ (after PAV) [37]	1501 kPa
Stiffness [38]	182 MPa
m-value [38]	0.32

Table 4. General property test results of the modified mixture.

	Asphalt Content (%)	Porosity (%)	Saturation (%)	Marshall Stability (N)	Flow Value(1/100)	Aggregate Porosity (%)	Effective Density	Theoretical Maximum Density (g/cm^3)
Standard	-	3~6	65~80	>7500	20~40	>13		
Average	5.3%	3.8	75.5	12,773	37.4	15.36	2.710	2.508

2.2. *Binder Tests*

2.2.1. Frequency Sweep Test

The Frequency Sweep Test (FST) is a very convenient test that can evaluate the rheological properties of a viscoelastic material according to time and temperature, and it is useful for checking the basic behavior of binders over a wide range of temperatures. In this study, FST was performed on the polymer-modified asphalt and the control mixture to create a master curve of the dynamic shear modulus, which can be used to predict the behavior of binders. The dynamic shear master curve can be used to compare the physical properties of asphalt binders, and it can be used as basic data for estimating the physical properties of asphalt mixtures in the future [39].

In this study, FST was conducted based on a dynamic shear rheometer based on ASTM D7552-22 [40]. Dynamic shear tests that use a dynamic shear rheometer (DSR) are generally used to analyze the viscous and elastic behavior characteristics of the asphalt binder that is in use, and it is possible to understand the physical property changes in the asphalt at the pavement's usage temperature by measuring the complex shear modulus (G*) and the phase angle (°) [41]. Methods for analyzing the rheological properties of asphalt using a DSR include the temperature sweep test (TST), which measures rheological behavior properties according to temperature, and the strain sweep test (SST), which measures the linear viscoelastic region of the asphalt and selects a strain rate [42]. These tests were performed at a strain level of 1.0% which was determined by the SST to create a master curve of the dynamic shear modulus, and tests were performed at 12 temperatures at 7 °C intervals from 13 °C to 90 °C to obtain the dynamic shear modulus for a wide range of temperatures. In the frequency sweep test, this research follows the popular standards as shown in related research [43,44]. Additionally, the number of testing temperatures was increased compared with the conventional method to better develop the dynamic shear modulus and quantify the behavior of the modified binder under various testing temperatures. As for the size of the binder specimen, a specimen with a thickness of 2 mm and a diameter of 8 mm was used at 48 °C and below, and a specimen with a thickness of 1 mm and a diameter of 25 mm was used at 48 °C and above to reduce error following ASTM standards [40]. The tests used a load control test method that repeatedly applies loads at varying frequencies within multiple temperature ranges, and the loading frequencies are outlined in Table 5.

Table 5. Each load cycle (Time Temperature Superposition).

Angular Frequency (rad/s)					
6.28	9.96	15.75	25.01	39.64	62.83

2.2.2. Multiple Stressed Creep and Recovery (MSCR) Test

The MSCR test is a testing method that was recently presented in ASTM D7405-10a [45] and AASHTO TP70 [46] for evaluating the high-temperature elastic recovery of asphalt binders, and it is mainly used to evaluate the high-temperature elastic recovery of asphalt binders. The existing ASTM D6373 [47] binder selection method recommends using a binder grade that is one or two steps above the grade determined by DSR tests if there is a high traffic volume or a low vehicle traveling speed in the area where the binder will be used. However, this method has a problem since it may lead to the use of unnecessary high-cost modified binders because this approach is used indiscriminately even when the area's maximum temperature is relatively low.

The MSCR test has the advantage of adequately reflecting the elastic and viscoelastic properties for evaluating the plastic resistance of asphalt binders by considering material properties in a relatively large strain rate range, and it performs evaluations after simulating the stress conditions that the mixture may experience due to the actual load. The loads that were used in MSCR were 0.1 kPa and 3.2 kPa, as suggested by ASTM D7405-10a [45] and AASHTO TP70 [46]. The 0.1 KPA load was applied for 1 s followed by a 9-s rest, and this process was repeated 10 times. Then, the load was increased to 3.2 kPa, and the process was repeated in the same way. Table 6 summarizes the load conditions that include the 2 load levels suggested by ASTM D7405-10a and AASHTO TP70, respectively. Figure 2 shows the stress curve and the resulting strain rate curve when loads of a certain size were applied and then a rest period was allowed. Figure 2A shows the applied shear stress, and Figure 2B shows the recoverable viscoelastic strain rate and the non-recoverable plastic deformation ($E_{recoverable}$ and $\varepsilon_{irrecoverable}$) that can occur due to shear stress, respectively [39].

Table 6. MSCR test process.

Number of Loads	**1**	**2**	**3**	**4**	**5**	**6**	**7**	**8**	**9**	**10**
Creep (kPa)	**0.1**	**0.2**	**0.4**	**0.8**	**1.6**	**3.2**	**6.4**	**12.8**	**25.6**	**61.2**
Time and Repetition	▪ Creep load time (seconds) = 1.0 ▪ Recovery time (seconds) = 9.0 ▪ Number of creep and recovery iterations = 10									

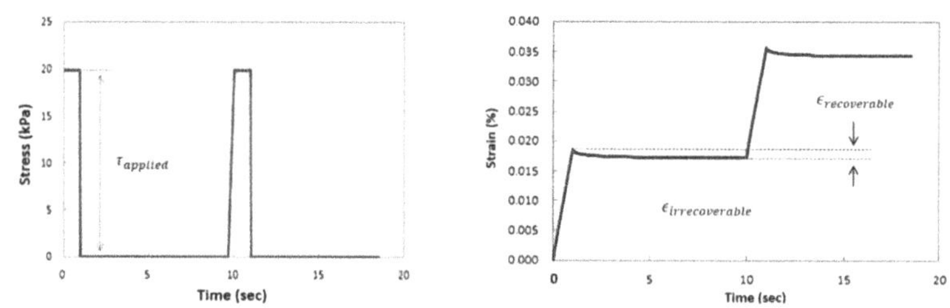

(**A**) (**B**)

Figure 2. (**A**) Example of stress curve of MSCR test; (**B**) Example of strain curve of MSCR test.

The MSCR test applies loads of a fixed level on specimens and then allows them for rest periods. The non-recoverable compliance J_{nr} can be obtained by dividing the unrecovered plastic strain rate by the level of the applied shear load, as shown in Equation (1) below. Generally, as the level of the load increases, the J_{nr} value increases, and at the same load, J_{nr} tends to increase as the binder's plastic deformation resistance decreases. Therefore, the maximum value of J_{nr} is suggested as the evaluation standard for the binder, as shown in the following Equation [45].

$$J_{nr} = \frac{\varepsilon_{irrecoverable}}{\tau_{applied}} \quad (1)$$

where J_{nr} is the non-recoverable compliance, $\varepsilon_{irrecoverable}$ is the non-recoverable shear strain rate, and $\tau_{applied}$ is the shear load.

In addition, the recoverable strain rate and the non-recoverable strain rate defined in the equation above can be used to determine the recovery percentage that is shown in Equation (2) below [45]. This can be used to effectively distinguish the elastic recovery properties of two different binders in which the same maximum strain rate occurs at the same load, and it can be used to effectively evaluate the properties of modified binders.

$$\text{Recovery}(\%) = \frac{\varepsilon_{recoverable}}{\varepsilon_{peak}} \quad (2)$$

where, recovery (%) is the recovery rate; $\varepsilon_{recoverable}$ is the recoverable shear strain rate; ε_{peak} is the maximum shear strain rate.

2.3. *Asphalt Concrete Mixture Tests*

2.3.1. Dynamic Modulus Test

The dynamic modulus can depict a variety of traffic conditions using various temperature conditions, loads, and speeds, and it changes according to the asphalt binder and the particle size of the aggregate that is used. It can be considered to be a physical property evaluation method that can depict the viscoelastic properties of asphalt mixtures particularly well. The viscoelastic behavior properties of asphalt mixtures can be identified based on the correlation between the deformation properties and the load that is measured when a sine-wave load is applied continuously without a rest period, and the parameter that is used here can be defined as the complex modulus (E*). The dynamic modulus values that are determined by the tests can be obtained based on the combination of test temperature and load. The master curve of the dynamic modulus and the transition function was calculated by using the superposition principle for the previously mentioned load time and temperature to the dynamic modulus values. The method that is recommended by AASHTO TP 62 [48] uses the viscosity properties of the asphalt binder, and it is called the viscosity-temperature susceptibility method. It uses the slope of the viscosity value according to the changes in the absolute temperature on a logarithmic graph. The relationship between viscosity and temperature susceptibility can be expressed in Equation (3) below [49].

$$\log(\log\eta) = \mathrm{A} + \mathrm{VTS}[\log(\mathrm{T_R})] \tag{3}$$

Here, η is the viscosity (c, Poise), T_R is the temperature(°R), A is the viscosity-temperature susceptibility curve intercept, and VTS is the slope of the relationship between viscosity and temperature susceptibility.

The transition function is developed by using viscosity-temperature susceptibility (VTS) to divide the viscosity of the center temperature by the viscosity of each temperature, as shown in Equation (4) below [49].

$$\log(\mathrm{T}) = \mathrm{c}\left(10^{\mathrm{A}+\mathrm{VTS}[\log(\mathrm{T_r})]} - 10^{\mathrm{A}+\mathrm{VTS}[\log(\mathrm{T_R})_0]}\right) \tag{4}$$

Here, c is a constant, T_R is the temperature to be converted (°R), and $(T_R)_0$ is the reference temperature (°R).

This study performed tests based on the AASHTO TP 62 [48], which was created by the United States FHWA, to evaluate asphalt mixtures that were created using ERCRP-modified binder and normal SBS-modified binder. A super-pave gyratory compactor (SGC) was used to create test specimens with a height of 178 mm and a diameter of 150 mm. Then, core work was performed to extract final samples having a size of 150 mm × 100 mm (Figure 3A). For the dynamic modulus tests, a UTM-25 universal testing machine was used (Figure 3B). The test temperature was set from −10 °C to 54 °C to understand the mixtures' physical properties at a variety of temperatures, and the loading frequency was divided into 6 levels from 25 Hz to 0.1 Hz. As in the aforementioned FST tests, a time-temperature shift factor was calculated, and the Excel program's Solver was used to fit the other elastic coefficient values horizontally based on a value of 20 °C. Ultimately, the master curve was acquired with an S-shaped sigmoidal function.

(A) (B)

Figure 3. (**A**) Specimen manufacturing process and (**B**) Process of measuring dynamic modulus.

2.3.2. Semi-Circular Bending (SCB) Test

In particular, fatigue cracking, which is one of several types of pavement damage, can quickly degrade pavement structures and drastically reduce driving quality once it occurs. Therefore, various test methods have been introduced to create mixtures that can effectively resist this kind of damage. The SCB test is similar to the indirect tensile test (IDT), which is an existing tensile failure test. Based on ASTM D8044-16 [50], the SCB tests used ERCRP-modified SBS binder and normal SBS-modified asphalt binder to fabricate the mixture. For the mixture, the tests used a normal-density 13 mm SMA that was created by the mix design process previously mentioned. For the test specimens, specimens with a diameter of 150 mm and a thickness of 50 mm were created using the gyratory compactor and cut into semi-circular shapes. Notches with a thickness of 1.5 and a length of 15 mm were added to the middle part of the specimens (Figure 4A). The tests were performed using a UTM-25 universal testing machine (Figure 4B), and the test temperature was set at −12 °C and −24 °C while the loading speed was set at 50 mm/min. The reaction force at the load point was monitored in real time by the data acquisition system in the UTM.

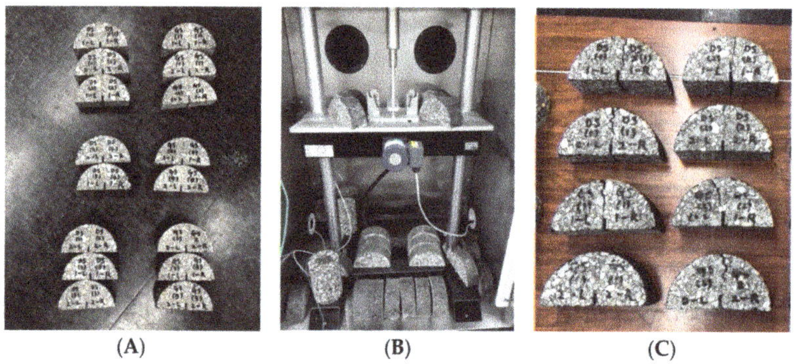

Figure 4. (**A**) SCE Specimens before testing, (**B**) SCB testing, and (**C**) Fractured specimens.

2.3.3. Cantabro Test

The cohesion of the asphalt binder is considered a very important element in allowing mixtures to resist damage such as potholes. As such, this study performed Cantabro tests using a Los Angeles abrasion loss tester (Figure 5A) to evaluate the aggregate shatter resistance of mixtures that use a modified binder based on the suggestion from ASTM D7064 [51] and Cox et al. [52]. In the Cantabro tests, Marshall test specimens were placed in a Los Angeles abrasion loss tester with the ball bearings removed, and the drum was rotated 30 times per minute following the KS F 2492 [53]. The mass loss ratio of the specimens after 300 rotations was used. The loss ratio was calculated using the following Equation (5) [53].

$$\text{Loss}(\%) = \frac{A - B}{A} \times 100 \quad (5)$$

Here, A: the mass of the specimen before the test (g), B: the mass of the specimen after the test (g).

Table 8. LFWD field test conditions.

LFWD Test Conditions			
Diameter of the Load Plate	**The Weight of the Load**	**Load Height**	**Poisson's Ratio**
150 mm	15 kg	27 inch	0.4

2.4.2. GPR (Ground Penetrating Radar) Exploration

GPR is an abbreviation of Ground Penetrating Radar, and it can be interpreted as radar that penetrates the ground following ASTM D4748-10 [55]. GPR survey is an electromagnetic wave survey that identifies the underground structure using electroporation pulses in the 10 MHz to several GHz frequency band. As a result, as many constructions are recently requiring large-scale and rapid construction, and the special law on underground safety management is enacted, the utilization of underground penetrating radar exploration, such as ground investigation, is gradually increasing. The GPR survey method is a physical exploration method that obtains underground information by propagating electromagnetic waves underground in the form of short-width pulses from an antenna and then receiving and analyzing electromagnetic waves reflected from the boundary of media with different physical properties during the propagation path.

The antenna frequency used in GPR exploration is proportional to the resolution and inversely proportional to the penetrating depth. The higher the frequency, the smaller the object can be detected, but the penetration depth of the wave becomes relatively shallow. On the other hand, the lower the frequency, the deeper the penetration depth, but the lower the ability to detect objects. Therefore, the size of the object to be searched for and the expected burial depth must be sufficiently investigated in advance to select an appropriate antenna. The minimum size of a detectable object is called 'resolution,' which varies depending on the soil and refers to half the length of one wavelength. Propagation speed differs depending on each medium, and since the site is an asphalt paved road, a general average propagation speed of 110 mm/ns was applied, and 3 cases in the longitudinal direction (50 m each) and 3 cases in the lateral direction (10 m each) were applied.

3. Results

3.1. Binder Tests

3.1.1. Frequency Sweep Test

In the asphalt mixture that is normally used in road pavement, low-temperature cracking due to tensile stress mainly occurs at temperatures of 0 °C and below, and as the temperature increases to 50 °C or above, plastic deformation or permanent deformation failure modes occur more often due to shear load. This is because the asphalt binder has the properties of a typical viscoelastic material in which the elastic modulus is high at low temperatures and low at high temperatures. To predict these results, a master curve of the dynamic shear modulus was created using TTS, which is a theory in which time and temperature variables related to viscoelastic material are expressed as a single curve. FST tests were conducted on the reference and the ERCRP-modified binder by performing a series of processes, and the results are shown in Figure 9 below. In general, the FST test shows a relatively identical shape between the two mixtures. Both mixtures share the common behavior of asphalt mixture since lower temperature results in a greater shear load value. Therefore, Figure 10 was conducted to further investigate the difference in mixtures.

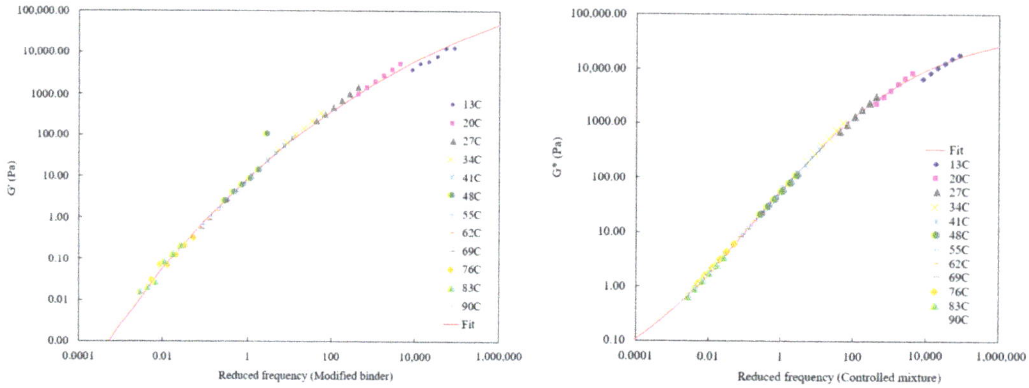

Figure 9. ERCRP-modified binder test results and controlled binder test results.

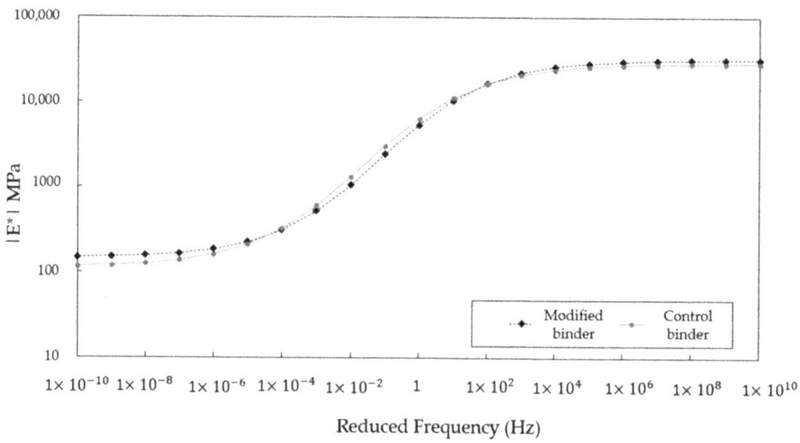

Figure 10. Dynamic modulus of test binder.

Based on the overlapping comparison of the normal SBS binder and the ERCRP-modified binder test results, although the ERCRP-modified binder slightly outperforms the control mixture at both critical stages (very slow or very high frequency), there was a neglectable performance difference concerning resistance to plastic deformation at high temperatures and resistance to temperature cracking at low temperatures.

3.1.2. MSCR Test

As mentioned before, the loads that were used in MSCR were 0.1 kPa and 3.2 kPa, as suggested by ASTM D7405-10a [45]. The 0.1 KPA load was applied for 1 s followed by a 9-s rest, and this process was repeated 10 times. Then, the load was increased to 3.2 kPa, and the process was repeated in the same way. In the MSCR test, the binder is exposed to increased amounts of stress and strain, thereby precisely simulating what would happen in a practical pavement. The behavior of the asphalt mixture in the MSCR test covers not only the hardening properties of the binder but also the elasticity through the use of greater degrees of stress in this test. Figure 11A,B present the MSCR test results of the modified and control SBS asphalt binder under the 0.1 and 3.2 KPa conditions, respectively. It can be seen that the cumulative strain rate due to the creep load was markedly lower in the ERCRP-modified binder than in the reference binder. In addition, the results of the comparison with the reference binders showed that the strain recovery slope of the ERCRP-modified binder was superior regardless of the creep load level. At both MSCR levels, the control binder shows a steep increase in strain while this magnitude was minor, as shown in the modified mix. For example, at MSCR of 0.1 kPa and 190 s, the strain value of the reference and the modified mix was 38.81 and 29.43, respectively. Meanwhile, the value of MSCR 3.2 kPa and 190 s was 808.98 and 630.49 in the former and latter binders, respectively.

This confirms the modifying effect of the CRP into the asphalt binder, which may add value to the stiffness of the mix. Considering the non-recoverable compliance at a shear creep load of 3.2 kPa, the MSCR test results indicate that both binder options share the equivalent shape at an early age ranging from 100 to 120 s. However, after this stage, the accumulated strain increases significantly in the reference binder, and the longer testing time leads to a higher gap between the two binders. As a result, the findings confirm that the performance of the ERCRP-modified binder was considerably improved compared with the normal binder. The addition of CRP into the mixture not only increases the resistance to deformation but also provides elastic behavior to the binder.

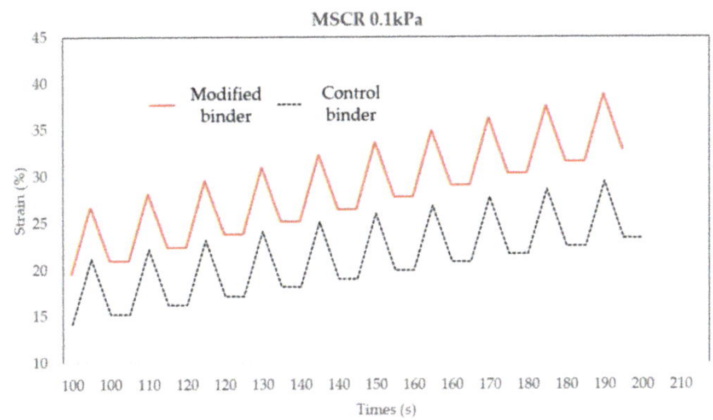

(**A**)

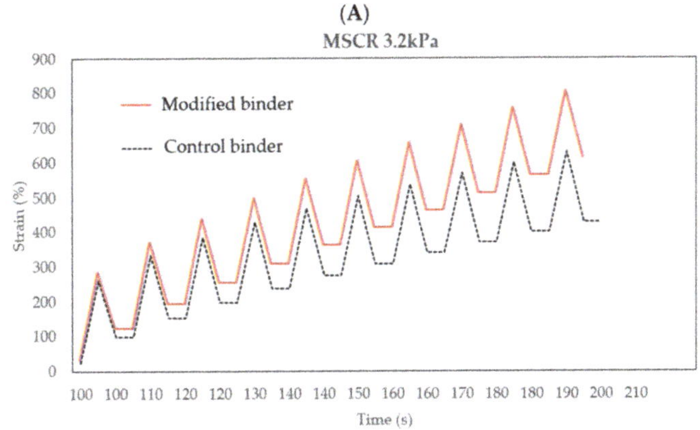

(**B**)

Figure 11. (**A**) MSCR test result (0.1 kPa) and (**B**) MSCR test result (3.2 kPa).

Additionally, the recovery (%) in Figure 12 exhibits the median elastic deformation values of the binders developed through this research. It can be observed that both SBSs mixtures show potential resistance against cracking by providing sufficient elastic recovery. The non-recovery (%) on the right side of Figure 12 indicates plastic deformation properties, and the findings notice that the modified binder had lower values than the conventional binder, suggesting a great plastic deformation.

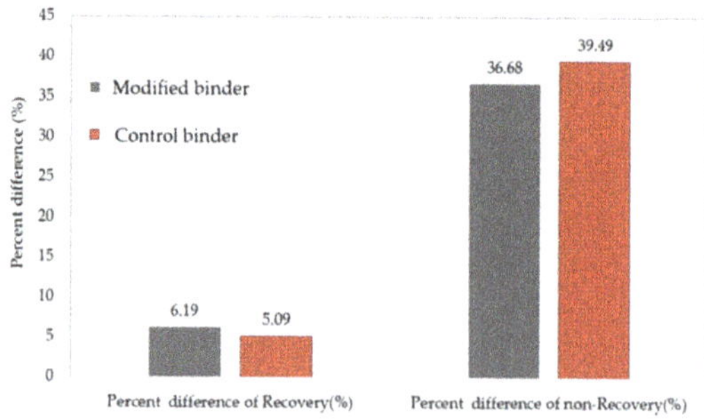

Figure 12. Recovery part (%) and non-recovery part (%).

3.2. *Asphalt Concrete Mixture Test Results*

3.2.1. Dynamic Modulus Test

As depicted in Figure 13 below, the findings suggest that the ERCRP-modified and control SBS mixtures exhibited relatively similar dynamic modulus curves for all load frequencies. However, after further investigation into the critical frequency range, the ERCRP-modified specimens showed higher dynamic modulus values than the control mix in low-temperature and high-temperature regions, but they showed similar behavior in normal medium temperatures. For instance, at the highest and lowest frequency, the dynamic modulus of the ERCRP-modified samples was approximately 32,267 and 189 MPa, while the value of the reference mixture was 28,730 and 105 MPa, respectively. This can be explained by the reinforcement effect of proper CRP content into the binder, which results in the stiffening effect of the mixture and, thereby, cultivating stronger resistance under repeated loads.

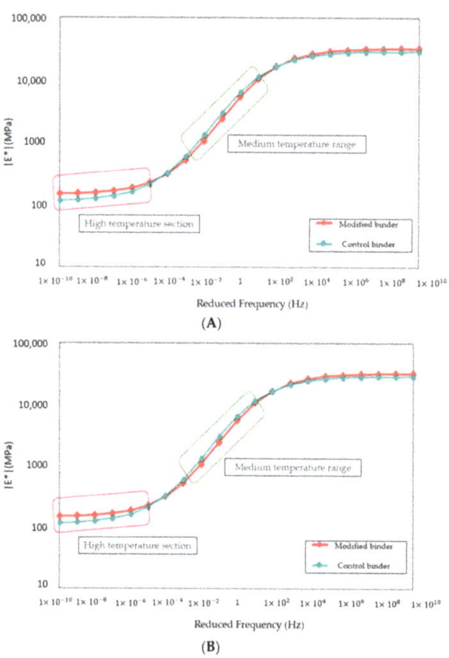

Figure 13. Master curve (HMA) of dynamic modulus test results: (**A**) log-log scale (high-temperature region) and (**B**) semi-log scale (low-temperature region).

3.2.2. SCB Test

Figure 14A,B demonstrate the results of SCB tests at −0 °C and −12 °C, respectively. The findings reveal that the vertical load applied to each specimen increased according to the gradual increase in displacement, which was controlled by an actuator. Once the specimens reached a fractured state at the maximum load value, the load value tended to decrease gradually until complete failure. The fracture behavior was different between testing conditions, and this is attributable to the hardened effect in the internal asphalt concrete mixture. The conditioned specimens at a lower temperature can achieve greater load, but abrupt fracture failure could be easily observed. At SCB tests conducted at −0 °C and −12 °C, the peak stress of the ERCRP-modified mixture was 4.75 and 5.4 MPa, respectively, while the value of the control mixture was only 4.2 and 5.3. Furthermore, the addition of CRP also contributes to the ductile behavior of asphalt mixture under applied load since the maximum displacement of the ERCRP-modified specimens was up to 3 mm. Meanwhile, the conventional samples can only reach 2.5 mm.

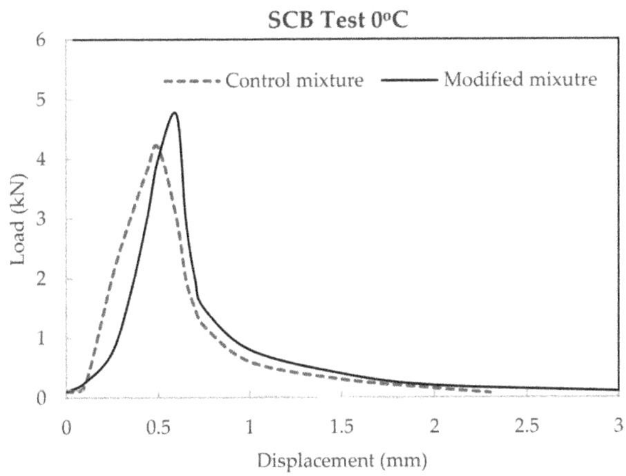

(A)

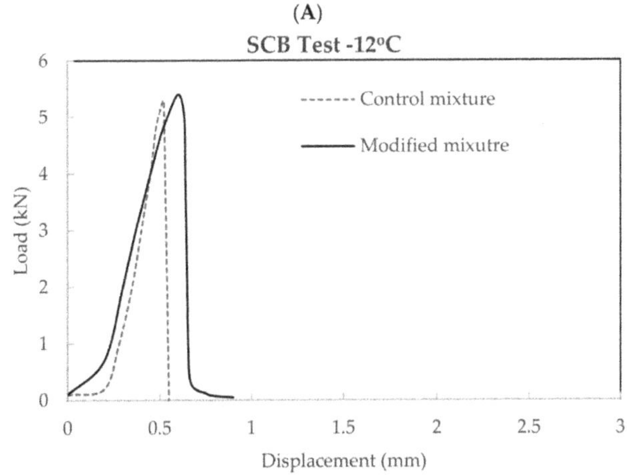

(B)

Figure 14. SCB test results (**A**) at 0 °C and (**B**) at −12 °C.

Table 9 below shows the fracture energy for each specimen. Fracture energy is a failure property that is used to evaluate resistance to cracking in asphalt mixtures, and it refers to the total energy that is applied to the specimen until the specimen fails. In the test results, the ERCRP-modified specimens had a superior fracture energy value of 820 J/m^2 at −0 °C, which was 8% higher than that of the reference specimens (766 J/m^2), indicating this mixture also acquires a higher resistance for practical application in cold regions. Considering the conditioning at −12 °C, the average fracture energy was 518 J/m^2, which was 23.75% larger than that of the control specimen.

Table 9. Average fracture energy (−12 °C) and average fracture energy (−24 °C).

Mix	0 °C Fracture Energy (J/m^2)	Mix	−12 °C Fracture Energy (J/m^2)
Control	766	Control	395
Modified	820	Modified	518

Table 9 also shows the rate of change in fracture energy according to temperature. The findings suggest that the ERCRP-modified specimen was the least affected by temperature, which suffered from a drop of about 36% in fracture energy. On the other hand, the controlled specimen showed lower values than the ERCRP-modified specimen, with degradation of more than 48%.

Based on the experiment results, it was determined that the mixture using the ERCRP-modified binder obtained feasible resistance against cracking. Due to the modification effect of using the proper content of CRP, it is expected to exhibit resilient characteristics and develop cracks gradually rather than fracturing suddenly.

3.2.3. Cantabro Durability Test

The Cantabro test results are summarized in Table 10. The original specimen weight was referred to as the dry weight of each Marshall test specimen, while the weight after testing was the final weight of the specimen subjecting to the tester 300 times at 30 times per minute. Overall, the Cantabro process caused critical surface damage to the test specimens, and the damage ratios (%) can be easily observed among all conditions.

Table 10. Summary of Cantabro test.

	Specimen Weight (g)	Weight after the Test (g)	Loss Rate (within 20% Based on Drainage)	Average Loss Rate (%)
Controlled mix 1	1191.0	1045.8	12.191	
Controlled mix 2	1193.4	1017.3	14.756	12.192
Controlled mix 3	1192.4	1077.6	9.628	
ERCRP-modified mix 1	1193.3	1112.4	6.780	
ERCRP-modified mix 2	1192.3	1033.0	13.361	10.375
ERCRP-modified mix 3	1193.4	1112.4	10.985	

Based on the test results, the ERCRP-modified specimen showed a loss rate of 10.375%, which was an improvement over the normal specimen, which had a loss rate of 12.192%. The improvement in effectiveness may be explained by the stiffening effect contributed by the incorporation of CRP. This finding confirms the results from the above dynamic modulus and SCB test, which indicate the ductile behavior of ERCRP-modified mixtures as well as the stronger bearing capacity.

3.3. Field Test Results

3.3.1. LFWD Test Results

Table 11 presents the LFWD test results between ERCRP-modified pavement and control pavement. In general, the applied falling weight leads to the average maximum loads of 4464 and 4151 N in the former and latter options. Although ERCRP-modified pavement received greater maximum loads compared with the control one, the maximum displacement among them showed comparable results, with the maximum displacement value ranging from 0.1 to 0.11 mm, respectively. As a result of the measured elastic modulus from a field test, the FLWD showed a slight improvement (6.75%) in the reinforcement effectiveness when using the CRP as a binder modifier in SBS pavements. This finding suggests that usage of CRP not only reduces the asphalt binder cost but also promotes the utilization of by-product materials in practice and thereby contributing to the development of sustainable pavements.

Table 11. LFWD test results.

No.	Maximum Load (N)		Maximum Displacement (mm)		Elastic Modulus (MPa)	
	ERCRP-Modified Pavement	Controlled Pavement	ERCRP-Modified Pavement	Controlled Pavement	ERCRP-Modified Pavement	Controlled Pavement
1	4649	4327	0.151	0.163	35.7	32.8
2	4495	4185	0.097	0.105	53.7	48.7
3	4470	4159	0.096	0.1	53.9	50.6
4	4466	4155	0.117	0.129	44.2	43.1
5	4454	4148	0.084	0.085	61.4	55.9
6	4478	4169	0.077	0.089	57.4	53.5
7	4370	4061	0.107	0.115	47.3	44.6
8	4417	4102	0.093	0.097	55	52.5
9	4376	4065	0.103	0.112	49.2	45.2
Average	4464	4152	0.1	0.11	50.87	47.43

3.3.2. GPR Test Results

As a result of GPR surveys conducted to investigate the overall construction condition (see Figure 15), the pavement thickness of the asphalt layer was generally uniform at around 0.25 m from the top of the pavement. In order words, the gap of GPR illustration lines from the top indication to the depth of 0.25 m was very consistent. Overall, both pavements showed homogeneous distribution of asphalt along the testing site, and there was no sign of a failure point in the surface layer. Hence, the test confirms the applicability of ER and CRP in modified SBS asphalt concrete mixture for actual construction. However, further long-term monitoring works should be conducted to verify the durability of pavement after prolonged traffic service life. In addition, the next stage of this research will focus on the cost-effectiveness of the proposed approach to promote this for large-scale applications.

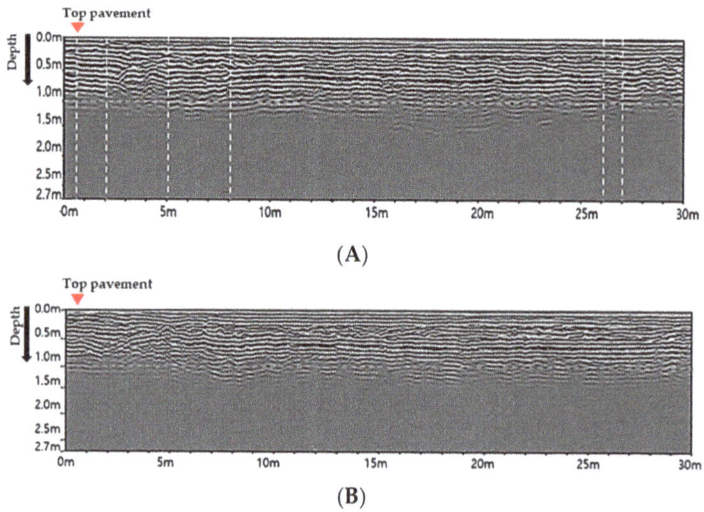

Figure 15. Sectional image of GPR exploration results. (**A**) Modified Sections. (**B**) Controlled Sections.

4. Conclusions

The proposed research aimed to incorporate epoxy resin (ER) and crumb rubber powder (CRP) contents to reduce the consumption of conventional asphalt binder and also promote the usage of recycled waste rubber in practice. To cope with this research objective, the ER and CRP were designed at 3% and 5% by weight of asphalt binder, respectively. In this research, various laboratory tests were performed to evaluate the rheological properties of modified asphalt binders as well as the performance of the proposed asphalt concrete mix, including Frequency Sweep Test, Multiple Stressed Creep and Recovery, Dynamic Modulus, Semi Circular Bending, Cantabro Durability Tests. Afterward, to verify the practical application of the newly developed material, a long-term performance review via field testbed was performed containing Falling Weight Deflectometer and Ground Penetrating Radar tests. The following conclusions can be drawn from the research:

- As a result of the frequency sweep test, the high-temperature plastic deformation resistance and low-temperature crack resistance showed almost similar results between the ERCRP-modified and reference mixtures.
- However, the improvement of ER and CRP in the ERCRP-modified binder was vital when the creep test was taken into account. The MSCR test result indicates that the cumulative strain of the ERCRP-modified binder due to creep load was significantly reduced compared with the reference mix. For instance, at MSCR of 0.1 kPa and 190 s, the strain value of the modified and reference mix was 29.43 and 38.81, respectively. On the other hand, the value of MSCR 3.2 kPa and 190 s was 630.49 and 808.98 in the former and latter binders, respectively. This confirms the contribution of the CRP to the deformation resistivity of the asphalt binder.
- In addition, the recovery (%) in the MSCR test also shows that both SBSs mixtures show potential resistance against cracking by providing sufficient elastic recovery. The plastic deformation (%) of the ERCRP-modified binder was lower than the conventional binder, suggesting a great plastic deformation resistance.
- Considering the behavior of asphalt concrete mixture at the critical frequency loads, the ERCRP-modified specimens obtained higher dynamic modulus values than the control mix in both low-temperature and high-temperature regions. For example, the dynamic modulus of the ERCRP-modified samples was approximately 32,267 and 189 MPa, while the value of the reference mixture was 28,730 and 105 MPa at the highest and lowest frequency, respectively.

- Regarding the SCB test results at −0 °C and −12 °C, the peak stress of the ERCRP-modified mixture was 4.75 and 5.4 MPa, respectively, while the value of the reference specimens was only 4.2 and 5.3 MPa, noticing the slight stress-bearing capacity improvement. Furthermore, the addition of CRP also enhances the ductile behavior of asphalt mixture under applied load since the maximum displacement of the ERCRP-modified specimens was up to 3 mm while this value of the conventional samples was 2.5 mm.
- Based on the test durability test results, the ERCRP-modified mix and reference samples present a loss rate of 10.375% and 12.192%, respectively, indicating a slight improvement in the durability of the modified mixture.
- Through in full-scale testbed, the FLWD elastic modulus of reinforced pavement shows a slight improvement (6.75%) compared with the control pavement. In general, based on the laboratory and field tests, the results confirm that usage of ERCRP not only diminishes the consumption of conventional asphalt binder but also promotes the utilization of by-product materials in practice and thereby developing the sustainable infrastructure objective.

Author Contributions: S.-Y.L., T.H.M.L.: conceptualization, methodology, writing—original draft. S.-Y.L. and T.H.M.L.: visualization, investigation, writing—review and editing. S.-Y.L. and T.H.M.L.: data curation, software. All authors have read and agreed to the published version of the manuscript.

Funding: Korea Ministry of SMEs and Startups.

Institutional Review Board Statement: Not applicable.

Informed Consent Statement: Not applicable.

Data Availability Statement: Data will be provided on request.

Acknowledgments: Research for this paper was carried out under the Korea Technology and Information Promotion Agency for SMEs Research Program (project no. S3044598, Research on development of large-scale utilization technology of waste plastic through road system multi application) funded by the Ministry of SMEs and Startups.

Conflicts of Interest: The authors declare no conflict of interest, financial or otherwise.

References

1. Yang, T.; Jia, Y.; Pan, Y.; Zhao, Y. Evaluation of the Low-Temperature Cracking Performance of Recycled Asphalt Mixture: A Development of Equivalent Fracture Temperature. *Buildings* **2022**, *12*, 1366. [CrossRef]
2. Wang, Y.; Lin, G.; Li, Z. Dynamic Response of Anisotropic Multilayered Road Structures Induced by Moving Loads Based on a Novel Spectral Element Method. *Buildings* **2022**, *12*, 1354. [CrossRef]
3. Li, M.; Luo, C.; Zhu, L.; Li, H.; Cong, P.; Feng, Y.; Yan, L. A novel epoxy-terminated polyethylene modified asphalt with low-viscosity and high storage stability. *Constr. Build. Mater.* **2022**, *335*, 127473. [CrossRef]
4. Huang, G.; Yang, T.; He, Z.; Yu, L.; Xiao, H. Polyurethane as a modifier for road asphalt: A literature review. *Constr. Build. Mater.* **2022**, *356*, 129058. [CrossRef]
5. Li, Y.; Chai, J.; Wang, R.; Zhou, Y.; Tong, X. A Review of the Durability-Related Features of Waste Tyre Rubber as a Partial Substitute for Natural Aggregate in Concrete. *Buildings* **2022**, *12*, 1975. [CrossRef]
6. Fan, Y.; Wu, Y.; Chen, H.; Liu, S.; Huang, W.; Wang, H.; Yang, J. Performance Evaluation and Structure Optimization of Low-Emission Mixed Epoxy Asphalt Pavement. *Materials* **2022**, *15*, 6472. [CrossRef]
7. Pstrowska, K.; Gunka, V.; Prysiazhnyi, Y.; Demchuk, Y.; Hrynchuk, Y.; Sidun, I.; Kułażyński, M.; Bratychak, M. Obtaining of Formaldehyde Modified Tars and Road Materials on Their Basis. *Materials* **2022**, *15*, 5693. [CrossRef]
8. Gamage, S.; Palitha, S.; Meddage, D.P.P.; Mendis, S.; Azamathulla, H.M.; Rathnayake, U. Influence of Crumb Rubber and Coconut Coir on Strength and Durability Characteristics of Interlocking Paving Blocks. *Buildings* **2022**, *12*, 1001. [CrossRef]
9. Guo, L.; Xu, W.; Zhang, Y.; Ji, W.; Wu, S. Selecting the Best Performing Modified Asphalt Based on Rheological Properties and Microscopic Analysis of RPP/SBS Modified Asphalt. *Materials* **2022**, *15*, 8616. [CrossRef]
10. Cui, S.; Guo, N.; Wang, L.; You, Z.; Tan, Y.; Guo, Z.; Luo, X.; Chen, Z. Effect of Freeze–Thaw cycles on the pavement performance of SBS modified and composite crumb rubber modified asphalt mixtures. *Constr. Build. Mater.* **2022**, *342*, 127799. [CrossRef]
11. Li, Y.; Yan, X.; Guo, J.; Wu, W.; Shi, W.; Xu, Q.; Ji, Z. Performance and Verification of High-Modulus Asphalt Modified by Styrene-Butadiene-Styrene Block Copolymer (SBS) and Rock Asphalt. *Coatings* **2022**, *13*, 38. [CrossRef]

12. Li, H.; Cui, C.; Temitope, A.A.; Feng, Z.; Zhao, G.; Guo, P. Effect of SBS and crumb rubber on asphalt modification: A review of the properties and practical application. *J. Traffic Transp. Eng.* **2022**, *9*, 836–863. [CrossRef]
13. Ma, Y.; Wang, S.; Zhang, M.; Jiang, X.; Polaczyk, P.; Huang, B. Weather aging effects on modified asphalt with rubber-polyethylene composites. *Sci. Total. Environ.* **2023**, *865*, 161089. [CrossRef]
14. Poulikakos, L.D.; Pasquini, E.; Tusar, M.; Hernando, D.; Wang, D.; Mikhailenko, P.; Pasetto, M.; Baliello, A.; Falchetto, A.C.; Miljković, M.; et al. RILEM interlaboratory study on the mechanical properties of asphalt mixtures modified with polyethylene waste. *J. Clean. Prod.* **2022**, *375*, 134124. [CrossRef]
15. Davoodi, A.; Esfahani, M.A.; Bayat, M.; Mohammadyan-Yasouj, S.E.; Rahman, A. Influence of nano-silica modified rubber mortar and EVA modified porous asphalt on the performance improvement of modified semi-flexible pavement. *Constr. Build. Mater.* **2022**, *337*, 127573. [CrossRef]
16. Chen, R.; Zhao, R.; Liu, Y.; Xi, Z.; Cai, J.; Zhang, J.; Wang, Q.; Xie, H. Development of eco-friendly fire-retarded warm-mix epoxy asphalt binders using reactive polymeric flame retardants for road tunnel pavements. *Constr. Build. Mater.* **2021**, *284*, 122752. [CrossRef]
17. Han, X.; Mao, S.; Zeng, S.; Duan, H.; Liu, Q.; Xue, L.; Yu, J. Effect of reactive flexible rejuvenators on thermal-oxidative aging resistance of regenerated SBS modified asphalt. *J. Clean. Prod.* **2022**, *380*, 135027. [CrossRef]
18. Zhang, Z.; Li, J.; Wang, Z.; Long, S.; Jiang, S.; Liu, G. Preparation and performance characterization of a novel high-performance epoxy resin modified reactive liquid asphalt. *Constr. Build. Mater.* **2020**, *263*, 120113. [CrossRef]
19. Saputra, R.; Walvekar, R.; Khalid, M.; Mubarak, N.M.; Sillanpää, M. Current progress in waste tire rubber devulcanization. *Chemosphere* **2021**, *265*, 129033. [CrossRef]
20. Boucher, J.; Friot, D. Primary Microplastics in the Oceans. *Mar. Environ. Res.* **2017**, *111*, 18–26.
21. Akbas, A.; Yuhana, N.Y. Recycling of Rubber Wastes as Fuel and Its Additives. *Recycling* **2021**, *6*, 78. [CrossRef]
22. Zheng, W.; Wang, H.; Chen, Y.; Ji, J.; You, Z.; Zhang, Y. A review on compatibility between crumb rubber and asphalt binder. *Constr. Build. Mater.* **2021**, *297*, 123820. [CrossRef]
23. Memon, N.A. Comparison between Superpave Gyratory and Marshall Laboratory Compaction Methods. Doctoral Dissertation, Universiti Teknologi Malaysia, Skudai, Malaysia, 2006.
24. *ASTM C127*; Standard Test Method for Density, Relative Density (Specific Gravity), and Absorption of Coarse Aggregate. Annual Book of ASTM Standard. ASTM International: West Conshohocken, PA, USA, 2004; pp. 1–5.
25. *ASTM Standard D5821*; Standard Test Method for Determining the Percentage of Fractured Particles in Coarse Aggregate. ASTM International: West Conshohocken, PA, USA, 2017; pp. 1–6.
26. *ASTM C131/C131M-14*; Standard Test Method for Resistance to Degradation of Small-Size Coarse Aggregate by Abrasion and Impact in the Los Angeles Machine; Annual Book of ASTM Standard. ASTM International: West Conshohocken, PA, USA, 2014; Volume 4, pp. 5–8.
27. *ASTM D 4791*; Flat Particles, Elongated Particles, or Flat and Elongated Particles in Coarse Aggregate. ASTM International: West Conshohocken, PA, USA, 1999; Volume 4, pp. 1–4.
28. *ASTM D24*; Standard Specification for Mineral Filler For Bituminous Paving Mixtures. ASTM International: West Conshohocken, PA, USA, 2001; Volume 4, pp. 1–2.
29. Gao, J.; Yao, Y.; Huang, J.; Yang, J.; Song, L.; Xu, J.; Lu, X. Effect of Hot Mixing Duration on Blending, Performance, and Environmental Impact of Central Plant Recycled Asphalt Mixture. *Buildings* **2022**, *12*, 1057. [CrossRef]
30. Cong, P.; Luo, W.; Xu, P.; Zhang, Y. Chemical and physical properties of hot mixing epoxy asphalt binders. *Constr. Build. Mater.* **2019**, *198*, 1–9. [CrossRef]
31. Zhang, H.; Zhang, Y.; Chen, J.; Liu, W.; Wang, W. Effect of Desulfurization Process Variables on the Properties of Crumb Rubber Modified Asphalt. *Polymers* **2022**, *14*, 1365. [CrossRef]
32. Ghafari, S.; Ranjbar, S.; Ehsani, M.; Nejad, F.M.; Paul, P. Sustainable crumb rubber modified asphalt mixtures based on low-temperature crack propagation characteristics using the response surface methodology. *Theor. Appl. Fract. Mech.* **2023**, *123*, 103718. [CrossRef]
33. *ASTM D5*; Standard Test Method for Penetration of Bituminous Materials. ASTM International: West Conshohocken, PA, USA, 2019.
34. *ASTM D36-16*; Standard Test Method for Softening Point of Bitumen (Ring-and-Ball Apparatus). ASTM International: West Conshohocken, PA, USA, 2016.
35. *ASTM D113-17*; Standard Test Method for Ductility of Asphalt Materials. ASTM International: West Conshohocken, PA, USA, 2017.
36. *ASTM D2872-22*; Standard Test Method for Effect of Heat and Air on a Moving Film of Asphalt (Rolling Rolling Thin-Film Oven Test) Annual Book of ASTM Standard. ASTM International: West Conshohocken, PA, USA, 2022; pp. 1–6.
37. *AASHTO R 28-09*; Accelerated Aging of Asphalt Binder Using a Pressurized Aging Vessel (PAV). Canadian Council of Independent Laboratories: Burlington, ON, USA, 2007; pp. 1–8.
38. *T313-10 A*; Determining the Flexural Creep Stiffness of Asphalt Binder Using the Bending Beam Rheometer (BBR). AASHTO: Washington, DC, USA, 2011; Standard Specifications for Transportation Materials and Methods of Sampling and Testing. pp. 1–21.
39. Cho, Y.-H.; Yun, T.; Kim, I.T.; Choi, N.R. The application of Recycled Concrete Aggregate (RCA) for Hot Mix Asphalt (HMA) base layer aggregate. *KSCE J. Civ. Eng.* **2011**, *15*, 473–478. [CrossRef]
40. *ASTM D7552-22*; Standard Test Method for Determining the Complex Shear Modulus (G *) of Bituminous Mixtures Using Dynamic Shear Rheometer. ASTM International: West Conshohocken, PA, USA, 2016; Volume I, pp. 1–11.

41. Lee, H.J.; Lee, J.H.; Park, H.M. Performance evaluation of high modulus asphalt mixtures for long life asphalt pavements. *Constr. Build. Mater.* **2007**, *21*, 1079–1087. [CrossRef]
42. Saboo, N.; Mudgal, A.; Singh, A. Using Frequency Sweep Test to Predict Creep and Recovery Response of Asphalt Binders. *J. Mater. Civ. Eng.* **2019**, *31*, 04019081. [CrossRef]
43. Walubita, L.F.; Alvarez, A.E.; Simate, G.S. Evaluating and comparing different methods and models for generating relaxation modulus master-curves for asphalt mixes. *Constr. Build. Mater.* **2011**, *25*, 2619–2626. [CrossRef]
44. Zhang, X.; Han, C.; Yang, J.; Xu, X.; Zhang, F. Evaluating the Rheological Properties of High-Modulus Asphalt Binders Modified with Rubber Polymer Composite Modifier. *Materials* **2021**, *14*, 7727. [CrossRef] [PubMed]
45. *ASTM D7405*; Standard Test Method for Multiple Stress Creep and Recovery (MSCR) of Asphalt Binder Using a Dynamic Shear Rheometer. ASTM International: West Conshohocken, PA, USA, 2020; pp. 1–4.
46. *AASHTO T 350*; Standard Method of Test for Multiple Stress Creep Recovery (MSCR) Test of Asphalt Binder Using a Dynamic Shear Rheometer (DSR). AASHTO: Washington, DC, USA, 2009; Volume 3, pp. 1–9.
47. *ASTM D6373-21a*; Standard Specification for Performance-Graded Asphalt Binder. ASTM International: West Conshohocken, PA, USA, 2021.
48. *AASHTO TP62-2017*; Standard Method of Test for Determining Dynamic Modulus of Hot Mix Asphalt (HMA). AASHTO: Washington, DC, USA, 2017.
49. Lee, K.; Kim, H.; Kim, N.; Kim, Y. Dynamic Modulus of Asphalt Mixtures for Development of Korean Pavement Design Guide. *J. Test. Eval.* **2007**, *35*, 143–150. [CrossRef]
50. *ASTM D8044-16*; Standard Test Method for Evaluation of Asphalt Mixture Cracking Resistance Using the Semi-Circular Bend Test (SCB) at Intermediate Temperatures. ASTM International: West Conshohocken, PA, USA, 2016; pp. 1–7.
51. *ASTM D 7064-08*; Standard Practice for Open-Graded Friction Course (OGFC) Mix Design. ASTM International: West Conshohocken, PA, USA, 2013; Volume 8, pp. 1–7.
52. Cox, B.C.; Smith, B.T.; Howard, I.L.; James, R.S. State of Knowledge for Cantabro Testing of Dense Graded Asphalt. *J. Mater. Civ. Eng.* **2017**, *29*, 0002020. [CrossRef]
53. *KS F 2492*; Cantabro Test Method for Drainage Asphalt Mixtures. Korean Standard: Seoul, Republic of Korea, 2017.
54. *ASTM E2835-07*; Standard Test Method for Measuring Deflections with a Light Weight Deflectometer (LWD). ASTM International: West Conshohocken, PA, USA, 2015; Volume 7, pp. 5–7.
55. *ASTM D4748-10*; Standard Test Method for Determining the Thickness of Bound Pavement Layers Using Short-Pulse Radar. Annual Book of ASTM Standard. ASTM International: West Conshohocken, PA, USA, 1998; Volume 10, p. 7.

buildings

MDPI

Article

Pavement Performance and Ice-Melting Characteristics of Asphalt Mixtures Incorporating Slow-Release Deicing Agent

Jiaqiang Zhang [1], Weicheng Wang [2], Jinzhou Liu [1], Shuyi Wang [3], Xiaochun Qin [4] and Bin Yu [1,*]

[1] School of Transportation, Southeast University, Nanjing 210089, China
[2] China Design Group Co., Ltd., Nanjing 210014, China
[3] College of Civil Engineering, Fuzhou University, Fuzhou 350116, China
[4] School of Civil Engineering, Beijing Jiaotong University, Beijing 100044, China
* Correspondence: yb@seu.edu.cn

Abstract: Icy pavement is one of the primary causes affecting driving safety in winter, and deicing asphalt mixture could effectively resist pavement icing. This study evaluated the effect of a slow-release deicing agent on pavement performance and ice-melting characteristics of the asphalt mixture. The asphalt mixture containing four different contents (0%, 30%, 50%, and 70% using the internal mixing method) by replacing mineral filler was designed. Pavement performance tests were used to investigate the effect of the deicing agent on the high-temperature stability, cracking resistance, and water stability of the asphalt mixture. Qualitative and quantitative tests were designed to compare the ice-melting characteristics and predict the ice-melting durability with different replacement amounts. The experimental results show that the high-temperature stability, low-temperature cracking resistance, and water stability of the asphalt mixture decrease with the increase inthe deicing agent. The mineral filler with a content of 50% deicing agent will enhance the stability of the mixture in the short term. Deicing asphalt mixture can significantly improve the ice and snow melting ability of the pavement, and the asphalt mixture with a content of 50% deicing agent will reduce the interface adhesion between mixture and ice by more than 55%. The slow-release deicing asphalt mixture can reach the maximum release concentration within two hours under rain and snow. The recommended replacement amount of slow-release deicing agent is 50%, and the predicted durability of deicing asphalt mixture is 5–8 years.

Keywords: road engineering; deicing mixture; pavement performance; ice-melting characteristics; durability prediction

Citation: Zhang, J.; Wang, W.; Liu, J.; Wang, S.; Qin, X.; Yu, B. Pavement Performance and Ice-Melting Characteristics of Asphalt Mixtures Incorporating Slow-Release Deicing Agent. *Buildings* **2023**, *13*, 306. https://doi.org/10.3390/buildings13020306

Academic Editors: Andrea Baliello and Di Wang

Received: 12 December 2022
Revised: 11 January 2023
Accepted: 17 January 2023
Published: 19 January 2023

1. Introduction

Snow and ice on asphalt pavement surfaces normally occur with snow or freezing rain weather in winter. A thin layer of ice will be created when snow and ice cover the pavement for a long time, leading to a reduction in the friction coefficient of the pavement surface and affecting the safety of vehicle driving [1]. Several methods have been proposed to remove the snow from the pavement surface, of which the common methods are mechanical plowing and manual sweeping. However, these kinds of methods require considerable specialized machinery and labor costs and may cause traffic congestion [2,3]. In addition, certain progress has also been made in active snow and ice removal methods, such as electrically conductive concrete, geothermal energy, microwave heating, and self-ice-melting asphalt pavement [4,5]. The self-ice-melting asphalt pavement is achieved by adding the deicing agent into the asphalt mixture. The active ingredient in deicing agents gradually migrating to the pavement surface under snow weather can reduce the freezing point, thus effectively reducing the icing situation of the pavement surface in winter to ensure traffic safety [6].

The deicing materials used for asphalt pavement were first studied as early as the 1960s. European countries such as Switzerland and Germany began developing deicing agents, which consist of various salts, including sodium chloride and sodium acetate [7,8]. Based on the successful experience of deicing agents in other countries, Chinese researchers have conducted numerous experiments to develop inorganic halides and organic salt deicing agents, such as Icebane and ZGHIT low freezing point filler [9]. Nevertheless, there is no unified standard to guide the deicing performance test. In the past decades, many scholars have designed different tests to evaluate the melting performance of deicing agents. Xia et al. [10] evaluated the ice resistance of the deicing agent by pulling and shearing test and assessed the long-term deicing performance after long-term environmental simulation Zhao et al. [11] measured the mass of the melted ice on the Marshall specimen to calculate the ice-melting rate and evaluate the ice-melting ability of different specimens. Chen et al. [12] developed test instruments, including a pulling force tester and shearing force tester, which could reflect the ice-pavement adhesion by normal and horizontal adhesion strength. Giuliani et al. [13] recorded the water-freezing process on the surface of asphalt film with a camera and identified the freezing point value of the deicing asphalt mixture. It was concluded that deicing filler significantly delayed the ice formation on the pavement surface, accelerated the ice-melting speed, and reduced the adhesion between ice and pavement. Ma et al. [14] replaced 25%, 50%, 75%, and 100% of the mineral filler with a deicing agent and indicated that the deicing effect was more significant with the increasing content of the deicing agent, as shown by the reduction of ice fracture strength, peel strength, and shear strength on the surface of the mixture.

The asphalt mixture with a deicing agent could avoid the occurrence of pavement icing to a certain extent. However, the replacement of mineral filler with a deicing agent will jeopardize the performance of the pavement. Therefore, the effect of the deicing mixture on the pavement performance should be noted [15,16]. Many studies have been conducted on the pavement performance of anti-icing asphalt mixture. Han et al. [17] studied the pavement performance of three types of deicing agents, Mafilon, calcium acetate, and magnesium acetate, at different dosing levels. The results concluded that the replacement of mineral filler by deicing agents impaired the pavement performance of the asphalt mixture. Liu et al. [18] tested the low-temperature properties and salt-releasing characteristics of antifreeze asphalt concrete and indicated that such properties were affected by salt content. The continuous immersion improved the bending strain at early ages. Ma et al. [19] added a polyester fiber modifier to the deicing asphalt mixture and found that the above mixture possesses much better engineering properties than the normal one. Min et al. [20] compared the engineering performance of a styrene-butadiene-styrene (SBS) modified mixture and an epoxy asphalt mixture. The results showed that the snow-melting agent has a lower influence on the engineering performance of the epoxy asphalt mixture. Yu et al. [21] prepared the asphalt mixture with snowmelt agent and high-elastic binder, which proved favorable pavement and de-icing performance, and the mixture can remove an ice layer less than 12 mm thick.

Meanwhile, the deicing effect of asphalt mixture relies on the migration of active ingredients in the deicing agent. The active ingredient will be gradually lost during the melting process. Thus, deicing asphalt mixture has a certain effective life [22]. Several methods have been used to predict the service life of deicing asphalt mixture, including the ion electrode method and freezing point method [23,24]. The cumulative and single-release concentrations are used as indicators in the ion electrode method, combined with the concept of critical effective ion concentration to predict the service life of the deicing agent [25]. Zhang et al. [26] established the prediction model of deicing longevity of self-ice-melting asphalt pavement based on mathematical regression fitting. Wu et al. [27] used Mafilon to replace the mineral filler in the mixture and derived a linear relationship between the specific gravity of the salt solution and the freezing point according to the freezing point test, combined with the water immersion test to quantitatively predict the deicing durability. However, there are few studies on the durability of deicing asphalt mixture. No unified test method has been proposed to predict the durability of deicing agents. Therefore, there remain issues in evaluating the pavement performance of the deicing mixture and the durability prediction of the deicing agent.

In conclusion, icy pavement surface is an important factor affecting driving safety in rain and snow weather. The active deicing technology of adding a deicing agent to the asphalt mixture can effectively solve this dilemma. A majority of the studies have been performedto evaluate the ice-melting ability and pavement performance of deicing asphalt mixture, while the corresponding behavior considering water stability requires to be clarified. Furthermore, the durability prediction of the deicing mixture is also significant. Therefore, this paper studied the pavement performance and melting characteristics of a slow-release deicing agent. The melting performance was evaluated in multiple dimensions by qualitative and quantitative methods. The immersion Marshall test after different immersion times and freeze-thaw splitting test were designed to verify the effect of the deicing agent on the water stability. Finally, the durability of the deicing agent was predicted based on the ice-melting ability. Figure 1 summarizes the experimental work and shows the flowchart of the methodology.

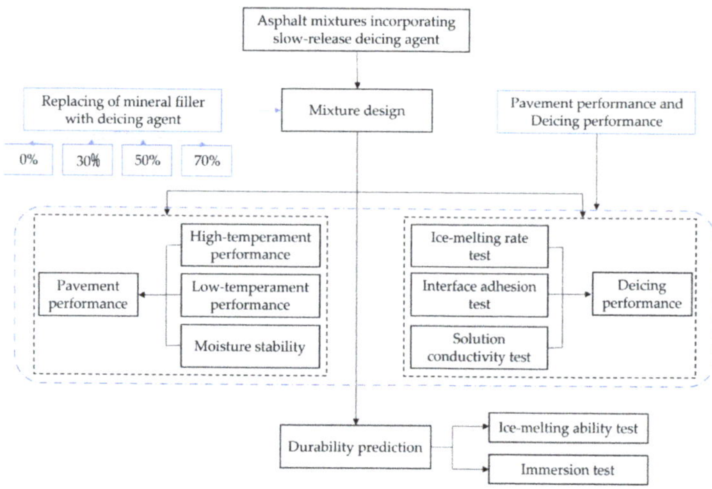

Figure 1. Experimental work flowchart.

2. Materials and Methods

2.1. Materials

Styrene-butadiene-styrene (SBS)-modified asphalt was used for this study. The basic properties of the SBS-modified asphalt are listed in Table 1. Basalt was used as the aggregate, and limestone powder was utilized as the mineral filler. The physical and mechanical characteristics of aggregate and mineral filler are shown in Table 2. It could be seen that all technical indices of asphalt and aggregate meet the requirements of the specification [28].

The deicing agent used in this study is ZGHIT, a product from Chenke Traffic Technology Co., Ltd. in Daqing, China, with a microscopic core-shell capsule structure of slow-release configuration. The maximum nominal particle size of the deicing agent is 0.075 mm, and the density is 2.20 g/cm^3, as shown in Figure 2. It consists of low surface energy water-repellent and low freezing point materials. The low freezing point materials could reduce the freezing point of the pavement, and the water-repellent materials could reduce water infiltration into the interior of the pavement, which can reduce the bonding ability of ice and pavement [11].

Table 1. Basic properties of SBS-modified asphalt.

Test Items	Unit	Test Value	Requirements
Penetration (25 °C,100 g, 5 s)	0.1 mm	63	40–80
Ductility (5 °C, 5 cm/min)	cm	34.2	≥20
Softening point	°C	74.5	≥60
Flash point	°C	304	≥230
Elastic recovery (25 °C)	%	80	≥75
	Rolling thin film ovens test (RTFOT)		
Quality variation	%	0.3	≤±1.0
Penetration ratio (25 °C)	%	76	≥65
Ductility (5 °C)	cm	23	≥15

Table 2. Engineering properties of aggregate with different particle sizes.

Aggregates	9–16 mm	4.75–9.5 mm	2.36–4.75 mm	0–2.36 mm	Requirements
Apparent specific gravity	2.862	2.879	2.846	2.787	≥2.60
Bulk specific gravity	2.713	2.722	2.709	2.683	≥2.50
Water absorption (%)	0.49	0.53	0.55	1.89	≤2.0

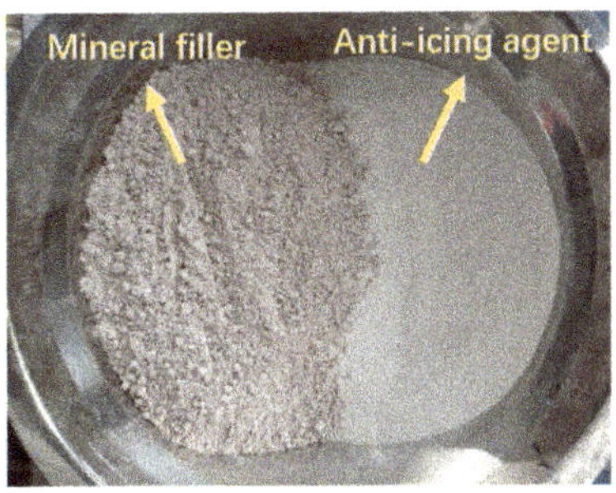

Figure 2. Mineral filler and ZGHIT deicing agent.

2.2. Mixture Design

The gradation of the deicing mixture was designed as a stone mastic asphalt (SMA) mixture with a maximum nominal particle size of 13 mm and a passing rate of 27.6% for the 4.75 mm sieve. The gradations of the mixture are illustrated in Figure 3. The lignin fiber with a content of 0.35% was blended in the mixture, and the optimum binder content was determined to be 6.2% by the standard [29]. ZGHIT deicing agent is a filler product, and it was incorporated into the mixture by internal mixing method, in which 0%, 30%, 50%, and 70% by volume of the mineral filler were replaced, respectively. The mixing process was the same as ordinary asphalt mixture, namely, adding aggregates and asphalt first and mixing for 90 s, then adding mineral filler and deicing agent for another mixing of 90 s.

2.3. Pavement Performance Tests

2.3.1. High-Temperament Performance Test

The high-temperature performance of the deicing asphalt mixture was evaluated via a wheel tracking test. The standard rutting specimen is a slab prepared by a roller compactor with a dimension of 300 mm × 300 mm × 50 mm according to the Chinese test specification [29]. Before the test, the rutting specimen was put into a constant temperature oven at 60 °C for more than 4 h. The test was carried out at 60 °C and a wheel pressure of 0.7 MPa. The dynamic stability (DS) was used to evaluate the high-temperature performance of the asphalt mixture, which was defined as the average loading cycles to form a 1 mm rutting depth. The number of specimens in each group was in accordance with the requirements of the specification, which is three parallel specimens [28,29]. If the error of a certain group does not meet the requirements, the specimens will be prepared and tested again until they meet the requirements.

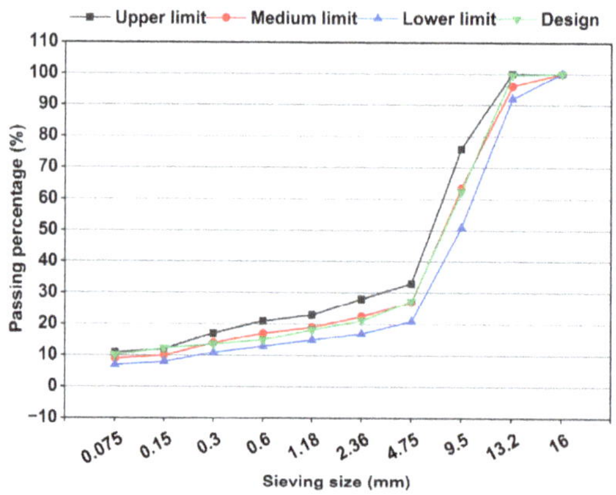

Figure 3. Grading curves of SMA-13 asphalt mixture.

2.3.2. Low-Temperament Performance Test

The low-temperature splitting test can determine the cracking resistance of the asphalt mixture at low temperatures. According to the test specification of the asphalt mixture splitting test in [29], firstly, the standard Marshall specimens (⌀101.6 mm × 63.5 mm) with different replacement amounts of the deicing agent were molded and then put into a −10 °C refrigerator for more than 4 h after demolding. Three parallel specimens were set in each group. During the splitting test, the specimens were loaded with a universal material testing machine at a constant rate of 1 mm/min until splitting damage under low-temperature conditions.

2.3.3. Moisture Stability Test

Moisture damage is one of the main forms of asphalt pavement distresses. The immersion Marshall test is usually used to assess the moisture damage resistance of the specimen. The specimens with different deicing agent replacement amounts were divided into two groups: the first group (MS_1) was tested for Marshall stability after 0.5 h in a water bath at 60 ± 1 °C; the second group (MS_2) was tested after 48 h in a water bath under the same conditions. In addition, the specification states that the residual stability after 48 h of water immersion normally meets the requirements in [30], while there may be additional cases where the residual stability is greater than 100%. Therefore, the Marshall stability of a constant temperature water bath for 10 days (MS_3) was tested for comparison.

Besides the immersion Marshall test, the indirect tensile strength ratio (TSR) test was also carried out to evaluate the moisture stability of the asphalt mixture under freezing and thawing conditions. According to the Chinese test specification [29], the test was performed via standard Marshall specimens, which were divided into two groups. The one group of specimens was maintained in a water bath at 25 °C for 2 h, and then the indirect tensile strength R_1 was measured. The other group was first vacuum-saturated with water at 0.09 MPa for 15 min and then saturated with water at atmospheric pressure for 0.5 h. Then the samples were placed in a −18 °C refrigerator for 16 h and then immersed in a water bath at 60 ± 0.5 °C for 24 h, and finally immersed in water at 25 °C for 2 h. After that, the indirect tensile strength R_2 was determined. The indirect tensile strength ratio was calculated by R_2/R_1. The number of parallel specimens used in the above tests was three.

2.4. Deicing Performance Tests

2.4.1. Ice-Melting Rate Test

The ice-melting test could be used to evaluate the ice-melting effect of deicing asphalt mixes visually. The Marshall specimens containing the deicing agent and 60 mL of the same size ice cylinders were prepared. The Marshall specimens were soaked in the water at a constant temperature of 5 °C to make them fully moistened and reduce the effect of water absorption on the ice melting rate of Marshall specimens. The ice cubes were placed over Marshall specimens, and the remaining mass of the ice cylinders was recorded at an interval of 10 min. In this test, the following parameter was obtained:

$$LR = \frac{m_0 - m_i}{m_i} \times 100 \tag{1}$$

where LR is the ice melting rate, %; m_0 is the mass of the ice cylinder before melting, g; m_i is the mass of the ice cylinder after melting weighed at i time, g.

2.4.2. Interface Adhesion Test

The interface adhesion test was designed regarding the adhesive binder bonding strength test (drawing method) in specification [29] to test the bonding force between the deicing asphalt mixture and the frozen sponge to reflect the de-icing effect of the deicing agent when the pavement is icy. The crossed nylon rope was fixed at the bottom of the sponge (120 mm × 100 mm × 60 mm) and then placed on the Marshall specimen after the sponge was filled with water. The combined specimen was then frozen in the refrigerator at −10 °C for 2 h until the water in the whole sponge was frozen into ice. The pointer-type tensiometer was hooked on one side of the nylon rope to slowly pull the frozen sponge and the Marshall specimen apart. Each group was set to three parallel specimens. The force during the pulling procedure was recorded as the interface adhesion, as shown in Figure 4. For the interface adhesion test, the smaller the pulling force between the icing sponge and the Marshall specimen means the weaker bond between the ice and the mixture.

Figure 4. Interface adhesion test.

2.4.3. Solution Conductivity Test

The deicing performance of the slow-release deicing asphalt mixture depends on its salt compounds. After the Marshall specimens containing the deicing agent were placed in water, the deicing agent released ions that made the solution appear to have conductive [31]. Thus, the conductivity meter can be used to test the conductivity of the immersion solution to evaluate the deicing characteristics of the asphalt mixture indirectly. The test water was pure water, and the conductivity meter was CT-3031 ranging from 0 to 19.99 mS/cm.

2.5. Durability Prediction of Deicing Mixture

2.5.1. Ice-Melting Ability Test

The melting ability of the deicing agent is closely related to the concentration of the active ingredient. When the concentration of the active ingredient is lower than the critical concentration, the melting ability would be lost. According to the method of durability prediction of the deicing agent in [8], the ice-melting ability test of the solution with gradient concentration was designed to determine the critical concentration and the corresponding critical conductivity of the deicing agent. Nine groups of ice samples with thesame volume were prepared by filling a plastic cup with 50 mL of pure water and freezing them in a refrigerator. Correspondingly, nine solutions of different concentrations ranging from 0.8 to 4 g/L were prepared according to the concentration gradient of 0.4 g/L. After that the conductivity of each group was measured. Before starting the test, the solutions were placed in a 0 °C environment to reduce the impact of temperature on the test results, as shown in Figure 5. The solutions were subsequently poured onto the ice samples, and the volume of melted liquid was measured after 30 min. The melting volume of the ice sample was used to express the ice melting ability of the gradient concentration solutions.

Figure 5. Ice-melting ability test of the solution with gradient concentration.

2.5.2. Immersion Test

The deicing components in asphalt pavement are released to the pavement surface through the connected pores in the mixture under the rainwater, thereby reducing the freezing point and playing the deicing role [32]. When the pavement structure is dry, it could be considered that the deicing agent is not released. At the same time, most of the rainwater flows into the rain well due to the road runoff in rain and snow weather, and the remaining rainwater penetrates the pores of the mixture under the action of external force, which could be approximately equivalent to the asphalt mixture immersed in water. Based on the above analysis and the results of the ice-melting ability test, the design of the immersion test used to predict ice-melting durability was as follows:

(1) Marshall specimen with 50% deicing agent content was placed in a container, and 600 mL pure water was injected to soak the top of the specimen slightly;
(2) After soaking for 24 h, the electrical conductivity of the solution was measured and recorded;
(3) The same volume of pure water was replaced and soaked again, and the cycle was repeated until the measured conductivity was lower than the critical value.

3. Results and Discussion

3.1. Analysis of Pavement Performance

3.1.1. High-Temperature Performance

As shown in Figure 6, the wheel tracking test results indicate that as the deicing agent content increases, the high-temperature performance of the asphalt mixture gradually decreases. The proportion of the decrease in DS is related to the amount of deicing agent replacement. The DS values of asphalt mixture with 30%, 50%, and 70% deicing agent replacement decreased by 10.98%, 25.10%, and 30.10%, respectively compared to the mixture with 0% deicing agent, while the DSs of those mixtures are still able to meet the specification requirements in [30]. The decrease in high-temperature performance may be contributed to the fact that the specific surface area of deicing agent filler is smaller than that of mineral filler, which plays a role in absorbing excess asphalt in the asphalt mix. The replacement of mineral filler by the deicing agent will lead to the increment of free asphalt content and the corresponding reduction of structural asphalt content that provides bond strength and mechanical strength [33]. Therefore, the dynamic stability of the asphalt mixture decreases.

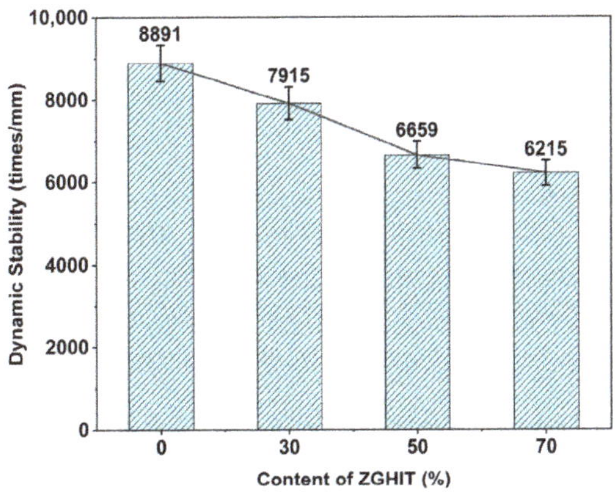

Figure 6. Results of wheel tracking test.

3.1.2. Low-Temperature Performance

Figure 7a shows the calculated splitting tensile strength of the deicing asphalt mixture, which reflects the stress required for splitting damage of the specimen. It was observed that replacing the mineral filler with the deicing agent would reduce the splitting tensile strength. The splitting tensile strength of the asphalt mixture without replacing the mineral filler is 3.64 MPa, while such strength with 30% replacement of deicing agent is 3.22 MPa, which is 11.6% lower than before. The line shows that the splitting tensile strength of asphalt mixture with a 50% replacement amount decreases by only 3.7% compared with that of 30%, and the splitting tensile strength of asphalt mixture with a 70% replacement amount decreases by 7.1% compared with that of 50%, which means that the splitting tensile strength of asphalt mixture with 50% deicing agent has a small decrease compared with that of the asphalt mixture with 30% deicing agent. In other words, the change in the content of the deicing agent from 30% to 50% has no evident influence on the low-temperature splitting tensile strength of the asphalt mixture. Figure 7b illustrates that the modulus of low-temperature failure strength of the asphalt mixture decreases with the increase inthe deicing agent content. This index is related to the flexibility of the asphalt mixture under low-temperature conditions. The increase in deicing agent replacement would lead to adecrease instructural asphalt in the mixture. In other words, the deicing agent would weaken the bonding effect between the aggregate and asphalt, which would reduce the low-temperature cracking resistance of the asphalt mixture [34].

3.1.3. Water Stability Performance

The results of Marshall stability values of asphalt mixture specimens with different soaking times are listed in Table 3. It could be seen that Marshall stability decreases with the increasing content of the deicing agent, and the residual stability of the mixture specimens with the deicing agent was higher than 100% after 48 h of soaking. The residual Marshall stability has a significant decrease when continues to soak up to 10 days. The reason is that the moisture absorption capacity of salt-based deicing agents is stronger than that of mineral filler, and the active ingredients are less released to accommodate the expansion of the specimens after 48 h soaking, which makes the specimens more structured and have a certain increase in strength. Interestingly, the residual stability of conventional specimens did not change significantly with soaking time, while the residual stability of those with deicing agent behaves a significant decrease after 10 days in contrast to 48 h soaking. The migration of salt solution from high concentration to low concentration, on the one hand, changes the stable skeleton structure and increases the gap of the asphalt mixture. As a result, more water enters the asphalt mixture. On the other hand, the capillary effect during the migration process weakens the bond strength of asphalt and aggregate, and the final residual stability decreases significantly [18].

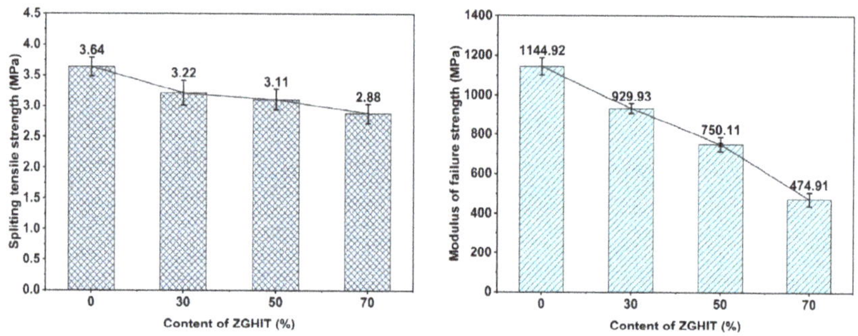

(a) (b)

Figure 7. Results of the low-temperature splitting test. (**a**) Splitting tensile strength. (**b**) Modulus of failure strength.

Table 3. Results of water immersion Marshall test.

Content of ZGHIT (%)	MS_1 (kN)	MS_2 (kN)	Residual Stability (%)	MS_3 (kN)	Residual Stability (%)
0	16.63	16.15	97.11	13.88	88.46
30	12.76	15.59	122.18	8.35	65.43
50	12.62	15.20	120.44	7.26	57.52
70	11.81	14.11	111.95	6.67	56.48

As shown in Figure 8, the histogram indicates the splitting strength of the asphalt mixture without freeze-thaw and after freeze-thaw, and the line indicates the freeze-thaw splitting strength ratio of the corresponding asphalt mixtures. The splitting strength values of the specimens without freeze-thaw were approximately equal, while they show a decreasing trend with the increasing content of the deicing agent after freeze-thaw. The strength ratio of freeze-thaw splitting also shows the same trend. The results are considered as the active ingredients in the deicing agent would cause water intrusion, and the water entering the specimen would freeze and expand at low temperatures, which accelerates the speed of the asphalt detaching from the aggregate surface. In addition, during the freeze-thaw cycle, the precipitated salts may re-crystallize and expand, resulting in potential stress concentration on the void surface, which also aggravates the water damage of the specimen to a certain extent.

Based on the synthesis of the water immersion Marshall test and freeze-thaw splitting test, it can be concluded that the deicing agent replacement of mineral filler would reduce the water stability of the asphalt mixture, and the degree of water stability reduction is related to the replacement amount.

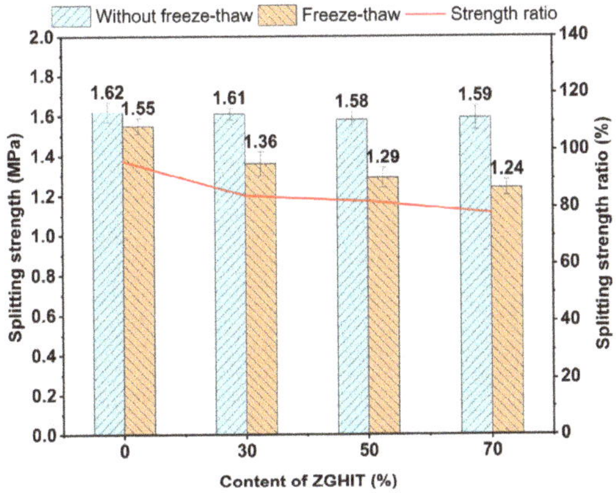

Figure 8. Results of the indirect tensile strength ratio test.

3.2. Analysis of Anti-Freezing Performance

3.2.1. Ice-Melting Rate

It could be visualized from Figure 9 that there was a significant difference in the ice melting effect of Marshall specimens with different deicing agent replacement amounts, and the deicing agent can improve the ice-melting ability of the asphalt mixture. The ice-melting rate of different specimens with time is shown in Table 4. When the melting time is 50 min, the specimens with 70% of the replacement amount of deicing agent melt all the ice, while the specimens without deicing agent melt only 18.28% of the ice. This suggests that the deicing agent has an obvious effect on melting ice. The ice-melting rates of the four groups of specimens were compared for 30 min, and it was found that the ice-melting rates of the specimens with 0, 30%, 50%, and 70% deicing agent replacement were 12.39%, 34.85%, 47.25%, and 63.34%, respectively, corresponding to the improvement of the ice melting performance of 1.81, 2.81 and 4.09. This means that the addition of a deicing agent can significantly improve the deicing ability of the specimens. However, the increase in the amount of deicing agent is limited to improve the ice-melting ability of the specimens.

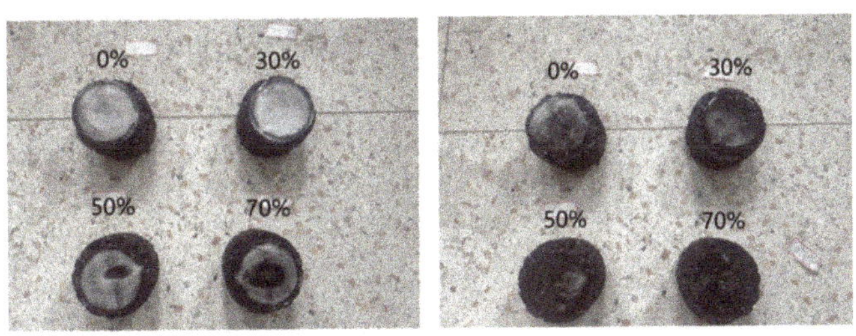

(a) Ice Melting in 5min (b) Ice Melting in 30min

Figure 9. Ice-melting rate of specimens with different content of deicing agent.

Table 4. Ice-melting rate of deicing specimens.

Time (min)		**0**	**10**	**20**	**30**	**40**	**50**
Content of ZGHIT (%)	0	0	2.10%	7.14%	12.39%	15.34%	18.28%
	30	0	10.17%	21.58%	34.85%	43.57%	55.39%
	50	0	15.89%	29.66%	47.25%	62.71%	79.24%
	70	0	21.26%	44.03%	63.34%	86.55%	100.00%

3.2.2. Interface Adhesion

As mentioned above, the interface adhesion test could reflect the deicing effect of the deicing agent. The test results are shown in Figure 10. The bonding force between the asphalt mixture with the deicing agent and the frozen sponge is smaller than that of the conventional asphalt mixture. Moreover, the bonding force decreases gradually with the increase inthe replacement ratio of the deicing agent. Interestingly, the reduction of the bonding force with 30%, 50%, and 70% deicing agents is gradually smaller, with the reduction being 44.4%, 55.5%, and 61.1% compared with the conventional asphalt mixture. The interface adhesion is related to the release of the active ingredient of the deicing agent. The active ingredients are released to the pavement surface to reduce the freezing point and promote the melting of the ice layer. The smaller the bonding force means the weaker bond between the ice layer and pavement, which signifies the better melting effect of the deicing asphalt mixture. Similarly, the icing resistance is more evident after incorporating the deicing agent into the mixture, while it cannot be further improved if merely increasing the content of the deicing agent. This observation of the interface adhesion reduction agreed with the findings of the previous studies [19].

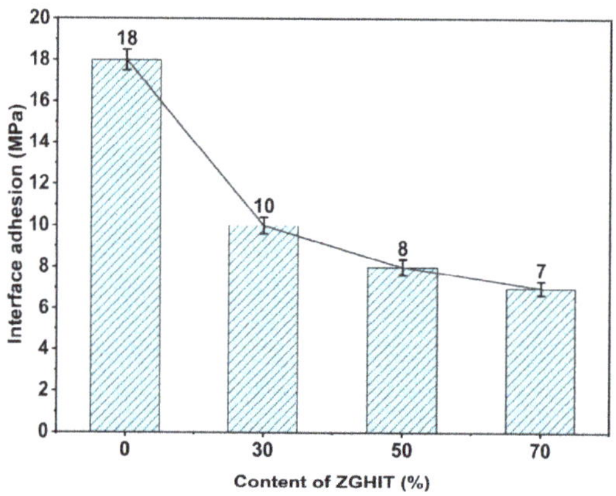

Figure 10. The bonding force between frozen sponge and deicing specimens.

3.2.3. Conductivity of Immersion Solution

Figure 11 demonstrates that the conductivity of the soaking solution after adding the deicing agent increases with the soaking time, especially within 2 h of soaking. The conductivity values of the soaking solution of different replacement mixtures are almost stable after 2 h. As a comparison, the conductivity of the solution of the conventional asphalt mixture maintains stability during the whole procedure. The conductivity curve of the soaking solution with 50% replacement of anti-icing agent asphalt mixture was between the conductivity curves of the soaking solution with 30% and 70% replacement of anti-icing agent asphalt mixture. Furthermore, the conductivity value of the soaking solution of asphalt mixture with 70% deicing agent replacement is about twice that of other deicing agent replacements of asphalt mixture. Undoubtedly, the conductivity of the soaked solution increases with the increasing content of the deicing agent.

According to the above results, the release rate of the slow-release deicing agent is positively correlated with its incorporating content in the mixture, and the deicing effect could be achieved more quickly with a larger replacement volume. Conductivity change with time could be divided into three stages including the rapid release stage, slow down release stage, and no longer release stage [17].

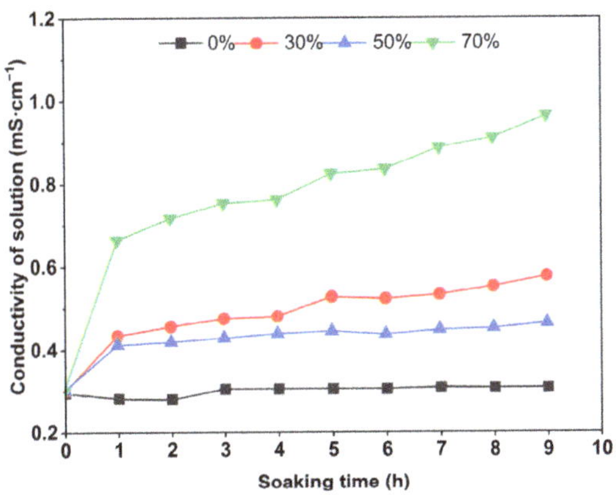

Figure 11. The release conductivity variation of deicing specimens at different times.

3.3. Durability Prediction

3.3.1. Ice-Melting Ability of Gradient Concentration Solution

Figure 12 shows the relationship between the melting amount of ice samples and the conductivity of the solution. It could be seen that there is a positive linear correlation between the concentration of the active ingredient solution and the conductivity, with a correlation coefficient greater than 0.99. The ice-melting ability increases with the increase inthe solution concentration. Furthermore, the ice-melting amount is approximately 0 when the solution concentration is 1.2 g/L, and the corresponding conductivity of the solution is 2.6 mS/cm. The freezing point of the solution needs to be reduced to make the ice sample melt, and the decrease inthe freezing point of the solution is related to the concentration of the active ingredient of the deicing agent. The ice sample would be melted only when the solution reaches a certain concentration, which could be defined as the critical ice-melting concentration [13]. Furthermore, the concentration of the solution could be indirectly characterized by measuring the conductivity. Therefore, it could be considered that the deicing mixture will lose the deicing capacity when the single-release concentration is lower than 1.2 g/L. That is, the solution conductivity is lower than 2.6 mS/cm. This critical ice-melting conductivity would be used to predict the durability of the deicing agent asphalt mixture in the immersion test.

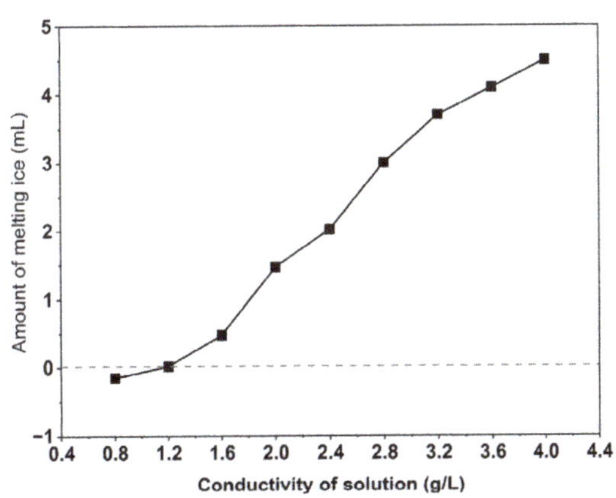

Figure 12. The amount of ice melting in the gradient concentration solution.

3.3.2. Durability Prediction of Deicing Agent

The durability prediction of the deicing agent was tested with an immersion test and, the experiment results are shown in Figure 13. It could be seen that the single-release concentration of the active ingredient in the Marshall specimen is gradually decreasing in all soaking cycles, and the changing rate in the conductivity of the solution at the beginning of soaking is low. This indicates that the single release of the deicing agent is generally stable and remains at a high level. At the middle stage of soaking, the decreased rate of conductivity increases obviously. This is due to the increasing porosity of the mixture resulting from the continuous release of the deicing agent, which in turn increases the precipitation rate of the active ingredients. By the end of the immersion cycle, most of the deicing filler in the mixture has been lost, resulting in a more stable decline in the rate of single releases. On the 15th day, the solution conductivity was down to the critical ice-melting conductivity of 2.6 mS/cm, at which point it is considered that the soaking sample has lost its deicing ability. The corresponding amount of water added to the test is 9000 mL.

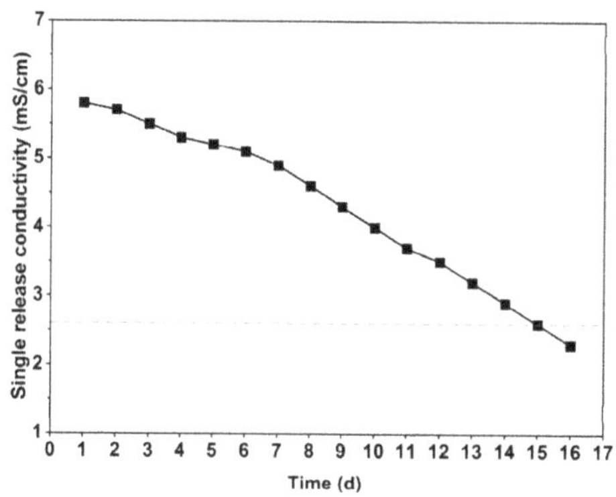

Figure 13. Cyclic soaking conductivity of deicing mixture.

The average annual rainfall data of five regions with different climate types in China were collected, and the runoff coefficient of urban asphalt pavement was considered to be0.8. The conversion of average annual rainfall could be calculated by Equation (2), and the ice-melting durability of slow-release deicing asphalt mixture is calculated by equivalent conversion, as shown in Table 5. The service life of the deicing mixture has a great variation in different regions and is contrary to the value of the average annual rainfall. It can be calculated that the predicted life range is 5–8 years for various regions. In other words, the ice-melting ability of the asphalt mixture containing the deicing agent would be lost after exceeding the durability, and winter maintenance will be required.

$$V = \pi \times r^2 \times H \times (1 - a)/1000 \quad (2)$$

where V is the conversion of average annual rainfall, mL; r is the radius of Marshall specimen, mm; H is the average annual rainfall, mm; and a is the runoff coefficient.

Table 5. Durability prediction of deicing agent in different regions.

Region	Average Annual Rainfall (mm)	Conversion of Average Annual Rainfall (mL)	Durability Prediction (Year)
Heilongjiang	753.1	1250.9	7.46
Beijing	644.2	1031.4	7.59
Shaanxi	740.4	1185.7	7.59
Jiangsu	1090.7	1746.8	5.15

4. Conclusions

To explore the effectiveness of adding a deicing agent to the asphalt mixture, the influence of the deicing agent on pavement performance and ice-melting characteristics were studied. Furthermore, the durability of the deicing asphalt mixture was predicted. Based on the entire investigation, the following conclusions can be summarized.

(1) The high-temperature performance and low-temperature cracking resistance of the asphalt mixture continuously decrease with the increasing content of the deicing agent. The replacement of mineral filler with a deicing agent can enhance the Marshall stability of the mixture in the short term, while the long-term residual stability after the release of the active ingredient is lower than that of the ordinary mixture.
(2) Deicing asphalt mixture can significantly improve the melting and snow removal ability of pavement. The interface adhesion can be reduced by more than 55% with a 50% replacement quantity. Considering the deicing performance and pavement performance of the mixture, the recommended replacement amount of slow-release deicing agent is found to be 50%.
(3) The slow-release deicing asphalt mixture can reach the maximum release concentration in 2 h under rain and snow, after which the solution maintains a stable concentration.
(4) The ice-melting capacity of the slow-release deicing agent is positively correlated with the released concentration. The effective life span of the deicing agent for this study is estimated to be 5–8 years after equivalent conversion.

Nonetheless, we also found that due to the lower surface area of the deicing agent than that of mineral fillers, adjusting the binder content of asphalt mixtures with different deicing agent contents might be beneficial to improving the pavement performance of the mixtures. It seems necessary to ensure the pavement performance of asphalt mixture containing the deicing agent and it is a promising research topic in the future. The influence of different binder content on the asphalt mixture with the deicing agent is also worthy of exploration.

Author Contributions: Conceptualization, J.Z. and B.Y.; Data curation, J.Z., W.W., J.L. and S.W.; Formal analysis, J.Z., W.W., X.Q. and S.W.; Funding acquisition, B.Y. and X.Q.; Methodology, J.Z.; Resources, W.W. and B.Y.; Validation, J.Z. and J.L.; Writing—original draft, J.Z.; Writing—review & editing, J.Z., J.L., X.Q. and S.W. All authors have read and agreed to the published version of the manuscript.

Funding: This research was funded by the Postgraduate Research & Practice Innovation Program of Jiangsu Province, grant number SJCX21_0057 and the Natural Science Foundation of China, grant numbers 51878163, 51878039 and 52078034.

Data Availability Statement: Not applicable.

Conflicts of Interest: The authors declare no conflict of interest.

References

1. Zheng, M.L.; Wu, S.J.; Wang, C.T.; Li, Y.F.; Ma, Z.H.; Peng, L. A Study on Evaluation and Application of Snowmelt Performance of Anti-Icing Asphalt Pavement. *Appl. Sci.* **2017**, *7*, 583. [CrossRef]
2. Zhao, W.K.; Li, L.; Wang, W.; Zhang, Y.N.; Su, W.T.; Chen, X.; Li, B.X. Thermal performances of porous snow by a hydronic heating system at different weather conditions. *J. Therm. Anal. Calorim.* **2020**, *141*, 1519–1528. [CrossRef]
3. Arvidsson, A.K. The Winter Model—A new way to calculate socio-economic costs depending on winter maintenance strategy. *Cold Reg. Sci. Technol.* **2017**, *136*, 30–36. [CrossRef]
4. Liu, K.; Fu, C.L.; Xie, H.Z.; Wang, F.; Wang, X.C.; Bai, H.J. Design of electric heat pipe embedding schemes for snow-melting pavement based on mechanical properties in cold regions. *Cold Reg. Sci. Technol.* **2019**, *165*, 102806. [CrossRef]
5. Xia, Y.F.; Jia, J.; Chen, Q. Road Performance and Ice-Melting Characteristics of Steel Wool Asphalt Mixture. *Adv. Civ. Eng.* **2022**, *2022*, 7029176. [CrossRef]
6. Wang, Z.J.; Zhang, T.; Shao, M.Y.; Ai, T.; Zhao, P. Investigation on snow-melting performance of asphalt mixtures incorporating with salt-storage aggregates. *Constr. Build. Mater.* **2017**, *142*, 187–198. [CrossRef]
7. Zhang, Z.; Luo, Y.; Zhao, F. Review of research on the effect of salt storage deicing material on the preformance of asphalt mixture. *Chem. Ind. Eng. Prog.* **2018**, *37*, 2282–2294.

8. Wu, S.Y.; Yang, J.; Yang, R.C.; Zhu, J.P.; Liu, S.; Wang, C.Y. Investigation of microscopic air void structure of anti-freezing asphalt pavement with X-ray CT and MIP. *Constr. Build. Mater.* **2018**, *178*, 473–483. [CrossRef]
9. Li, F.; Wang, Z. Experiment of Road Performance of Asphalt Mixture with Automatic Long-term Snowmelt Agent. *J. Highw. Transp. Res. Dev.* **2012**, *29*, 7. [CrossRef]
10. Xia, H.Y.; Zhao, X.; Wu, Y.C.; Yuan, T.; Song, L.F.; Yan, M.J.; Wang, F.Y.; Chen, H.X. Preparation and performance of antifreeze adhesive materials for asphalt pavement. *Constr. Build. Mater.* **2020**, *258*, 119554. [CrossRef]
11. Zhao, Y.; Chen, C.; Xiang, Y.K.; Wang, J. Preparation, Characterization, and Anti-Icing Properties of Sustained-Release Low-Freezing-Point Asphalt Mixture. *J. Mater. Civ. Eng.* **2022**, *34*, 04022171. [CrossRef]
12. Chen, H.X.; Wu, Y.C.; Xia, H.Y.; Zhang, Z.; Yuan, T. Anti-freezing asphalt concrete: Ice-adhesion performance. *J. Mater. Sci.* **2018**, *53*, 4781–4795. [CrossRef]
13. Giuliani, F.; Merusi, F.; Polacco, G.; Filippi, S.; Paci, M. Effectiveness of of sodium chloride-based anti-icing filler in asphalt mixtures. *Constr. Build. Mater.* **2012**, *30*, 174–179. [CrossRef]
14. Ma, T.; Geng, L.; Ding, X.H.; Zhang, D.Y.; Huang, X.M. Experimental study of deicing asphalt mixture with anti-icing additives. *Constr. Build. Mater.* **2016**, *127*, 653–662. [CrossRef]
15. Xu, O.M.; Han, S.; Zhang, C.L.; Liu, Y.M.; Xiao, F.P.; Xu, J. Laboratory investigation of andesite and limestone asphalt mixtures containing sodium chloride-based anti-icing filler. *Constr. Build. Mater.* **2015**, *98*, 671–677. [CrossRef]
16. Tan, Y.; Zhang, C.; Xu, H.; Tian, D. Snow Melting and Deicing Characteristics and Pavement Performance of Active Deicing and Snow Melting Pavement. *China J. Highw. Transp.* **2019**, *32*, 1–17.
17. Han, S.; Yin, Y.Y.; Peng, B.A.; Dong, S.A.; Wu, S. Experimental Study of Asphalt Mixture with Acetate Anti-Icing Filler. *Arab. J. Sci. Eng.* **2022**, *47*, 4225–4237. [CrossRef]
18. Liu, Z.Z.; Sha, A.M.; Xing, M.L.; Li, Z.Z. Low temperature property and salt releasing characteristics of antifreeze asphalt concrete under static and dynamic conditions. *Cold Reg. Sci. Technol.* **2015**, *114*, 9–14. [CrossRef]
19. Ma, T.; Ding, X.H.; Wang, H.; Zhang, W.G. Experimental Study of High-Performance Deicing Asphalt Mixture for Mechanical Performance and Anti-Icing Effectiveness. *J. Mater. Civ. Eng.* **2018**, *30*, 04018180. [CrossRef]
20. Min, Z.H.; Xia, Y.Y.; Li, X.C.; Tao, Z. Performances evaluation of epoxy asphalt mixture containing snow-melting agent. *Constr. Build. Mater.* **2017**, *155*, 762–769. [CrossRef]
21. Yu, W.; Zhang, X.; Zhong, K. De-icing performance of asphalt mixture with snowmeltagent and high-elastic modified asphalt binder. *J. China Univ. Min. Technol.* **2015**, *44*, 912–916.
22. Zhong, K.; Sun, M.Z.; Chang, R.H. Performance evaluation of high-elastic/salt-storage asphalt mixture modified with Mafilon and rubber particles. *Constr. Build. Mater.* **2018**, *193*, 153–161. [CrossRef]
23. Zhou, J.; Li, J.; Liu, G.Q.; Yang, T.; Zhao, Y.L. Long-Term Performance and Deicing Effect of Sustained-Release Snow Melting Asphalt Mixture. *Adv. Civ. Eng.* **2019**, *2019*, 1940692. [CrossRef]
24. Dan, H.C.; Tan, J.W.; Du, Y.F.; Cai, J.M. Simulation and optimization of road deicing salt usage based on Water-Ice-Salt Model. *Cold Reg. Sci. Technol.* **2020**, *169*, 102917. [CrossRef]
25. Tian, M.M.; Zhu, B.L.; Cao, Y.H.; Jiang, R.N.; Yan, S.X. Anti-icing Capacity and Service Life Research on the Controlled Released Anti-icing Asphalt Modifier. In Proceedings of the 7th International Conference on Mechanical and Electronics Engineering (ICMEE), Dalian, China, 26–27 September 2015.
26. Zhang, H.H.; Guo, R.H. Investigation of Long-Term Performance and Deicing Longevity Prediction of Self-Ice-Melting Asphalt Pavement. *Materials* **2022**, *15*, 6026. [CrossRef]
27. Wu, S.Y.; Yang, J.; Sun, X.Y.; Wang, C.Y.; Yang, R.C.; Zhu, J.P. Preparation and characterization of anti-freezing asphalt pavement. *Constr. Build. Mater.* **2020**, *236*, 117579. [CrossRef]
28. *JTG F40-2019*; Technical Specifications for Construction of Highway Asphalt Pavement. China Ministry of Transport: Beijing, China, 2019.
29. *JTG E20-2019*; Standard Test Methods of Bitumen and Bituminous Mixtures for Highway Engineering. China Ministry of Transport: Beijing, China, 2019.
30. *JTG D50-2017*; Specifications for Desigen of Highway Asphalt Pavement. China Ministry of Transport: Beijing, China, 2017.
31. Li, L.M.; Li, H.Z.; Tan, Y.Q.; Yang, X.; Xu, H.N. Investigation of chloride release characteristic of Chlorine-Based Anti-Icing asphalt mixture. *Constr. Build. Mater.* **2021**, *312*, 125410. [CrossRef]
32. Qin, K.; Ma, Q.; Wu, J. Research on Performance of Asphalt Concrete Under the Action of Temperature and Salt Corrosion. *Bull. Chin. Ceram. Soc.* **2013**, *32*, 952–956.
33. Zhang, J.P.; Li, X.Q.; Liu, G.Q.; Pei, J.Z. Effects of material characteristics on asphalt and filler interaction ability. *Int. J. Pavement Eng.* **2019**, *20*, 928–937. [CrossRef]
34. Liu, Z.Z.; Xing, M.L.; Chen, S.F.; He, R.; Cong, P.L. Influnence of the chloride-based anti-freeze filler on the properties of asphalt mixture. *Constr. Build. Mater.* **2014**, *51*, 133–140. [CrossRef]

MDPI AG
Grosspeteranlage 5
4052 Basel
Switzerland
Tel.: +41 61 683 77 34

Buildings Editorial Office
E-mail: buildings@mdpi.com
www.mdpi.com/journal/buildings

www.ingramcontent.com/pod-product-compliance
Lightning Source LLC
Chambersburg PA
CBHW061506210926
51850CB00008B/1838
9783725819416